LECTURE NOTES ON
CALCULUS OF VARIATIONS

PEKING UNIVERSITY SERIES IN MATHEMATICS

Series Editors: Kung-Ching Chang, Pingwen Zhang, Bin Liu,
and Jiping Zhang *(Peking University, China)*

Peking University ⬡ Series in Mathematics — Vol. 6

LECTURE NOTES ON
CALCULUS OF
VARIATIONS

Kung Ching Chang
Peking University, China

Translated by
Tan Zhang
Murray State University, USA

World Scientific

NEW JERSEY · LONDON · SINGAPORE · BEIJING · SHANGHAI · HONG KONG · TAIPEI · CHENNAI

Published by

World Scientific Publishing Co. Pte. Ltd.

5 Toh Tuck Link, Singapore 596224

USA office: 27 Warren Street, Suite 401-402, Hackensack, NJ 07601

UK office: 57 Shelton Street, Covent Garden, London WC2H 9HE

Library of Congress Cataloging-in-Publication Data
Names: Zhang, Gongqing. | Zhang, Tan, 1969–
Title: Lecture notes on calculus of variations / by Kung Ching Chang
    (Peking University, China) ; translated by: Tan Zhang (Murray State University, USA).
Other titles: Calculus of variations
Description: New Jersey : World Scientific, 2016. | Series: Peking University
    series in mathematics ; volume 6 | Includes bibliographical references and index.
Identifiers: LCCN 2016025413| ISBN 9789813144682 (hardcover : alk. paper) |
    ISBN 9789813146235 (softcover : alk. paper)
Subjects: LCSH: Calculus of variations. | Mathematical analysis. | Functionals.
Classification: LCC QA315 .Z434 2016 | DDC 515/.64--dc23
LC record available at https://lccn.loc.gov/2016025413

British Library Cataloguing-in-Publication Data
A catalogue record for this book is available from the British Library.

Printed in Singapore

Preface

Calculus of variations first appeared around the time when calculus was invented, over 300 years ago. It used to be a mandatory course for undergraduate students majoring in mathematics, following calculus and ordinary differential equations. The main content was to turn variational problems into problems of solving differential equations. However, since only a few differential equations have explicit solutions, this limitation hindered the study. Since the second half of the 20th century, as a course of its own, calculus of variations has been condensed or merged into other related courses gradually.

Nevertheless calculus of variations has close connections to many other branches of mathematics. Among the 23 problems proposed by Hilbert, three of which deal with variational problems, its importance is evident. Variational problems arise naturally from mechanics, physics, economics, operation research, and engineering, etc. In particular, since the 1970s, finite element methods and optimization techniques have provided numerical solutions to variational problems, thereby elevating its status in the realm of applied mathematics.

In the past few decades, great development has taken place in both theoretical and applied aspects of calculus of variations. It is noted by the mathematical community that calculus of variations is quickly becoming a necessity in undergraduate mathematics education, without it, students would struggle with new demands of modern science and technology. However, there is no unanimous agreement so far on how to remedy this shortcoming, it will take some time to explore. This book attempts to bring the readers up to date in this subject area.

In the academic years between 2006 and 2010, the author taught the course entitled "Calculus of Variations" to both advanced undergraduate students and beginning level graduate students in mathematics at the School of Mathematical Sciences, Peking University. The organization of the course content was based on the following three principles:

1. The lectures should not only introduce the classical theory but also the modern development of calculus of variations. Furthermore, more in-depth studies were conducted stemming from various research problems.

2. The course should put its main emphasis on the most frequently used theorems as well as techniques.

3. The course should welcome a large audience, including students in the area of pure mathematics, numerical mathematics, mathematical statistics, information science, and financial mathematics.

The prerequisites for this course were Mathematical Analysis, Modern Algebra, and Analytic Geometry; in addition, students were expected to be somewhat familiar with Ordinary Differential Equations, Real Analysis, Functional Analysis, Differential Geometry, and Mathematical Physics.

The entire course was divided into three sections: classical theory of calculus of variations, the existence and regularity of solutions, and special topics; whereas the latter played a role in the development of modern day calculus of variation. This book not only introduces fundamental concepts, basic theorems, and techniques used in calculus of variations, but reinforces them with abundant examples and counterexamples as well. In particular, it sheds new light on the definitions, their interplays and compatibilities with existing theorems and methods. Exercises are given at the end of most lectures.

This book is based on these lectures notes.

Due to the experimental nature of the book, I sincerely welcome any input and critique on how to improve it.

I am very grateful to Tian Fu Zhao and Shan Nian Lu from the Higher Education Publishing Company. Their careful proofreading and comments are greatly appreciated.

Kung Ching Chang
Peking University
December 2010

Contents

Lecture 1

The theory and problems of calculus of variations

1.1 Introduction

Calculus of variations is an important part of mathematical analysis, it is closely related to many other branches of mathematics and has numerous applications; some examples include:

⋄ Many important equations in mathematical physics, differential equations in elastic and plastic mechanics, biomembrane equation, and differential equations arising from geometry are some particular Euler equations for certain functionals.

⋄ Optimal control problems are often different kinds of variational problems with constraints, which appear in both engineering and economics.

Variational problems also occur frequently in intelligence material, image processing, and optimal designs.

⋄ Variational method is the main tool in establishing the existence of solutions for elliptic partial differential equations, it has since become an integral part of the study of partial differential equations. The intimate connections between calculus of variations and partial differential equations are readily seen by Hilbert's 19th and 20th problems.

⋄ Numerical methods used in partial differential equations, in particular, the finite element method comes directly from variational structures. The rapid development of optimization techniques has made it feasible to numerical solutions for the extremal values of variational problems.

⋄ The interplay between topology and the calculus of variations has led to a new brach in mathematics — global analysis. It has vastly accelerated to the advancement of critical point theory. In particular, Morse theory reveals the interplay between analysis and topology, it has since become a core topic in differential topology. In the same vein, Floer homology has also emerged as their heir.

◇ Calculus of variations has seeped through to many other areas, such as Riemannian Geometry, Finsler Geometry, Symplectic Geometry, and Conformal Geometry, etc. Many variational problems with rich geometric background, such as geodesics, minimal surfaces, harmonic maps, etc. have stimulated research interests in finding new theory (for example, geometric measure theory), new methods, and new techniques.

◇ Variational methods play an important role in the study of periodic orbits of Hamiltonian dynamical systems, Mather set, and chaos.

◇ Malliavin Calculus, the stochastic calculus of variations is an interplay between differential calculus and probability theory. It has become an essential part of financial mathematics.

It is clear that variational problems, theory, and methods have profound influence on various areas of modern mathematics, including pure mathematics, applied mathematics, numerical mathematics, information science, and mathematical economics, etc. No doubt, calculus of variations has taken the center stage of modern mathematics.

Comparing to some classical textbooks (for examples [LL], [Ka], [GF]), this volume offers the following unique features:

◇ While introducing the classical theory of calculus of variations, emphasis is put on the first order as well as second order conditions of a minimum.

Since many examples in partial differential equations, differential geometry, and mathematical physics involve several variables, we give more extensive discussions on multivariate cases.

◇ In the classical theory of calculus of variations, we strengthen the Hamilton–Jacobi theory and the conservation law, since they are very useful content in physics and geometry.

◇ Aside from the classical theory of calculus of variations, we also emphasize the direct methods and their applications. The direct methods are the main body of modern calculus of variations; it is the foundation for establishing the existence of solutions of differential equations and their numerical solutions. This approach is accessible to students with previous exposure to functional analysis, and it constitutes for nearly half of the material in this volume.

◇ Eigenvalue problem is one of the central problems in analysis. Our treatment is to present them as an application of solving constrained optimization problems, which mirrors nicely with corresponding topics in functional analysis.

Furthermore, we conduct careful investigation on some special topics, which may be considered optional.

◇ Critical point theory is the fastest growing branch of calculus of variations in the past few decades, it has wide applications. It is particularly important in proving the existence of solutions of differential equations. This is a very rich area, we will only introduce one of the simplest results — the Mountain Pass Theorem, as the first introductory step to this exciting subject.

◇ Periodic solutions of Hamiltonian systems, homoclinic orbits, and hetero-clinic orbits are heated topics in dynamical systems and symplectic geometry. Under certain conditions, the existence of solutions may be obtained via variational methods.

◇ Both geodesics and minimal surfaces are simple geometric examples of variational nature, this lecture may be regarded as an introduction to geometric analysis.

◇ Finite element methods and optimization techniques are two commonly used numerical solutions to variational problems. However, it is worth noting that the theoretical background of finite element method is built upon calculus of variations.

◇ Some additional topics with real life applications are presented toward the end of the book, for instance, some optimal control problems and problems from image processing. They are independent of each other and yet help the readers to appreciate this subject in our modern society.

There are a total of 20 lectures. The first eight lectures are considered classical calculus of variations; Lectures 9 through 14 introduce direct methods. Together they are the main focus of the book. Lectures 15 through 20 are special topics, they may be optional material. Topics with * may be omitted during the first reading of this volume.

1.2 Functionals

Calculus of variations examines the extremal values (or more generally, the critical values) of functionals.

Generally speaking, a functional is a mapping from any set M to the field of real numbers $\mathbb{R}$ or the field of complex numbers $\mathbb{C}$. However, in calculus of variations, a functional will only take values in $\mathbb{R}$, whose domain M is a set of functions, i.e. $I : M \to \mathbb{R}$.

For example, let $\Omega \subset \mathbb{R}^n$ be an open set, $x_0 \in \Omega$ be a fixed point, $F \in C(\bar{\Omega})$, and $M = C^1(\bar{\Omega})$.

$$I_1(u) = \max_{x \in \Omega} |u(x)|,$$

$$I_2(u) = u(x_0),$$

$$I_3(u) = \int_\Omega [|\nabla u(x)|^2 - F(u(x))]dx$$

are all functionals. However, regardless of the choices of M and the single variable function f, the composite function

$$I_4(u) = f(u(x))$$

is not a functional.

Given a function $L \in C^1(\Omega \times \mathbb{R}^N \times \mathbb{R}^{nN})$, one mainly considers the following functional:

$$I(u) = \int_\Omega L(x, u(x), \nabla u(x))dx,$$

where M is a subset of $C^1(\bar{\Omega})$, the set of continuously differentiable functions on $\bar{\Omega}$. Sometimes, M could also be a subset of differentiable functions in some generalized sense, given by certain prescribed constraints (such as integral form boundary conditions, pointwise boundary conditions, and boundary conditions with or without differentials).

Occasionally, the integral expression of I may also contain higher order derivatives terms, the set M should be modified accordingly.

1.3 Typical examples

Example 1.1 (The line of steepest descent) Given two points $A = (x_1, y_1)$ and $B = (x_2, y_2)$ in the xy-plane, where $x_1 < x_2$ and $y_1 > y_2$. A particle is free falling along a smooth curve joining A and B. Assuming the initial velocity of the particle is zero, what trajectory would be the fastest to travel from A to B? (see Figure 1.1)

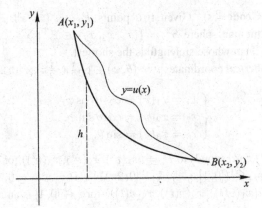

Fig. 1.1

Assume $u \in C^1([x_1, x_2])$, let $\{(x, u(x)) \mid x \in [x_1, x_2], u(x_i) = y_i, i = 1, 2\}$ be a curve connecting A and B. Since

$$\begin{cases} \dfrac{1}{2}mv^2 = mgh, \\[2mm] v = \dfrac{ds}{dt}, \end{cases}$$

we have

$$v = \sqrt{2g(y_1 - u(x))}$$

and

$$dt = \frac{ds}{v} = \sqrt{\frac{1 + |u'(x)|^2}{2g(y_1 - u(x))}} dx.$$

The total time is

$$T = \int_{x_0}^{x_1} dt = \frac{1}{\sqrt{2g}} \int_{x_0}^{x_1} \sqrt{\frac{1 + |u'(x)|^2}{(y_1 - u(x))}} dx.$$

Let

$$M = \{u \in C^1([x_1, x_2])) \mid u(x_i) = y_i, i = 1, 2\},$$

then the mapping $M \to \mathbb{R}$ via $u \mapsto T$ is a "functional". Here u is the independent variable, $T = T(u)$ is the dependent variable, and we wish to find $u \in M$ such that T attains its minimum.

Example 1.2 (Geodesics) Given two points $P_0 = (x_1^0, x_2^0, x_3^0)$ and $P_1 = (x_1^1, x_2^1, x_3^1)$ on the unit sphere $S^2 = \{(x_1, x_2, x_3) \in \mathbb{R}^3 \mid \sum_{i=1}^3 x_i^2 = 1\}$, find the path joining them whose arclength is the shortest.

We adopt spherical coordinates $v = (\theta, \varphi) \in [-\frac{1}{2}\pi, \frac{1}{2}\pi] \times [0, 2\pi)$ on S^2 such that

$$\begin{cases} x_1 = x_1(v) = \cos\theta\cos\varphi, \\ x_2 = x_2(v) = \cos\theta\sin\varphi, \\ x_3 = x_3(v) = \sin\theta, \end{cases}$$

and $v^i = (\theta^i, \varphi^i)$ corresponds to $P_i = (x_1(v^i), x_2(v^i), x_3(v^i))$ for $i = 0, 1$.

Let $M = \{v \in C^1([0,1], [-\frac{\pi}{2}, \frac{\pi}{2}] \times [0, 2\pi)) \mid v(i) = v^i, i = 0, 1\}$, then for all $v \in M$, $u(t) = (x_1(v(t)), x_2(v(t)), x_3(v(t)))$ for $t \in [0,1]$ is an arc connecting P_0 and P_1.

The square of the line element is

$$ds^2 = dx_1^2 + dx_2^2 + dx_3^2 = d\theta^2 + \cos^2\theta \, d\varphi^2 = (\theta'(t)^2 + \cos^2\theta(t)\varphi'(t)^2)dt^2.$$

Hence, the arclength $L : M \to \mathbb{R}$ is given by

$$L(u) = \int_0^1 |ds| = \int_0^1 \sqrt{(\theta'(t)^2 + \cos^2\theta(t)\varphi'(t)^2)}dt.$$

The arclength L is now a functional of the parametric function $v(t) = (\theta(t), \varphi(t))$, and we wish to find $v \in M$ such that $L(v)$ is a minimum.

Example 1.3 (Minimal surfaces) Given a Jordan curve Γ in $\mathbb{R}^3$, is it possible to find a surface S bounded by Γ whose area is a minimum?

We give the following parametrization of $S : (u, v) \mapsto Z = (x, y, z)$ as a map from $\bar{D} \to \mathbb{R}^3$, where $D \subset \mathbb{R}^2$ is the unit circle, $u^2 + v^2 \leq 1$ such that

$$\begin{cases} x = x(u, v), \\ y = y(u, v), \\ z = z(u, v). \end{cases}$$

The area $A(Z)$ of S is given by

$$A(Z) = \int_D |Z_u \times Z_v| du dv$$

$$= \int_D \sqrt{(x_u y_v - y_u x_v)^2 + (y_u z_v - z_u y_v)^2 + (z_u x_v - x_u z_v)^2} du dv.$$

The area A is therefore a functional of the parametric function Z, and we wish to minimize A under the condition that $Z|_{\partial D}$ is homeomorphic to ($\simeq$) Γ. Namely, we want to find the vector-valued function $Z(u, v) = (x(u, v), y(u, v), z(u, v)) \in M$, where $M = \{Z \in C^1(\bar{D}, \mathbb{R}^3) \mid Z|_{\partial D} \simeq \Gamma\}$ such that $A(Z)$ is a minimum.

Example 1.4 (Eigenvalue problems and inequalities) Given a bounded region $\Omega \in \mathbb{R}^n$. For all $u \in H_0^1(\Omega)$, where $H_0^1(\Omega)$ denotes the Sobolev space with zero boundary condition (for details, see Lecture 10), we define the energy

$$E(u) = \int_\Omega |\nabla u(x)|^2 dx$$

and the constraint

$$G(u) = \int_\Omega |u(x)|^2 dx = 1.$$

We define

$$M = \{u \in H_0^1(\Omega) \mid G(u) = 1\}$$

and we wish to find a function u_1 such that the functional $E : M \to \mathbb{R}$ attains its minimum at u_1. Furthermore,

$$\lambda_1 = \min\{E(u) \mid G(u) = 1\}$$

is called the first eigenvalue with the corresponding first eigenfunction u_1. The eigenpair has great importance in geometry, physics, and engineering.

Often times, inequalities in analysis and geometry can be posed as a variational problem. For instance, the Sobolev's inequality asserts

$$\int_{\mathbb{R}^N} |\nabla u(x)|^2 dx \geq S_N \left(\int_{\mathbb{R}^N} |u(x)|^{\frac{2N}{N-2}} dx \right)^{\frac{n-2}{N}},$$

where

$$S_N = N(N-2)\pi \left(\frac{\Gamma(N)}{\Gamma(N/2)} \right)^{-2/N}.$$

We may turn this into a variational problem as follows. Let

$$M = \left\{ u \in H_0^1(\mathbb{R}^N) \; \middle| \; \int_{\mathbb{R}^N} |u(x)|^{\frac{2N}{N-2}} dx = 1 \right\}$$

and

$$E(u) = \int_{\mathbb{R}^N} |\nabla u(x)|^2 dx,$$

we now find $u \in M$ such that the functional E has a minimum. If the minimal value is S_N, then S_N is the best constant for the above inequality.

Example 1.5 (Vibrations of thin plates) To study the vibrations of a thin elastic plate (thin means the ratio between the thickness h and the minimal span a of the plate satisfies $h/a \ll 1$; the plate is assumed to be homogeneous and isotropic) under exterior force. In continuum mechanics, Kirchnoff proposed the so-called "assumption of straight line", i.e. straight lines normal to the plate remains straight and no stretch nor strain on the plate occurs during deformation.

Suppose a plane region Ω represents a thin plate, its density function is $\rho(x, y)$. Let $w(x, y)$ be the displacement of $(x, y) \in \Omega$. The stress–strain relation can then be expressed via the potential energy density, which depends on the Hessian matrix of $w(x, y)$:

$$\begin{pmatrix} w_{xx} & w_{xy} \\ w_{yx} & w_{yy} \end{pmatrix}.$$

Since all physical quantities remain independent of the coordinates chosen, the potential energy density can only depend on the quantities

$$w_{xx} + w_{yy} \quad \text{and} \quad w_{xx}w_{yy} - w_{xy}^2.$$

Let $f(x, y)$ be the density of the stress put on Ω, the resulting potential energy density is only considered exterior work done. If we ignore the stress on $\partial\Omega$, the boundary of Ω and the bending on the boundary, the total potential energy is then given by

$$U(w) = \int_\Omega \left\{ \frac{1}{2}[(w_{xx}+w_{yy})^2 - 2(1-\mu)(w_{xx}w_{yy}-w_{xy}^2)] + f(x, y)w(x, y) \right\} dxdy,$$

where μ is determined by the material of the plate itself, known as the Poisson ratio.

Suppose now we fix the boundary of the thin plate, i.e. $w|_{\partial\Omega} = \varphi(x)$ with φ a given function on $\partial\Omega$. Let $M = \{w \in C^2(\bar{\Omega}) \mid w|_{\partial\Omega} = \varphi(x)\}$, then U is a functional of the displacement function $w \in M$.

From the principles of mechanics, we know that the equilibrium of the plate must obey variational principle. In other words, the potential energy functional U of the plate achieves its minimum at the displacement w where the plate is balanced.

1.4 More examples

Unlike the above tranditional examples, certain problems may not appear to be related to problems of finding extremal values of functionals at first, however, by proper modifications, they can be transformed into variational problems.

Example 1.6 (Reinvestment) We assume the production rate of certain goods at time t is $q = q(t)$, the growth rate $\dot{q}$ is proportional to the percentage of reinvestment $u(t)$ at time t, i.e.

$$\dot{q} = \alpha u q,$$

where $\alpha > 0$ is a constant.

Consequently, the total number of goods produced on the time interval $[0, T]$ is given by

$$J(u, q) = \int_0^T (1 - u(t)) q(t)\, dt.$$

Given the initial value $q(0) = q_0$, the question to consider is: how to choose the reinvestment percentage $u(t)$ such that the total product J is a maximum?

Once we set

$$M = \{(u, q) \in C[0, T] \times C^1[0, T] \mid 0 \le u(t) \le 1, \dot{q} = \alpha u q, q(0) = q_0\},$$

the mapping $(u, q) \mapsto J(u, q)$ becomes a functional on M. The question is to find $(u, q) \in M$ such that J achieves its maximum. This is an optimal control problem, where the roles of u and q are unequal. We call u the control variable, q the state variable and J the target functional.

In dealing with variational problems, we often refer to the variables as "functions", but the meaning of "functions" can be quite broad.

Example 1.7 (Image segmentation) Imagine the following scenario: a picture of someone (with possible damage) is presented to us and we are asked to detect the edge of the image of the person.

Suppose the background of the picture occupies a plane region $\Omega \subset \mathbb{R}^2$ and $g : \Omega \to \mathbb{R}^1$ is its image. We wish to find a function $u : \Omega \to \mathbb{R}^1$ such that u is the best fit for g along the edge of the image, without leaving special marks elsewhere.

In order to describe the edge of the image, we introduce a closed subset K of $\bar{\Omega}$ with finite one dimensional Hausdorff measure $H^1(K)$. We define

$$I(K, u) = \int_{\Omega \setminus K} |\nabla u|^2 dx dy + \mu \int_{\Omega \setminus K} |u - g|^2 dx dy + \lambda H^1(K),$$

where $\lambda, \mu > 0$ are adjustable parameters.

It is worth noting that I not only depends on the function u, but also on the closed subset K. Although the variable K is not a function, we may replace it by its characteristic function

$$\chi_K(x) = \begin{cases} 1, & x \in K \\ 0, & x \in \Omega \setminus K \end{cases}$$

and I becomes a functional depending on χ_K and u. We may henceforth use K and χ_K interchangeably.

We can thus turn this image segmentation problem into an extremal problem for the functional I. For this reason, we define the set

$$M = \left\{ (K, u) \mid K \subset \Omega \text{ is closed}, H^1(K) < \infty, \int_\Omega (|\nabla u|^2 + |u|^2) dxdy < \infty \right\},$$

where ∇u should be understood in the sense of distribution. Our goal is to find $(K, u) \in M$ such that $I : M \to \mathbb{R}$ achieves its minimum.

Example 1.8 (Harmonic mappings) Assume (M, g) and (N, h) are two compact Riemannian manifolds. Given a mapping $u \in C^1(M, N)$, its differential $du \in \Gamma(T^*M \times u^{-1}TN)$ is a cross section of the product bundle, where $u^{-1}TN$ is a bundle over M with metric $h \circ u$ and T^*M is the cotangent bundle on M. In local coordinates: $x = (x^1, \ldots, x^m)$ and $u = (u^1, \ldots, u^n)$, we have:

$$du = \frac{\partial u^i}{\partial x^\alpha} dx^\alpha \otimes \frac{\partial}{\partial u^i}.$$

We define the energy density to be

$$L(u, du) = \frac{1}{2}|du(x)|_h^2 = \frac{1}{2} \sum_{i,j=1}^n \sum_{\alpha,\beta=1}^m h_{ij}(u(x)) g^{\alpha\beta}(x) \frac{\partial u^i}{\partial x^\alpha} \frac{\partial u^j}{\partial x^\beta}.$$

Let dV_g denote the volume element of M, then the total energy

$$E(u) = \int_M L(u(x), du(x)) dV_g$$

is a functional of u.

The mapping u itself can also be regarded as a "function". If we take $M = C^1(M, N)$, then $E : M \to \mathbb{R}$ is a functional. Those mappings $u \in M$ for which E attains its minimal value are called harmonic mappings. They play an important role in differential geometry.

Example 1.9 (Hamiltonian system) Given a function $H \in C^1(\mathbb{R}^1 \times \mathbb{R}^n \times \mathbb{R}^n)$. We introduce the following notation: $x = (x_1, \ldots, x_n)$, $p = (p_1, \ldots, p_n)$, $(t, x, p) \in \mathbb{R}^1 \times \mathbb{R}^n \times \mathbb{R}^n$, and $(x, p)_{\mathbb{R}^n} = \sum_{i=1}^n x_i p_i$.

When both x and p are (vector-valued) functions of time t, the system of ordinary differential equations

$$\begin{cases} \dot{x} = H_p(t, x, p), \\ \dot{p} = -H_x(t, x, p) \end{cases}$$

is called a Hamiltonian system.

Let M be the collection of all functions $H \in C^1(\mathbb{R}^1 \times \mathbb{R}^n \times \mathbb{R}^n)$ such that the functions as well as their derivative functions are integrable on $\mathbb{R}^1$, we impose certain conditions on H such that

$$I(x,p) = \int_{\mathbb{R}^1} [(\dot{x}, p)_{\mathbb{R}^n} - H(t, x, p)]dt$$

is a functional on M. Clearly, I is neither bounded above nor below, hence it has no extremal values. However, in our subsequent lectures, we shall find out for certain choices of M, those pairs $(x(t), p(t))$ for which I attains its critical values (or stationary values) are precisely solutions of the Hamiltonian system.

Example 1.10 (Einstein field equation) In the theory of general relativity, it is customary to use the signature $(1, 3)$ Minkowski metric (g_{ij}) on $\mathbb{R}^4$ to describe the gravitational field. The Minkowski metric is a non-degenerate symmetric bilinear form with sign convention $(1, -1, -1, -1)$ and whose line element is given by

$$ds^2 = \sum_{i,j} g_{ij} dx^i dx^j.$$

We designate the Euclidean coordinate by (x, y, z) and the time coordinate by t, then $(x^0, x^1, x^2, x^3) = (ct, x, y, z)$, where c is the speed of light. Using an inertia frame, we have:

$$ds^2 = (dx^0)^2 - (dx^1)^2 - (dx^2)^2 - (dx^3)^2.$$

We use the Christoffel symbols Γ^i_{jk} to compute the curvature tensor to be

$$R_{iklm} = \frac{1}{2}\left(\frac{\partial^2 g_{im}}{\partial x^k \partial x^l} + \frac{\partial^2 g_{kl}}{\partial x^i \partial x^m} - \frac{\partial^2 g_{il}}{\partial x^k \partial x^m} - \frac{\partial^2 g_{km}}{\partial x^i \partial x^l}\right) + \sum_{n,p} g_{np}(\Gamma^n_{kl}\Gamma^p_{im} - \Gamma^n_{km}\Gamma^p_{il}),$$

the Ricci tensor is given by

$$R_{ik} = \sum_l \left(\frac{\partial \Gamma^l_{ik}}{\partial x^l} - \frac{\partial \Gamma^l_{il}}{\partial x^k}\right) + \sum_{l,m}(\Gamma^l_{ik}\Gamma^m_{lm} - \Gamma^m_{il}\Gamma^l_{km})$$

and the scalar curvature is

$$R = \sum_{i,k} g^{ik} R_{ik} = \sum_{i,k}\sum_{l,m} g^{il} g^{km} R_{iklm}.$$

Einstein introduced the following decomposition of $S = S(g)$:

$$S = S_g + S_m,$$

where

$$S_g = \int R d\Omega, \qquad d\Omega = \sqrt{|\det(g_{ij})|} d^4 x$$

is called the Einstein–Hilbert action field, it represents the contribution of the gravitational field without any matter. The quantity S_m is given by

$$S_m = \frac{1}{c} \int \Lambda \sqrt{-\det(g_{ij})} d^4 x,$$

where Λ is a function determined by the matter and metric, it measures the contribution of the gravitational field.

Note that the metric g is a variable, also a function taking its values in non-degenerate real symmetric 4×4 matrices of signature $(1, 3)$, whereas $S = S(g)$ is regarded as a functional of g.

The motion of matter obeys variational principles, hence the gravitational field-Minkowski metric helps to stabilize S.

In summary, variational problems have very rich content. From classical mechanics to gauge theory, the laws of motion follow variational principles. The reflection of a light beam, the bending surface of fluid in capillary, and soap bubbles are also examples of natural phenomena which abide by variational principles. Furthermore, from engineering designs to social-economic life, we are continuously challenged by problems seeking for maximal speed, maximal distance, minimal consumption, optimal shape, the lightest weight, optimal revenue, and the most crisp image, etc. all of which ultimately lead to solving extremal value problems of functionals.

Lecture 2

The Euler–Lagrange equation

2.1 The necessary condition for the extremal values of functions — a review

Before we begin our study on the necessary condition for the extremal values of a functional, we shall first review the necessary condition for the extremal values of a real-valued function. Let $\Omega \subset \mathbb{R}^n$ be an open set. Suppose the function $f \in C^1(\Omega)$ attains its (local) minimum at some point $x_0 \in \Omega$, i.e. there exists an open neighborhood $U \subset \Omega$ of x_0 such that

$$f(x) \geq f(x_0), \quad \forall\, x \in U.$$

Hence, $\forall\, h \in \mathbb{R}^n \setminus \{\theta\}$, where θ denotes the zero vector in $\mathbb{R}^n$, $\exists\, \varepsilon(h) > 0$ such that whenever $0 < |\varepsilon| < \varepsilon(h)$, $x_0 + \varepsilon h \in U$ and

$$f(x_0 + \varepsilon h) \geq f(x_0),$$

i.e.

$$\frac{1}{\varepsilon}[f(x_0 + \varepsilon h) - f(x_0)] \geq 0.$$

Letting $\varepsilon \to 0$, it yields

$$(\nabla f(x_0), h)_{\mathbb{R}^n} = 0,$$

where $(\cdot, \cdot)$ denotes the standard inner product on $\mathbb{R}^n$. Since $h \in \mathbb{R}^n \setminus \{\theta\}$ is arbitrary, it follows that $\nabla f(x_0) = \theta$; a necessary condition for x_0 to be a local minimum of f.

2.2 The derivation of the Euler–Lagrange equation

In this section, we only consider the case of a functional depending on a single variable function (it may be vector-valued). For the multivariate cases, we give detailed discussions in Lecture 6.

Given an interval $J = [t_0, t_1] \subset \mathbb{R}^1$ and an open subset $\Omega \subset \mathbb{R}^N$. For a given continuously differentiable function $L = L(x, u, p)$, $L \in C^1(J \times \Omega \times \mathbb{R}^N, \mathbb{R}^1)$ together with two points $P_0, P_1 \in \Omega$, we define the set

$$M = \{u \in C^1(J, \Omega) \mid u(t_i) = P_i, \ i = 0, 1\}$$

and the functional I on M via

$$I(u) = \int_J L(t, u(t), \dot{u}(t)) dt.$$

We call u^* a minimum of I on M if there exists an open neighborhood U of x^* in M with respect to the $C^1(J, \Omega)$ topology such that

$$I(u) \geq I(u^*), \quad \forall u \in M \cap U.$$

Assuming the existence of u^*, we shall determine the necessary condition for which I attains its minimum at u^*.

Similar to the extremal value problems of functions, $\forall \ \varphi \in C_0^1(J, \mathbb{R}^N)$ ($C_0^1(J, \mathbb{R}^N)$ is the closure of $C_0^\infty(J, \mathbb{R}^N)$ with respect to the $C^1(J, \Omega)$ topology), $\exists \ \varepsilon = \varepsilon(\varphi) > 0$ such that whenever $0 < |\varepsilon| < \varepsilon(\varphi)$, $u^* + \varepsilon\varphi \in U$ and

$$I(u^* + \varepsilon\varphi) - I(u^*) \geq 0.$$

It follows that

$$\delta I(u^*, \varphi) = \lim_{\varepsilon \to 0} \frac{1}{\varepsilon}(I(u^* + \varepsilon\varphi) - I(u^*)$$

$$= \int_J \sum_{i=1}^N [L_{u^i}(t, u^*(t), \dot{u}^*(t))\varphi^i(t) + L_{p^i}(t, u^*(t), \dot{u}^*(t))\dot{\varphi}^i(t)] dt$$

$$= \int_J (\sum_{i=1}^N \int_{t_0}^t L_{u^i}(s, u^*(s), \dot{u}^*(s)) ds - L_{p^i}(t, u^*(t), \dot{u}^*(t))) \dot{\varphi}^i(t) dt$$

$$\geq 0,$$

$\forall \varphi \in C_0^1(J, \mathbb{R}^N)$. We call $\delta I(u^*, \varphi)$ the *first order variation* of I with respect to φ.

If we replace $\varepsilon > 0$ by $\varepsilon < 0$, it is equivalent to replacing φ by $-\varphi$ in the above equation. Hence, $\forall \ \varphi \in C_0^1(J, \mathbb{R}^N)$, we have:

$$\boxed{\int_J \left(\sum_{i=1}^N \int_{t_0}^t L_{u^i}(s, u^*(s), \dot{u}^*(s)) ds - L_{p^i}(t, u^*(t), \dot{u}^*(t)) \right) \dot{\varphi}^i(t) dt = 0.}$$

In order to better understand this integral in relation to u^*, it is desirable to remove the arbitrary function φ. To do so, we shall need the following lemma.

Lemma 2.1 (du Bois–Reymond) *If $\psi \in C[t_0, t_1]$ satisfies*

$$\int_J \psi(t) \cdot \dot{\lambda}(t) dt = 0, \qquad \forall \lambda \in C_0^1(J),$$

where $C_0^1(J) = C_0^1(J, \mathbb{R}^1)$, which is equal to $\{u \in C^1(J) \,|\, u(t_0) = u(t_1) = 0\}$, then ψ is a constant.

Proof Let $c = \frac{1}{|J|} \int_J \psi(t) dt$ and $\lambda(t) = \int_{t_0}^t (\psi(s) - c) ds$, then $\lambda \in C_0^1(J)$. Thus,

$$\int_J (\psi(t) - c)^2 dt = \int_J \psi(t)(\psi(t) - c) dt = \int_J \psi(t) \cdot \dot{\lambda}(t) dt = 0.$$

By continuity, ψ must be a constant. □

This leads to

Theorem 2.1 *Assume $u^* \in M$ is a minimizer of the functional I, it must satisfy the integral form of the Euler–Lagrange equation (for short E-L equation henceforth):*

$$\boxed{\int_{t_0}^t L_{u^i}(s, u^*(s), \dot{u}^*(s)) ds - L_{p^i}(t, u^*(t), \dot{u}^*(t)) = \text{const.}, \quad 1 \le i \le N, \ \forall t.}$$

Note that E-L equation is a necessary condition for u^* to be a minimizer of I on M, it needs not be sufficient. Moreover, the solutions of the E-L equation correspond to critical points of the functional I, they may be maxima, minima, or other extrema.

Remark 2.1 When $L \in C^1$ and $u \in C^1$, using the theory of distributions, the integral form of the E-L equation may also be written as

$$\boxed{-DL_p(t, u(t), \dot{u}(t)) + L_u(t, u(t), \dot{u}(t)) = 0,}$$

where D is the generalized derivative. We can define the Euler–Lagrange operator E_L as follows:

$$(E_L u)(t) = -DL_p(t, u(t), \dot{u}(t)) + L_u(t, u(t), \dot{u}(t)).$$

In particular, if $L \in C^2$ and $u \in C^2$, then the above expression is valid pointwise, hence we may simply replace D by the usual $\frac{d}{dt}$.

Remark 2.2 We can also relax the $C^1(J)$ requirement in Theorem 2.1. For example, we may consider Lipschitz functions on J, $\text{Lip}(J)$. Since a Lipschitz function

$u(t)$ is absolutely continuous on J, it has a derivative function $\dot{u}(t)$ almost everywhere on J. The integral used in the functional I can therefore be interpreted in the sense of a Lebesgue integral, and we let

$$M = \{u \in \text{Lip}(J, \Omega) \mid u(t_i) = P_i,\ i = 0, 1\}.$$

Note that in M, $\forall\, \delta > 0$, $U = \{v \in \text{Lip}(J) \mid |v(t) - u^*(t)| + |\dot{v}(t) - \dot{u}^*(t)| < \delta \text{ a.e.}, t \in J\}$ is a neighborhood of u^*.

The E-L equation still holds in the sense of Lebesgue integral for almost all $t \in J$:

$$\int_{t_0}^{t} L_{u^i}(s, u(s), \dot{u}(s))ds - L_{p^i}(t, u(t), \dot{u}(t)) = \text{const.}, \quad 1 \le i \le N.$$

In fact, since $\dot{u}$ is bounded almost everywhere (a.e.) on J, there exists a compact neighborhood $W \subset \Omega \times \mathbb{R}^N$ of $(u(t), \dot{u}(t))$ such that the derivative of L is bounded a.e. on $J \times W$. Hence, we have:

$$\delta I(u^*, \varphi) = \lim_{s \to 0} \frac{1}{s}(I(u^* + s\varphi) - I(u^*))$$

$$= \lim_{s \to 0} \frac{1}{s} \int_J [(L(t, u^*(t) + s\varphi(t), \dot{u}^*(x) + s\dot{\varphi}(x)) - L(t, u^*(t), \dot{u}^*(t))]dt$$

$$= \int_J \sum_{i=1}^{n} [L_{u^i}(t, u^*(t), \dot{u}^*(t))\varphi^i(t) + L_{p^i}(t, u^*(t), \dot{u}^*(t))\dot{\varphi}^i(t)]dt.$$

To see this, we use the fact that the difference inside the above integral is uniformly bounded, hence by the Lebesgue's Dominant Convergence Theorem, we can pass the limit inside the integral.

Aside from this, in the du Bois–Reymond Lemma, we may replace the requirement $\psi \in C^1(J)$ by $\psi \in L^\infty(J)$ and $\lambda \in C_0^1(J)$ by $\lambda \in \text{AC}_0(J)$, the space of all absolutely continuous functions on J which vanish on the boundary of J.

Remark 2.3 A piecewise C^1 continuous function is a Lipschitz function. By a piecewise C^1 continuous function u, we mean there exists a finite set $D = \{a_1, \ldots, a_k\}$ such that $u \in C^1(J \setminus D)$ and $\dot{u}(a_i \pm 0)$ exists for $1 \le i \le k$. In this case, the integral form of the E-L equation continues to hold for the class of piecewise C^1 continuous functions.

Many fundamental equations in mechanics and geometry are E-L equations. We have the following examples:

Example 2.1 (The displacement of a moving particle) A force F is put on a particle of mass m. Assume the particle's displacement coordinate is given by $x = (x_1, x_2, x_3) \in \mathbb{R}^3$ with $|x|^2 = x_1^2 + x_2^2 + x_3^2$, the velocity $v = \dot{x}$ and kinetic energy $T = \frac{1}{2}mv^2$. Suppose F is of a potential, i.e. there exists a function V such that $-\nabla V = F$, we then call the function

$$L = T - V = \frac{1}{2}mv^2 - V = \frac{1}{2}m|\dot{x}|^2 - V(x)$$

the Lagrangian. On a properly chosen domain M, we consider the functional

$$I(x) = \int_{t_1}^{t_2} L(x(t), \dot{x}(t))dt.$$

A minimizer $x(t)$ of I satisfies the E-L equation

$$F = m\ddot{x},$$

which is exactly the orbit governed by Newton's second law of motion.

Given a collection of particles with degree of freedom n, we denote the displacement coordinate by $q = (q_1, \ldots, q_n)$, it follows that
- the kinetic energy is given by $T = \frac{1}{2}\sum_{i,j=1}^{n} a_{ij}\dot{q}_i\dot{q}_j$, where (a_{ij}) is a positive definite matrix,
- the potential is given by $V = V(q_1, \ldots, q_n)$,
- the Lagrangian is given by $L = T - V$,
- the functional is given by

$$I(q) = \int_{t_0}^{t_1} L(q(t), \dot{q}(t))dt = \int_{t_0}^{t_1}\left(\frac{1}{2}\sum_{i,j=1}^{n} a_{ij}\dot{q}_i\dot{q}_j - V(q_1, \ldots, q_n)\right)dt.$$

The derived E-L equations are

$$\sum_{j=1}^{n} a_{ij}\ddot{q}_j = -\frac{\partial V}{\partial q_i} \quad (i = 1, 2, \ldots, n),$$

which are also Newtonian equations.

Example 2.2 (Geodesics) Suppose $(\tilde{M}, g)$ is a Riemannian manifold equipped with a Riemannian metric $g_{ik}(u)$, where (g_{ik}) is an $N \times N$ positive definite matrix. Given two points P_1 and P_2 belonging to the same coordinate chart $U \subset \tilde{M}$. U is homeomorphic to an open subset in $\mathbb{R}^N$. We then define $L : \mathbb{R}^1 \times \mathbb{R}^N \to \mathbb{R}^1$ by

$$L(u, p) = \sum_{i,j=1}^{N} g_{ij}(u)p_i p_j.$$

Let $M = \{u \in C^1(J, U) \mid u(i) = P_i\}$, for the functional

$$I(u) = \int_J L(u(x), \dot{u}(x))dx,$$

its E-L equation looks like

$$\frac{d}{dt}\sum_j\{g_{ij}(u)\dot{u}^j\} = \frac{1}{2}\sum_{j,k}g_{jk,i}(u)\dot{u}^j\dot{u}^k,$$

where

$$g_{jk,i}(u) = \frac{\partial}{\partial u^i}g_{jk}(u).$$

Consequently,

$$\sum_j g_{ij}(u)\ddot{u}^j + \sum_{j,k}(g_{ij,k}(u)\dot{u}^j\dot{u}^k - \frac{1}{2}g_{kj,i}\dot{u}^j\dot{u}^k) = 0.$$

Note that

$$\sum_{j,k}g_{lj,k}(u)\dot{u}^k\dot{u}^j = \sum_j\frac{d}{dt}g_{lj}(u)\dot{u}^j = \sum_k\frac{d}{dt}g_{lk}\dot{u}^k = \sum_{j,k}g_{kl,j}(u)\dot{u}^k\dot{u}^j.$$

By means of Christoffel symbols of the first kind

$$\Gamma_{jlk}(u) = \frac{1}{2}\{g_{lj,k} + g_{kl,j} - g_{jk,l}\},$$

we see that

$$\sum_j g_{ij}(u)\ddot{u}^j + \sum_{jk}\Gamma_{jik}(u)\dot{u}^k\dot{u}^j = 0.$$

Let (g^{ik}) be the inverse matrix of (g_{ik}), or using the Christoffel symbols of the second kind

$$\Gamma^i_{jk} = \sum_l g^{il}\Gamma_{jlk},$$

the E-L equation becomes

$$\ddot{u}^i + \sum_{j,k}\Gamma^i_{jk}(u)\dot{u}^j\dot{u}^k = 0 \quad \forall\, i.$$

This coincides with the geodesic equations in differential geometry.

In the following lectures, we will also use the E-L equation to derive many fundamental equations in physics.

• The variational derivative

Our earlier derivation of the E-L equation took place on the whole interval J. However, $\forall\, c \in \text{int}(J)$, the pointwise E-L equation only depends on the behavior of L near c. This means, it only depends on any open interval $(c - h, c + h) \subset \text{int}(J)$. Since every test function φ with support inside $(c - h, c + h)$ belongs to

$C_0^1(J, \mathbb{R}^N)$, by the arbitrariness of φ, it yields the E-L equation on $(c - h, c + h)$ (see Figure 2.1)

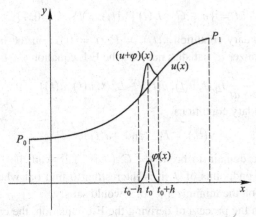

Fig. 2.1

Next, we examine such local behavior via limits. For simplicity, we shall take $N = 1$ and assume $L \in C^2$, $u \in C^2$. Note that

$$\lim \frac{I(u + \varphi) - I(u)}{\Delta \sigma} = -\lim \frac{\int_{c-h}^{c+h} \int_{t_0}^{t} E_L(u)(s)ds \dot{\varphi}(t)dt}{\Delta \sigma}$$

$$= \lim \frac{\int_{c-h}^{c+h} E_L(u)(t)\varphi(t)dt}{\Delta \sigma}$$

$$= E_L(u)(c),$$

where $\theta = \theta(t) \in (0, 1)$, the support of φ is contained in $(c - h, c + h)$, and $\Delta \sigma = \int_{c-h}^{c+h} \varphi(t)dt$ is the area of the sector bounded by the curves $u(t) + \varphi(t)$ and $u(t)$ for $t \in (c - h, c + h)$. It should also be noted that the limiting process is given by

$$h \to 0, \quad \sup_{t \in [c-h, c+h]} |\dot{\varphi}(t)| \to 0.$$

In light of the above calculations, we call the Euler–Lagrange operator of u at t

$$E_L(u)(t) = -\frac{d}{dt}L_p(t, u(t), \dot{u}(t)) + L_u(t, u(t), \dot{u}(t))$$

the variational derivative of I at t.

2.3 Boundary conditions

Recall in §2.1, a function u from the domain

$$M = \{u \in C^1(J, \Omega) \mid u(t_i) = P_i, \ i = 0, 1\}$$

satisfies the boundary condition $u(t_i) = P_i$ ($i = 0, 1$) on the interval J. This implies the minimizer u^* satisfies not only the E-L equation

$$-\frac{d}{dt} L_p(t, u(t), \dot{u}(t)) + L_u(t, u(t), \dot{u}(t)) = 0,$$

but also the boundary conditions

$$u(t_0) = P_0 \quad \text{and} \quad u(t_1) = P_1.$$

If we change the domain to be $M = C^1(J, \mathbb{R}^N)$ instead, i.e. we impose no condition on the endpoints of J, it is interesting to find out what equation and boundary condition the minimizer u^* of I would satisfy.

From §2.1, in the process of deriving the E-L equation, the crucial step is to choose a nearby function u of u^* and compare their functional values. To be more specific, we choose

$$u = u^* + \varepsilon\varphi$$

for $\varphi \in C_0^1(J, \mathbb{R}^N)$. Since φ vanishes at the endpoints of J, u and u^* share the same boundary values.

In our current setting, since there is no need to impose any boundary condition on φ, any $\varphi \in C^1(J, \mathbb{R}^N)$ would work. Suppose $u^* \in C^2(J, \mathbb{R}^N)$, using integration by parts, we see that

$$\delta I(u^*, \varphi)$$

$$= \int_J \left(\sum_{i=1}^n \int_{t_0}^{t_1} [L_{u^i}(s, u^*(s), \dot{u}^*(s)) ds - L_{p^i}(t, u^*(t), \dot{u}^*(t)) \right) \dot{\varphi}^i(t) dt$$

$$= -\int_J \left(\sum_{i=1}^n \int_{t_0}^{t_1} [L_{u^i}(t, u^*(t), \dot{u}^*(t)) - \frac{d}{dt} L_{p^i}(t, u^*(t), \dot{u}^*(t)) \right) \dot{\varphi}^i(t) dt$$

$$- \sum_{i=1}^N [L_{u^i}(t, u^*(t), \dot{u}^*(t)) \varphi^i(t)|_{t_0}^{t_1}].$$

Since $C_0^1(J, \mathbb{R}^N) \subset C^1(J, \mathbb{R}^N)$, we first choose an arbitrary $\varphi \in C_0^1(J, \mathbb{R}^N)$ to obtain the same E-L equation and then choose $\varphi \in C^1(J, \mathbb{R}^N)$ (with arbitrary boundary value). Since the first term of the right-hand side disappears and $\varphi^i(t_j)$ ($j = 0, 1, i = 1, \ldots, N$) in the second term is arbitrary, it must be the case where

$$L_{u^i}(t_j, u^*(t_j), \dot{u}^*(t_j)) = 0, \quad i = 1, \ldots, N, \ j = 0, 1.$$

Of course, there are many other choices for the set M; for example, one may choose to fix only one endpoint and leave the other one free; we may also impose different boundary conditions on the different components of a vector-valued function.

In some of our latter discussions, we will also encounter other types of boundary conditions, such as periodic and free boundary conditions.

Caution: In all of our previous discussions, we have always assumed all functions $u \in M$ are continuously differentiable. If we replace the continuously differentiable functions by piecewise C^1 functions, although the E-L equation remains the same (locally), it is however necessary to insert corner conditions at those points where the derivative function has a jump discontinuity (see Exercise 2.4).

2.4 Examples of solving the Euler–Lagrange equations

For $N = 1$, we shall consider the following special cases where the E-L equation can be simplified.

• **Case 1.** Suppose u is absent from L, then $L = L(t, p)$ and

$$\frac{d}{dt} L_p(t, \dot{u}(t)) = 0.$$

Since $L_p(t, \dot{u}(t)) = c$ is a first order equation without u, assuming we can solve for $\dot{u}$ (e.g. $L_{pp}(t, p) \neq 0$) to get

$$\dot{u}(t) = g(t, c),$$

then by integration, $u(t)$ is also solved.

Example 2.3 Let $M = \{u \in C^1([1, 2]) \mid u(1) = 0, u(2) = 1\}$ and

$$I(u) = \int_1^2 \sqrt{1 + \dot{u}^2} \frac{dt}{t}.$$

Find u such that it minimizes the functional I.

Solution Since $L = t^{-1}\sqrt{1 + p^2}$,

$$L_p = \frac{p}{t\sqrt{1 + p^2}} = C.$$

It follows that

$$\dot{u}^2(1 - C^2 t^2) = C^2 t^2 \quad \text{or} \quad \dot{u} = \pm \frac{Ct}{\sqrt{1 - C^2 t^2}}.$$

Taking into account of the sign as part of the constant C, we integrate again to obtain

$$u = \frac{1}{C}\sqrt{1 - C^2 t^2} + C_1.$$

Using the above boundary conditions, we deduce that $C = \frac{1}{\sqrt{5}}$ and $C_1 = 2$. Therefore,

$$(u-2)^2 + t^2 = 5.$$

• **Case 2. (Autonomous systems)** Suppose L is independent of t, then $L = L(u,p)$. We introduce the Hamiltonian

$$H(u,p) = pL_p(u,p) - L(u,p).$$

Theorem 2.2 *Assume $L \in C^2$ and it is independent of t. If $u \in C^2(J, \mathbb{R}^1)$ is a solution of the E-L equation, then*

$$H(u(t), u'(t)) = \text{const.}, \quad \forall t.$$

Proof By direct calculation, we have:

$$\frac{d}{dt}H(u(t), u'(t)) = u'(t) \cdot E_L(u)(t) = 0.$$

$\square$

Example 2.4 (The line of steepest descent) Here we have:

$$L(u,p) = \frac{1}{\sqrt{2g}} \frac{\sqrt{1+p^2}}{\sqrt{y_1 - u}}.$$

By Theorem 2.2, it implies

$$pL_p - L|_{(u,u')} = \text{const.},$$

i.e. $\exists$ a constant c such that

$$-\frac{\sqrt{1+u'^2}}{\sqrt{y_1 - u}} + \frac{u'^2}{\sqrt{(1+u'^2)(y_1 - u)}} = c.$$

Hence,

$$c^2(1+u'^2)(y_1 - u) = 1.$$

Let k be a positive constant yet to be determined. We make the following coordinate transformation: let θ be a parameter such that

$$\begin{cases} x = x(\theta), \\ u = u(\theta), \end{cases}$$

and

$$u(\theta) = y_1 - k(1 - \cos\theta).$$

Then

$$c^2\left(1 + k^2\frac{\sin^2\theta}{\dot{x}(\theta)^2}\right)k(1-\cos\theta) = 1.$$

Taking $c = \sqrt{\frac{1}{2k}}$, we then have:

$$\dot{x}(\theta) = k(1 - \cos\theta).$$

Thus, we have:

$$\begin{cases} x(\theta) = x_1 + k(\theta - \sin\theta), \\ u(\theta) = y_1 - k(1 - \cos\theta), \\ \theta \in [0, \Theta]. \end{cases}$$

Lastly, we use

$$\begin{cases} x(\Theta) = x_2, \\ u(\Theta) = y_2, \end{cases}$$

to determine the k and Θ.

Example 2.5 (A minimal surface generated by a surface of revolution) Given two points $P_1 = (x_1, y_1)$ and $P_2 = (x_2, y_2)$ in the xy-plane, where $x_1 < x_2$ and $y_1, y_2 > 0$, we are to find a function $u \in C^1([x_1, x_2])$ whose graph passing through these points such that the surface generated by revolving the graph of u about the x-axis has minimal area.

Without loss of generality, we may assume $u(x_i) = y_i$, $i = 1, 2$ and $u(x) > 0$. The area of the surface of revolution is given by

$$I(u) = 2\pi \int_{x_1}^{x_2} u(x)\sqrt{(1 + u'(x)^2)}\,dx.$$

We now find u to minimize I.

Since the Hamiltonian is conservative, we have:

$$L(u(t), p(t)) - \dot{u}(t)L_p(u(t), p(t)) = C,$$

hence,

$$u\sqrt{1 + \dot{u}^2} - \frac{u\dot{u}^2}{\sqrt{1 + \dot{u}^2}} = C, \quad \forall\, t,$$

or equivalently,

$$\dot{u} = C^{-1}\sqrt{u^2 - C^2}.$$

After integration, we arrive at

$$C\ln\frac{u + \sqrt{u^2 - C^2}}{C} = x + C_1,$$

or the equivalent form of

$$u = C\cosh\frac{x + C_1}{C},$$

which are the standard catenary equations.

Example 2.6 (Geodesics on a sphere) As in Example 1.2, we adopt the parameter φ. By Theorem 2.2, we can rewrite the E-L equation as

$$\frac{\theta'^2(\varphi)}{\sqrt{\cos^2\theta(\varphi)+\theta'(\varphi)^2}}-\sqrt{\cos^2\theta(\varphi)+\theta'(\varphi)^2}=c,$$

where c is a constant. By definition, $-1\le c<1$. Hence,

$$-\cos^2\theta(\varphi)=c\sqrt{\cos^2\theta(\varphi)+\theta'(\varphi)^2},$$

namely,

$$c^2\theta'^2=\cos^4\theta-c^2\cos^2\theta.$$

By substituting $t=\tan\theta$ in the above equation, it yields that

$$\pm\varphi+\varphi_0=\int\frac{cd\theta}{\cos\theta\sqrt{\cos^2\theta-c^2}}=\int\frac{cdt}{\sqrt{(1-c^2)-c^2t^2}}=\arcsin\frac{ct}{\sqrt{1-c^2}},$$

for $0<c^2<1$. This gives us

$$\tan\theta(\varphi)=\frac{\sqrt{1-c^2}}{c}\sin\left(\pm\varphi+\varphi_0\right)$$

or

$$\theta(\varphi)=\arctan\frac{\sqrt{1-c^2}}{c}\sin\left(\pm\varphi+\varphi_0\right).$$

The constants c and φ_0 are determined by P_0 and P_1.

It turns out that this corresponds to the "great circles" on the sphere. When
$c=0$, $\theta=\pm\frac{\pi}{2}$, which corresponds to the north and south pole, a degenerate case
as this is not a curve. When $c=-1$, it corresponds to the equator.

• **Case 3.** Suppose p is absent from L, i.e. $L=L(t,u)$, then a solution of the
E-L equation

$$L_u(t,u)=0$$

is a single curve or several curves.

Example 2.7 For the functional $I(u)=\int_a^b(t-u)^2dt$, its E-L equation is

$$t-u=0;$$

which are lines with equation $u=t\ \forall\,t\in[a,b]$.

• **Coordinate transformations** In the following, we use the variational
derivative to prove that E-L equation is invariant under coordinate transformations. Let

$$\begin{cases}s=s(t,u),\\v=v(t,u),\end{cases}$$

whose inverse is

$$\begin{cases} t = t(s, v), \\ u = u(s, v). \end{cases}$$

The Lagrangian L is now changed to

$$\tilde{L}(s, v, q) = L\left(t(s, v), u(s, v), \frac{u_s + u_v q}{t_s + t_v q} \right)(t_s + t_v q).$$

Suppose the image of $t \in [t_0, t_1]$ under the transformation is $s \in [s_0, s_1]$, we then have:

$$\int_{t_0}^{t_1} L(t, u(t), \dot{u}(t)) dt = \int_{s_0}^{s_1} \tilde{L}(s, v(s), \dot{v}(s)) ds.$$

Hence, the E-L equation is now of the form

$$\tilde{L}_v - \frac{d}{ds} \tilde{L}_q = 0.$$

We can solve the latter equation first in the new coordinates and then convert it back under the inverse transformation.

Example 2.8 Consider the extremal values of the functional:

$$I(r) = \int_{\varphi_0}^{\varphi_1} \sqrt{r^2 + \dot{r}^2} d\theta,$$

where $r = r(\theta)$. Its corresponding E-L equation is

$$\frac{r}{\sqrt{r^2 + \dot{r}^2}} - \frac{d}{d\theta} \frac{\dot{r}}{\sqrt{r^2 + \dot{r}^2}} = 0.$$

Using polar coordinates:

$$x = r \cos\theta, u = r \sin\theta,$$

the functional I is of the form

$$I(u) = \int_{x_0}^{x_1} \sqrt{1 + \dot{u}^2} dx,$$

whose E-L equation becomes

$$\ddot{u} = 0.$$

A general solution to this second order equation is of the form

$$u = ax + b.$$

Substituting back to the original variables, it follows that

$$r \sin\theta = ar \cos\theta + b.$$

Exercises

1. Given an interval $J = [t_0, t_1] \subset \mathbb{R}^1$ and an open subset $\Omega \subset \mathbb{R}^N$. Let $L = L(x, u, p) \in C^1(J \times \Omega \times \mathbb{R}^N, \mathbb{R}^1)$ be a continuously differentiable function. For two vectors $\xi_0, \xi_1 \in \mathbb{R}^N$, we define

$$M_1 = \{u \in C^1(J, \mathbb{R}^N) \mid \dot{u}(t_i) = \xi_i, \ i = 0, 1\}$$

and the functional I on M_1 by

$$I(u) = \int_J L(t, u(t), \dot{u}(t))dt.$$

Find the necessary condition for which $u_0 \in M_1$ is a minimizer of I.

2. (The first Erdmann corner condition) Under the assumptions of Exercise 1, we further choose two points $P_0, P_1 \in \Omega$. Assume

$$u_0 \in M_2 = \{u \in PWC^1(J, \mathbb{R}^N) \mid u(t_i) = P_i, \ i = 0, 1\}$$

is a minimizer of I, where PWC^1 denotes the set of piecewise C^1 continuous functions. If there exists $t^* \in (t_0, t_1)$ such that $\dot{u}_0(t^* - 0) \neq \dot{u}_0(t^* + 0)$, prove that

$$L_P(t^*, u_0(t^*), \dot{u}_0(t^* - 0)) = L_P(t^*, u_0(t^*), \dot{u}_0(t^* + 0)), \quad i = 1, \ldots, N.$$

3. Let $L \in C^2(\mathbb{R}^N \times \mathbb{R}^N)$ and J be a closed interval. Assume $u \in C^2(J, \mathbb{R}^N)$ is a solution of the E-L equation of the functional $I(u) = \int_J L(u(t), \dot{u}(t))dt.$ Define the Hamiltonian to be

$$H(u, p) = \sum_{i=1}^{N} p_i L_{p_i}(u, p) - L(u, p),$$

prove that

$$H(u(t), \dot{u}(t)) = \text{const.}$$

Given a collection of particles with degree of freedom n, we denote the displacement coordinate by $q = (q_1, \ldots, q_n)$, the kinetic energy by

$$T = \frac{1}{2} \sum_{i,j=1}^{n} a_{ij} \dot{q}_i \dot{q}_j,$$

where (a_{ij}) is a positive definite matrix, and the potential by

$$V = V(q_1, \ldots, q_n).$$

Let the Lagrangian be $L = T - V$, we may ask the following questions: what is the Hamiltonian H in the case? What physical meaning does it have?

4. (The second Erdmann corner condition) Under the assumptions of Exercise 2, prove that

$$L(t^*, u_0(t^*), \dot{u}_0(t^*-0)) - \sum_{i=1}^{N} L_{p_i}(t^*, u_0(t^*), \dot{u}_0(t^*-0))$$

$$= L(t^*, u_0(t^*), \dot{u}_0(t^*+0)) - \sum_{i=1}^{N} L_{p_i}(t^*, u_0(t^*), \dot{u}_0(t^*+0)), \quad i=1,\ldots,N.$$

Hint: Introduce the following coordinate transformation

$$t = v_{N+1}(s), \quad u_i(t) = v_i(s), \quad 1 \le i \le N, \quad s \in \Lambda,$$

where $v_{N+1} : \Lambda \to J$ is a homeomorphism. Choose a function $F \in C^1(\mathbb{R}^{N+1} \times \mathbb{R}^N)$ such that

$$F(y_1, \ldots, y_{N+1}, q_1, \ldots, q_{N+1})$$

$$= L\left(y_{N+1}, y_1, \ldots, y_N, \frac{q_1}{q_{N+1}}, \ldots, \frac{q_N}{q_{N+1}}\right) q_{N+1},$$

show that

(1) The functional $K(v) = \int_\Lambda F(v(s), \dot{v}(s)) ds$ and the functional $I(u) = \int_J L(t, u(t), \dot{u}(t)) dt$ have the same set of extremal values. Furthermore, their extrema can be derived from one another via the above coordinate transformation.

(2)

$$\forall \lambda > 0, \quad F(y, \lambda q) = \lambda F(y, q).$$

(3) Using positive homogeniety, show that the Euler's equation holds:

$$F(y, q) = \sum_{i=1}^{N+1} F_{q_i}(y, q) q_i.$$

Lecture 3

The necessary condition and the sufficient condition on extremal values of functionals

3.1 The extremal values of functions — a revisit

Assume $f \in C^2(\Omega, \mathbb{R}^1)$, $\Omega \subset \mathbb{R}^n$ is an open set, $x_0 \in \Omega$ is such that $\nabla f(x_0) = 0$, we may ask the question what are the necessary condition and the sufficient condition for x_0 to be a (local) minimum of f.

Suppose x_0 is a minimum of f, there must be a neighborhood $U \subset \Omega$ of x_0 satisfying

$$f(x) - f(x_0) \geq 0, \quad \forall x \in U.$$

This means $\exists \, \epsilon_0 > 0$ such that when $0 < |\epsilon| < \epsilon_0$, for all $h \in \mathbb{R}^n \setminus \{0\}$, $x_0 + \epsilon h \in U$ and

$$f(x_0 + \epsilon h) \geq f(x_0), \quad |\epsilon| < \epsilon_0.$$

This implies the one variable function $\epsilon \mapsto f(x_0 + \epsilon h)$ has 0 as its minimum, hence,

$$\frac{d^2}{d\epsilon^2} f(x_0 + \epsilon h)|_{\epsilon=0} \geq 0.$$

Namely,

$$(d^2 f(x_0)h, h) = \sum_{i,j=1}^n \frac{\partial^2 f}{\partial x_i \partial x_j}(x_0) h_i h_j \geq 0.$$

Therefore, the matrix $d^2 f(x_0) = (\frac{\partial^2 f}{\partial x_i \partial x_j})(x_0)$ is positive semi-definite.

Conversely, suppose the matrix $d^2 f(x_0)$ is positive definite, then x_0 must be a strict local minimum of f.

3.2 Second order variations

We now return to the discussion on the extremal values of functionals. We have shown that the E-L equation is a first order variation; it serves only as a necessary condition for the minimizer but not sufficient. From both the functional analysis and differential topology points of view, a solution satisfying the E-L equation is only a critical point of the functional. Just like its counterpart in the finite dimensional case, we also need the second order variation to determine whether it is a (local) minimum.

Let $L \in C^2(J \times \mathbb{R}^N \times \mathbb{R}^N)$ and

$$I(u) = \int_J L(t, u(t), \dot{u}(t)) dt.$$

We assume that $u_0 \in M$ is a solution of the E-L equation $E_L(u_0) = 0$ of the functional I. For all $\varphi \in C_0^\infty(J, \mathbb{R}^N)$, let

$$g(s) = I(u_0 + s\varphi),$$

then the one variable function $s \mapsto g(s)$ has 0 as its minimum. We call the following expression

$$\delta^2 I(u_0, \varphi) = \ddot{g}(0)$$

$$= \frac{d^2}{ds^2} I(u_0 + s\varphi)|_{s=0}$$

$$= \frac{d^2}{ds^2} \int_J L(t, u_0(t) + s\varphi(t), \dot{u}_0(t) + s\dot{\varphi}(t)) dt|_{s=0}$$

$$= \sum_{i,j} \int_J [L_{u^i u^j}(t, u_0(t), \dot{u}_0(t)) \varphi^i(t) \varphi^j(t)$$

$$+ 2L_{u^i p^j}(t, u_0(t), \dot{u}_0(t)) \varphi^i(t) \dot{\varphi}^j(t)$$

$$+ L_{p^i p^j}(t, u_0(t), \dot{u}_0(t)) \dot{\varphi}^i(t) \dot{\varphi}^j(t)] dt$$

the second order variation of I along φ at u_0.

On one hand, suppose u_0 is a minimizer, then $\ddot{g}(0) \geq 0$, so

$$\delta^2 I(u_0, \varphi) \geq 0, \quad \forall \, \varphi \in C_0^1(J, \mathbb{R}^N). \tag{3.1}$$

On the other hand, suppose $u_0 \in C_0^1(J, \mathbb{R}^N)$ satisfies the E-L equation and suppose $\exists \, \lambda > 0$ such that

$$\delta^2 I(u_0, \varphi) \geq \lambda \int_J \{|\varphi|^2 + |\dot{\varphi}|^2\} dt, \quad \forall \, \varphi \in C_0^1(J, \mathbb{R}^N), \tag{3.2}$$

then u_0 must be a strict minimum of I. To see this, consider

$$g(s) - g(0) = g(s) - g(0) - \dot{g}(0)s = \frac{s^2}{2}\ddot{g}(\theta s) = \frac{s^2}{2}[\ddot{g}(\theta s) - \ddot{g}(0)] + \frac{s^2}{2}\ddot{g}(0),$$

for $\theta \in (0,1)$ depending only on φ. We introduce the following function-valued matrices:

$$A = (L_{p_i p_j}(t, u, p)),$$
$$B = (L_{p_i u_j}(t, u, p)),$$
$$C = (L_{u_i u_j}(t, u, p)),$$

together with their restrains along the function $u_0(t)$:

$$A_{u_0} = (L_{p_i p_j}(t, u_0(t), \dot{u}_0(t))),$$
$$B_{u_0} = (L_{p_i u_j}(t, u_0(t), \dot{u}_0(t))),$$
$$C_{u_0} = (L_{u_i u_j}(t, u_0(t), \dot{u}_0(t))).$$

We then have:

$$\delta^2 I(u_0, \varphi) = \int_J [(A_{u_0}\dot{\varphi}, \dot{\varphi}) + 2(B_{u_0}\dot{\varphi}, \varphi) + (C_{u_0}\varphi, \varphi)]dt.$$

Since

$$\ddot{g}(s) = \int_J [(A_{u_0+s\varphi}\dot{\varphi}, \varphi) + 2(B_{u_0+s\varphi}\dot{\varphi}, \varphi) + (C_{u_0+s\varphi}\dot{\varphi}, \varphi)] \, dt$$

and $L \in C^2$, for all $\|\varphi\|_{C^1(J)} \le 1$, as $s \to 0$, we have the uniform estimate:

$$|A_{u_0+s\varphi} - A_{u_0}| + |B_{u_0+s\varphi} - B_{u_0}| + |C_{u_0+s\varphi} - C_{u_0}| = o(1),$$

which yields

$$\ddot{g}(s) - \ddot{g}(0) = o(s^2) \int_J (|\nabla\varphi|^2 + |\varphi|^2) \, dt.$$

Hence, $\forall \varphi \in C_0^1(J, \mathbb{R}^N)$, as long as $|s|$ is sufficiently small, there exists $\epsilon < \lambda$ such that

$$I(u_0 + \epsilon\varphi) - I(u_0) \ge (\lambda - \epsilon) \int_J (|\nabla\varphi|^2 + |\varphi|^2) \, dt \qquad \square$$

Although (3.1) and (3.2) give us the necessary and sufficient conditions respectively on u_0 such that it minimizes I, its dependence on the arbitrary function φ is nevertheless unsatisfying. We shall continue our study in the subsequent section.

3.3 The Legendre–Hadamard condition

In our previous setting, notice that the roles of the three matrices A_0, B_0, and C_0 in determining whether u_0 is a minimum are not all equal.

In fact, $\forall \tau \in \text{int}(J)$, $\forall \xi \in \mathbb{R}^N$, $\forall \mu > 0$ sufficiently small, one may choose $v \in C^1(\mathbb{R}^1)$ with $v(s) = 0$ satisfying for $|s| \geq 1$, $\int_{\mathbb{R}^1} \dot{v}(s)^2 ds = 1$. Let

$$\varphi(t) = \xi \mu v\left(\frac{t - \tau}{\mu}\right),$$

then

$$\dot{\varphi}(t) = \xi \dot{v}\left(\frac{t - \tau}{\mu}\right).$$

For all $\mu > 0$ sufficiently small, we have:

$$\int_J \dot{\varphi}_i \dot{\varphi}_j dt = \xi_i \xi_j \mu,$$

$$\int_J \dot{\varphi}_i \varphi_j dt = \xi_i \xi_j \mu^2 \int_{R^1} v(t)\dot{v}(t) dt,$$

$$\int_J \varphi_i \varphi_j dt = \xi_i \xi_j \mu^3 \int_{R^1} v(t)^2 dt.$$

Substituting into (3.1) and letting $\mu \to 0$, it shows that

$$\delta^2 I(u_0, \varphi) = \mu(A_{u_0}\xi, \xi) + o(\mu).$$

We introduce the *Legendre–Hadamard* condition as follows:

$$(A_{u_0}\xi, \xi) = \sum_{i,j=1}^{N} L_{p_i p_j}(\tau, u_0(\tau), \dot{u}_0(\tau))\xi^i \xi^j \geq 0, \ \forall \tau \in J, \ \forall \xi \in \mathbb{R}^N. \quad (3.3)$$

Suppose $\exists \lambda > 0$ such that

$$\sum_{i,j=1}^{N} L_{p_i p_j}(\tau, u(\tau), \dot{u}(\tau))\xi^i \xi^j \geq \lambda |\xi|^2, \ \forall \tau \in J, \ \forall \xi \in \mathbb{R}^N, \quad (3.4)$$

we then call it the *strict Legendre–Hadamard condition*.

Theorem 3.1 *Let* $L \in C^2(J \times \mathbb{R}^N \times \mathbb{R}^N)$. *Suppose* $u_0 \in M$ *is a minimizer of* I, *then the Legendre–Hadamard condition (3.1) holds. Conversely, if* $u_0 \in M$ *satisfies the E-L equation, and if there exists* $\lambda > 0$ *such that (3.2) holds, then* u_0 *is a strict minimizer of* I.

We have stated (3.2) involves an arbitrary function φ, in order to remove the influence of φ, we must establish its relation with the strict Legendre–Hadamard condition (3.4).

It turns out, as seen by the next lemma, we can remove the $|\varphi|^2$ term from the integral on the right-hand side of (3.2).

Lemma 3.1 (Poincaré) *Let* $\varphi \in C_0^1(J, \mathbb{R}^N)$, *then we have:*

$$\int_J |\varphi|^2 dt \le \frac{(t_2 - t_1)^2}{2} \int_J |\dot{\varphi}|^2 dt.$$

Proof Since

$$\varphi(t) = \int_{t_0}^t \dot{\varphi}(s) ds,$$

by the Cauchy–Schwarz inequality, we have:

$$|\varphi(t)|^2 \le \left(\int_{t_0}^t |\dot{\varphi}(s)| ds \right)^2 \le (t - t_0) \int_J |\dot{\varphi}(s)|^2 ds.$$

After integrating, it gives that

$$\int_J |\varphi(t)|^2 dt \le \frac{(t_1 - t_0)^2}{2} \int_J |\dot{\varphi}|^2 dt.$$

If we remove the $|\varphi|^2$ term from the integral on the right-hand side of (3.2) and replacing it by

$$\delta^2 I(u, \varphi) > \lambda \int_J |\dot{\varphi}|^2 dt, \quad \forall \, \varphi \in C_0^1(J, \mathbb{R}^N), \tag{3.2}'$$

for some $\lambda \ge 0$, then Theorem 3.1 remains valid.

Example 3.1 Let $L = \sqrt{1 + p^2}$ and $M = \{u \in C^1([0, b]) \mid u(0) = u(b) = 0\}$. The E-L equation of the functional $I(u) = \int_0^b \sqrt{1 + \dot{u}^2(t)} \, dt$ is given by

$$\frac{d}{dt} \frac{\dot{u}}{\sqrt{1 + \dot{u}^2}} = 0,$$

which has a solution $u = 0 \in M$.

Since

$$L_{uu} = L_{up} = L_{pu} = 0, \quad L_{pp} = \frac{1}{(1 + p^2)^{\frac{3}{2}}},$$

$$\delta^2 I(0, \varphi) = \int_0^b \dot{\varphi}^2 dt.$$

By Theorem 3.1, $u = 0$ is a minimum. $\square$

On one hand, from the second order variation, it is not difficult to see if the matrix

$$\begin{pmatrix} A_{u_0} & B_{u_0} \\ B_{u_0} & C_{u_0} \end{pmatrix}$$

is positive definite, then the solution u_0 of the E-L equation must be a minimum. However, from the next example, we see that the positive definiteness of the matrix is not necessary for u_0 to be a minimum.

Example 3.2 Let $I(u) = \int_J (\dot{u}(t)^2 - u(t)^2)dt$, then for $u = 0$,

$$\begin{pmatrix} A & B \\ B & C \end{pmatrix} = \begin{pmatrix} 1 & 0 \\ 0 & -1 \end{pmatrix}$$

is not positive semi-definite.

However, when $|J| = t_1 - t_0$ is sufficiently small, from Poincaré's inequality, we still have;

$$\delta^2 I(0, \varphi) = \int_J (\dot{\varphi}^2 - \varphi^2)dt \geq \left(1 - \frac{(t_1 - t_0)^2}{2}\right)\int_J \dot{\varphi}^2 dt.$$

So $u = 0$ is still the minimum. □

On the other hand, it is not difficult to check (3.4) is not a sufficient condition for u_0 to be a minimum. In the following, we will investigate the additional requirement needed for (3.4) to be sufficient.

3.4 The Jacobi field

In this section, we introduce the notion of a Jacobi field.

Let $L \in C^3$ and assume u_0 is a solution of the E-L equation, along u_0, we define

$$\Phi_{u_0}(t, \xi, \eta) = (A_{u_0}\eta, \eta) + 2(B_{u_0}\xi, \eta) + (C_{u_0}\xi, \xi), \quad \forall\, (\xi, \eta) \in \mathbb{R}^N \times \mathbb{R}^N$$

to be the (accessory) Lagrangian.

Suppose u_0 is a minimum, we examine the following integral based on the accessory Lagrangian:

$$Q_{u_0}(\varphi) = \int_J \Phi_{u_0}(t, \varphi(t), \dot{\varphi}(t))\, dt, \quad \forall\, \varphi \in C_0^1(J, \mathbb{R}^N).$$

Since $Q_{u_0}(\varphi) = \delta^2 I(u_0, \varphi) \geq 0, \forall\, \varphi \in C_0^1(J, \mathbb{R}^N)$ and $Q_{u_0}(\theta) = 0$, θ must be a minimum.

We extend the domain of functional Q_{u_0} to be $\mathrm{Lip}_0(J, \mathbb{R}^N)$ (Lipschitz functions with vanishing boundary values), its integral form of the E-L equation looks like

$$A_{u_0}\dot\varphi(t) + B_{u_0}^\top \varphi(t) - \int_{t_0}^t (B_{u_0}\dot\varphi(t) + C_{u_0}\varphi(t)) = \mathrm{const.}$$

If L along u_0 satisfies the strict Legendre–Hadamard condition, i.e. A_{u_0} is positive definite, then using the above integral form of the E-L equation, the solution $\varphi \in C^2(J, \mathbb{R}^N)$. Furthermore, φ must satisfy the homogeneous second order ordinary differential equation:

$$J_{u_0}(\varphi) = \frac{d}{dt}[A_{u_0}\dot\varphi(t) + B_{u_0}^\top(t)\varphi] - [B_{u_0}\dot\varphi(t) + C_{u_0}\varphi] = 0, \quad \forall t \in J.$$

We call this equation the Jacobi equation and the operator J_{u_0} the Jacobi operator along u_0 (a solution of the E-L equation).

The Jacobi operator is a linear ordinary differential operator of second order, and it plays a similar role in variational problems as that of the Hessian matrix in extremal problems of functions.

We call a C^2-solution of the Jacobi equation a Jacobi field along the orbit $u_0(t)$. All Jacobi fields together constitute a linear space of dimension $2N$.

Theorem 3.2 *If φ_0 is a Jacobi field along u_0, then $Q_{u_0}(\varphi_0) = 0$. Conversely, if $\varphi \in \mathrm{Lip}_0(J, \mathbb{R}^N)$ satisfies $Q_{u_0}(\varphi_0) = 0$ and $Q_{u_0}(\varphi) \geq 0$ for all $\varphi \in C_0^1(J, \mathbb{R}^N)$, then φ_0 is a Jacobi field along u_0.*

Proof "$\Rightarrow$" Since Φ_{u_0} is homogeneous of degree two with respect to (ξ, η), by Euler's identity, we have:

$$2\Phi_{u_0}(t, \xi, \eta) = (\Phi_{u_0})_\xi(t, \xi, \eta)\xi + (\Phi_{u_0})_\eta(t, \xi, \eta)\eta.$$

Suppose $\varphi_0 \in C_0^1([a, b], \mathbb{R}^N)$, $[a, b] \subset \mathrm{int}(J)$ is a Jacobi field along u_0, then

$$2\int_a^b (\Phi_{u_0})(t, \varphi_0(t), \dot\varphi_0(t))dt$$
$$= \int_a^b [(\Phi_{u_0})_\xi(t, \varphi_0(t), \dot\varphi_0(t))\varphi_0(t) + (\Phi_{u_0})_\eta(t, \varphi_0(t), \dot\varphi_0(t))\dot\varphi_0(t)]dt$$
$$= \int_a^b [(\Phi_{u_0})_\xi(t, \varphi_0(t), \dot\varphi_0(t)) - \frac{d}{dt}(\Phi_{u_0})_\eta(t, \varphi_0(t), \dot\varphi_0(t))]\varphi_0(t)dt$$
$$= -\int_a^b J_{u_0}(\varphi)\, dt = 0.$$

Since a and b are arbitrary, it follows that $Q_{u_0}(\varphi_0) = 0$.

"$\Leftarrow$" Using smooth function approximation, we have:

$$Q_{u_0}(\varphi) \geq 0, \quad \forall\, \varphi \in \mathrm{Lip}_0(J, \mathbb{R}^N).$$

Thus, φ_0 is a minimum of Q_{u_0}. From our previous argument, it must satisfy the integral form of the E-L equation, hence it also satisfies the differential form of the E-L equation $J_{u_0}(\varphi_0) = 0$. □

Lemma 3.2 *Given a sufficiently smooth Lagrangian L, suppose it satisfies the strict Legendre–Hadamard condition along a solution u_0 of the E-L equation, namely, A_{u_0} is positive definite. Suppose there exists $\mu > 0$ such that*

$$Q_{u_0}(\varphi) \geq \mu \int_J |\varphi|^2 dt,$$

then there exists $\lambda > 0$ such that

$$Q_{u_0}(\varphi) \geq \lambda \int_J (|\dot{\varphi}|^2 + |\varphi|^2) dt.$$

Consequently, u_0 is a strict minimum of the functional

$$I(u) = \int_J L(t, u(t), \dot{u}(t)) \, dt.$$

Proof For any two continuous functions ϕ, ψ on J, let $\langle \phi, \psi \rangle = \int_J \phi(t)\psi(t)dt$. Since A_{u_0} is positive definite, there exists $\alpha > 0$ such that

$$\langle A_{u_0}\dot{\varphi}, \dot{\varphi} \rangle \geq \alpha \int_J |\dot{\varphi}|^2 dt.$$

From

$$Q_{u_0}(\varphi) = \langle A_{u_0}\dot{\varphi}, \dot{\varphi} \rangle + 2\langle B_{u_0}\dot{\varphi}, \varphi \rangle + \langle C_{u_0}\varphi, \varphi \rangle,$$

we can find two positive constants C_1 and C_2 such that

$$\alpha \int_J |\dot{\varphi}|^2 \, dt$$
$$\leq Q_{u_0}(\varphi) + C_1 \left(\left(\int_J |\dot{\varphi}|^2 \, dt \right)^{\frac{1}{2}} \left(\int_J |\varphi|^2 \, dt \right)^{\frac{1}{2}} + \int_J |\varphi|^2 \, dt \right)$$
$$\leq \frac{\alpha}{2} \int_J |\dot{\varphi}|^2 \, dt + Q_u(\varphi) + C_2 \int_J |\varphi|^2 \, dt.$$

Using the assumption

$$\int_J |\varphi|^2 dt \leq \mu^{-1} Q_{u_0}(\varphi),$$

we find that

$$\int_J |\dot{\varphi}|^2 dt \leq \frac{2}{\alpha}(1 + C_2\mu^{-1})Q_{u_0}(\varphi).$$

According to Theorem 3.1 and Poincaré's inequality, u_0 is a strict minimum of I.

□

3.5 Conjugate points

Definition 3.1 (Conjugate points) Let u_0 be a solution of the E-L equation of the functional $I(u) = \int_J L(t, u(t), \dot{u}(t))\, dt$. We call $(a, u_0(a))$ and $(b, u_0(b))$ a pair of conjugate points along the orbit $(t, u_0(t))$, if there exists a nonzero Jacobi field $\varphi \in C_0^1([a, b], \mathbb{R}^N)$ along $u_0(t)$ (see Figure 3.1).

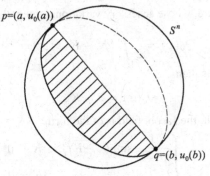

Fig. 3.1

Sometimes, if there are no conjugate points on the orbit $\{(t, u_0(t)) \mid t \in (t_0, t_1)\}$, we simply say that u_0 has no conjugate points.

Example 3.3 Given the metric

$$e(x, y)dx^2 + 2f(x, y)dxdy + g(x, y)dy^2$$

on a surface S in $\mathbb{R}^3$. We choose a geodesic γ in S and without loss of generality, we may assume that it is the x-axis $(y = 0)$ and the curves $x = $ const are perpendicular to γ. We furnish S with an orthonormal frame, under which, the square of the line element of the curve $y = u(x)$ is given by

$$ds^2 = e(x, y)dx^2 + dy^2,$$

where $e > 0$, $e(x, 0) = 1$, and $e_y(x, 0) = 0$. The arclength functional is

$$I(u) = \int_a^b \sqrt{e(x, u) + \dot{u}^2}dx,$$

i.e.

$$L(t, u) = \sqrt{e(x, u) + p^2}.$$

Hence,

$$L_{pp} = \frac{e}{(e + p^2)^{\frac{3}{2}}}, \quad L_{up} = L_{pu} = 0, \quad L_{uu} = \frac{2e_{uu}(e + p^2) - e_u^2}{4(e + p^2)^{\frac{3}{2}}}.$$

Along the geodesic γ: $y = 0$, we have:

$$\begin{pmatrix} A & B \\ B & C \end{pmatrix} = \begin{pmatrix} 1 & 0 \\ 0 & \frac{1}{2}e_{uu} \end{pmatrix}$$

In differential geometry, we call the quantity

$$K(x) = -\frac{1}{2}e_{uu}(x,0)$$

the Gaussian curvature, whose accessory variational integral is

$$Q_0(\varphi) = \frac{1}{2}\int_a^b [\dot\varphi^2 - K(x)\varphi^2]dx.$$

The Jacobi operator is then

$$J_0(\varphi) = \ddot\varphi + K\varphi.$$

When K is a constant, the Jacobi field is of the form

$$\varphi(t) = \begin{cases} \dfrac{1}{\sqrt{-K}}\sinh(\sqrt{-K}t), & K < 0 \\ t, & K = 0, \\ \dfrac{1}{\sqrt{K}}\sin(\sqrt{K}t), & K > 0. \end{cases}$$

It follows that if $K \leq 0$, then there are no conjugate points. However, when $K > 0$, the first conjugate point of $(0,0)$ along γ is $(\pi/\sqrt{K}, 0)$.

Remark 3.1 For a general Riemannian manifold (M, g), g is a Riemannian metric on M, the Lagrangian of a geodesic is

$$L(u,p) = \sum g_{ij}(u)p_i p_j,$$

the corresponding Jacobi equation is

$$\frac{d^2\varphi}{dt^2} + R\left(\dot u(t), \varphi(t)\right)\dot u(t) = 0,$$

where $R(\cdot,\cdot)$ denotes the Riemann curvature operator.

Theorem 3.3 *Suppose u_0 is a solution of the E-L equation of the functional $I(u) = \int_J L(t, u(t), \dot u(t))\, dt$ and A_{u_0} is positive definite. If $\delta^2 I(u_0, \varphi) \geq 0$ for all $\varphi \in C_0^1(J, \mathbb{R}^N)$, then there is no $a \in (t_0, t_1)$ such that $(a, u_0(a))$ is conjugate to $(t_0, u_0(t_0))$.*

Proof Suppose not, then $\exists\, a \in (t_0, t_1)$ such that $(a, u_0(a))$ and $(t_0, u_0(t_0))$ are conjugate points, i.e. there exists a nonzero Jacobi field $\xi \in C^2([t_0, a], \mathbb{R}^N)$ along $u_0(t)$ satisfying: $J_{u_0}(\xi) = 0$ and $\xi(t_0) = \xi(a) = 0$. Let

$$\tilde\xi(t) = \begin{cases} \xi(t) & t \in [t_0, a], \\ 0 & t \in [a, t_1], \end{cases}$$

then $\tilde{\xi} \in \text{Lip}(J, \mathbb{R}^N)$ with $\tilde{\xi}(t_0) = \tilde{\xi}(t_1) = 0$ and

$$Q_{u_0}(\tilde{\xi}) = \int_{t_0}^{a} \Phi_{u_0}(t, \xi(t), \dot{\xi}(t)) dt = 0.$$

By Theorem 3.2, $\tilde{\xi} \in C^2(J, \mathbb{R}^N)$ must satisfy the Jacobi equation $J_{u_0}(\tilde{\xi}) = 0$. By the uniqueness of the solution of the initial value problem of a second order ordinary differential equation, $\tilde{\xi} \equiv 0$, contradictory to $\xi \neq 0$, hence completes the proof. $\qquad \square$

For the special case $N = 1$, we can also show that the converse of the above theorem is also true.

We shall henceforth assume u_0 is a solution of the E-L equation. Notice that if u_0 has no conjugate points on $(t_0, t_1]$, then there exists a positive Jacobi field $\psi > 0, \forall t \in J$.

To see this, suppose λ is a Jacobi field with the initial conditions $\lambda(t_0) = 0$ and $\dot{\lambda}(t_0) = 1$. By assumption, the next root a satisfies $a > t_1$. Since the solution of an ordinary differential equation varies continuously dependent on the initial values, there exist $\epsilon > 0$ and a Jacobi field ψ such that $\psi(t_0 - \epsilon) = 0$, $\dot{\psi}(t_0 - \epsilon) = 1$, and $\psi(t) > 0, \forall t \in J$.

Lemma 3.3 *Suppose $\psi(t) > 0, \forall t \in J$ is a Jacobi field along u_0, then for all $\varphi \in C_0^1(J)$, we have:*

$$Q_{u_0}(\varphi) = \int_J A_{u_0}(t) \psi^2(t) \left(\left(\frac{\varphi}{\psi} \right)'(t) \right)^2 dt.$$

Proof Let $\lambda = \frac{\varphi}{\psi}$, then $\varphi = \lambda \psi$ and $\varphi' = \lambda' \psi + \lambda \psi'$. Hence,

$$A_{u_0} \varphi'^2 + 2B_{u_0} \varphi' \varphi + C_{u_0} \varphi^2$$
$$= \lambda^2 (A_{u_0} \psi'^2 + 2B_{u_0} \psi' \psi + C_{u_0} \psi^2) + 2\lambda' \lambda \psi (A_{u_0} \psi' + B_{u_0} \psi) + \lambda'^2 A_{u_0} \psi^2.$$

Now since ψ satisfies the Jacobi equation, we then have:

$$\int_J (A_{u_0} \varphi'^2 + 2B_{u_0} \varphi' \varphi + C_{u_0} \varphi^2) \, dt$$
$$= \int_J \left\{ \left[\psi' \lambda^2 (A_{u_0} \psi' + B_{u_0} \psi) + \frac{d(A_{u_0} \psi' + B_{u_0} \psi)}{dt} \psi \lambda^2 \right. \right.$$
$$\left. \left. + 2\lambda' \lambda \psi (A_{u_0} \psi' + B_{u_0} \psi) \right] + A_{u_0} \lambda'^2 \psi^2 \right\} dt$$
$$= \int_J (A_{u_0} \lambda'^2 \psi^2) dt + \psi \lambda^2 (A_{u_0} \psi' + B_{u_0} \psi)|_{t_0}^{t_1}$$
$$= \int_J (A_{u_0} \lambda'^2 \psi^2) dt. \qquad \square$$

Theorem 3.4 *Let $N = 1$ and assume $u_0 \in C^1(J)$ is a solution of the E-L equation. If $\exists \lambda > 0$ such that $A_{u_0}(t) \geq \lambda, \forall t \in J$ and there exists a Jacobi field $\psi > 0$ on J, then u_0 is a strict minimum.*

Proof Denote

$$\alpha = \inf_J (A_{u_0}(t)\psi^2(t)) > 0.$$

$\forall \varphi \in C_0^1(J)$, we use Lemma 3.3 and Poincaré's inequality to obtain:

$$Q_{u_0}(\varphi) = \int_J A_{u_0} \psi^2 \left(\frac{\varphi}{\psi}\right)'^2 dt$$

$$\geq \alpha \int_J \left(\frac{\varphi}{\psi}\right)'^2 dt$$

$$\geq \alpha \frac{1}{|J|^2} \int_J \left(\frac{\varphi}{\psi}\right)^2 dt$$

$$\geq \alpha \inf_J \left(\frac{1}{\psi^2}\right) \frac{1}{|J|^2} \int_J \varphi^2 dt.$$

Thus, there exists $\mu > 0$ such that

$$Q_{u_0}(\varphi) \geq \mu \int_J |\varphi|^2 dt.$$

The assertion now follows from Lemma 3.2. $\qquad\square$

Example 3.4 Let $M = \{u \in C^1([0,1]) \mid u(0) = a, u(1) = b\}$ and consider the functional

$$I(u) = \int_0^1 (t\dot{u} + \dot{u}^2) dt.$$

Since $L_u = 0$, $L_p = 2p + t$, and $L_{pp} = 2$, its E-L equation

$$2\ddot{u}(t) + 1 = 0$$

has solutions

$$u(t) = -\frac{t^2}{4} + \left(b + \frac{1}{4}\right)t + a.$$

The accessory variational integral is

$$Q_u(\varphi) = \int_0^1 \dot{\varphi}^2 dt,$$

with the corresponding Jacobi equation

$$\ddot{\varphi} = 0.$$

Using the initial conditions $(\varphi(0), \dot{\varphi}) = (0, 1)$, we see that $\varphi = t$. This Jacobi field has no conjugate point and hence u is a strict minimum.

Exercises

1. Find the minimum of each of the following functionals:

   (1)

   $$I(u) = \int_0^1 (t\dot{u} + \dot{u}^2)\, dt, \quad M = C_0^1(0,1).$$

   (2)

   $$I(u) = \int_0^1 u\sqrt{1+\dot{u}^2}\, dt, \quad M = \{v \in C_0^1[a,b] \mid v(a) = \cosh a, v(b) = \cosh b\},$$

   $$\text{for } 0 < a < b.$$

   (3)

   $$I(u) = \int_0^1 (u^2 + \dot{u}^2)\, dt, \quad M = \{v \in C^1[0,b] \mid v(0) = 0, v(b) = B\}.$$

2. Assume that φ is an absolutely continuous function on $[a,b]$ whose a.e. derivative function φ' is square integrable on $[a,b]$. If $\varphi(a) = 0$, prove the following Poincaré's inequality:

   $$\int_a^b \varphi^2(x)\, dx \le \frac{(b-a)^2}{2} \int_a^b [\varphi'(x)]^2\, dx.$$

3. In $\mathbb{R}^3$, consider the surface of revolution generated by revolving the curve $r = r(z) > 0$ about the z-axis:

   $$S : r = \sqrt{x^2 + y^2}.$$

   (1) Find the metric on S.
   (2) Write the equation of geodesics in S.
   (3) First write the geodesic equations for the cylinder $r =$ const. and the circular cone $r = z$ respectively, then determine whether they are minima.

4. Assume $L = L(t,u,p)$ is a differentiable function which is bounded from below. Furthermore, assume L is strictly convex with respect to (u,p). Let $M = C_0^1(J)$. Show that the solution $u \in M$ of the E-L equation must be a strict minimum of the functional.

Lecture 4

Strong minima and extremal fields

4.1 Strong minima and weak minima

Similar to a minimum of a given function, a minimum of a given functional is also a local minimum. The notion of 'local' is determined by neighborhoods. In calculus of variations, the space M is often an infinite dimensional function space and an infinite dimensional space is usually equipped with many distinct topologies; it is therefore crucial to specify what topology we are considering in our investigations.

Definition 4.1 Let $J = [t_0, t_1]$, $u \in C^1(J, \mathbb{R}^N)$ is called a strong (weak) minimum of the functional

$$I = \int_J L(t, u(t), \dot{u}(t)) \, dt$$

if there exists $\epsilon > 0$ such that for all $\varphi \in C_0^1(\Omega, \mathbb{R}^N)$ with

$$\|\varphi\|_{C^0(J)} < \epsilon \quad (\|\varphi\|_{C^1(J)} < \epsilon),$$

we have:

$$I(u + \varphi) \geq I(u).$$

The C^1 requirement can be replaced by Lip and using the Lip-norm instead of the C^1-norm. Without any confusion, we still call this a weak minimum. In Lecture 3, the notion of a minimum agrees with a weak local minimum (cf. §3.2).

It is clear that a strong minimum is always a weak minimum and a weak minimum in the sense of Lipschitz is also a weak minimum under the C^1-topology. We have the following example.

Example 4.1 Let

$$I(u) = \int_0^1 (u'^2 + u'^3) dx$$

43

and

$$M = \{u \in \mathrm{Lip}([0,1], \mathbb{R}^1) \mid u(0) = u(1) = 0\},$$

then $u = 0$ is a weak minimum.

In fact, for $\|u\|_{\mathrm{Lip}} < \frac{1}{2}$, we have:

$$I(u) - I(0) = \int_0^1 (u'^2 + u'^3)dx \geq \frac{1}{2}\int_0^1 u'^2 dx \geq 0.$$

Moreover, from

$$\delta^2 I(0, \varphi) = \int_0^1 \dot{\varphi}^2 \, dt, \quad \forall \, \varphi \in C_0^1([0,1])$$

and Poincaré's inequality, we see that $u = 0$ must be a strict (weak) local minimum.

On the other hand, we claim $u = 0$ is not a strong minimum. $\forall \, 0 < h < 1 - h^2$, let

$$u_h(x) = \begin{cases} -\dfrac{x}{h}, & x \in [0, h^2], \\[2mm] \dfrac{h(x-1)}{1-h^2}, & x \in [h^2, 1], \end{cases}$$

then

$$(u_h'^2 + u_h'^3)(x) = \begin{cases} \dfrac{1}{h^2} - \dfrac{1}{h^3} \leq -\dfrac{1}{2h^3}, & x \in [0, h^2], \\[3mm] \left(\dfrac{h}{1-h^2}\right)^2 + \left(\dfrac{h}{1-h^2}\right)^3 \leq 2, & x \in [h^2, 1]. \end{cases}$$

Thus, $\|u_h\|_{C^0} \leq h$ and

$$I(u_h) - I(0) = I(u_h) \leq 2 - \frac{1}{2h} \longrightarrow -\infty \quad (h \to 0).$$

4.2 A necessary condition for strong minimal value and the Weierstrass excess function

Let $L \in C^1(J \times \mathbb{R}^N \times \mathbb{R}^N)$ be a Lagrangian and suppose u is a solution of the E-L equation of the functional

$$I(u) = \int_J L(t, u(t), \dot{u}(t)) \, dt,$$

we seek a necessary condition for u to be a strong minimum of I. On that note, we compare $I(u)$ to the values of I in a C^0-neighborhood of u.

Suppose u is a strong minimum, then $\forall \, \varphi \in C_0^1(J, \mathbb{R}^N)$ with $\|\varphi\|_{C^0} < \epsilon$, $I(u + \epsilon) \geq I(u)$.

We now construct such a function φ. $\forall \, \xi \in \mathbb{R}^N$, $\forall \, \tau \in (t_0, t_1)$, we choose $\lambda > 0$ sufficiently small so that $[\tau - \lambda^2, \tau + \lambda] \subset (t_0, t_1)$. Let

$$\psi_\lambda(s) = \begin{cases} 0, & s \in (-\infty, -\lambda^2] \cup [\lambda, \infty) \\ \lambda^2 + s, & s \in [-\lambda^2, 0] \\ \lambda^2 - \lambda s, & s \in [0, \lambda], \end{cases}$$

then

$$\psi_\lambda'(s) = \begin{cases} 0, & s \in (-\infty, -\lambda^2] \cup [\lambda, \infty) \\ 1, & s \in [-\lambda^2, 0] \\ -\lambda, & s \in [0, \lambda]. \end{cases}$$

We define

$$\varphi_\lambda(t) = \xi \psi_\lambda(t - \tau),$$

it is easy to check that $\|\varphi_\lambda\|_{C^0} = O(\lambda^2)$, $\|\dot{\varphi}_\lambda\|_{C^0} = \|\xi\|_{\mathbb{R}^N}$, and $\|\varphi_\lambda\|_{C^1([\tau, \tau + \lambda])} = O(\lambda)$. In particular, if we choose $\varphi = \varphi_\lambda$, then

$$I(u + \varphi_\lambda) - I(u) \geq 0.$$

It follows from the E-L equation that

$$\int_J F(t) \, dt := \int_J \{L(t, u(t) + \varphi_\lambda(t), \dot{u}(t) + \dot{\varphi}_\lambda(t)) - L(t, u(t), \dot{u}(t))$$
$$- \varphi_\lambda(t) L_u(t, u(t), \dot{u}(t)) - \dot{\varphi}_\lambda(t) L_p(t, u(t), \dot{u}(t)) \} dt$$
$$\geq 0$$

and

$$\int_J F(t) \, dt = \left[\int_{t_0}^{\tau - \lambda^2} + \int_{\tau - \lambda^2}^{\tau} + \int_\tau^{\tau + \lambda} + \int_{\tau + \lambda}^{t_1} \right] F(s) \, ds.$$

Note the first and fourth integrals are both equal to zero, whereas the integrand of the third integral is $o(\lambda)$, whence

$$\lim_{\lambda \to 0} \frac{1}{\lambda^2} \int_\tau^{\tau + \lambda} F(s) \, ds = 0.$$

Lastly, the second integral yields:

$$\lim_{\lambda \to 0} \frac{1}{\lambda^2} \int_{\tau - \lambda^2}^{\tau} F(s) \, ds$$
$$= L(\tau, u(\tau), \dot{u}(\tau) + \xi) - L(\tau, u(\tau), \dot{u}(\tau)) - \xi L_p(\tau, u(\tau), \dot{u}(\tau)).$$

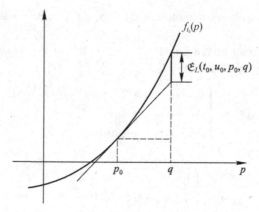

Fig. 4.1

We call the following

$$\mathfrak{E}_L(t, u, p, q) = L(t, u, q) - L(t, u, p) - (q - p) \cdot L_p(t, u, p)$$

the Weierstrass excess function. Figure 4.1 illustrates its geometric meaning.

As seen in the graph, $t_0 \in J$, $u_0 = u(t_0)$, $p_0 = \dot{u}(t_0)$. For $f_{t_0}(p) = L(t_0, u_0, p)$, $\mathfrak{E}_L(t_0, u_0, p_0, q)$ is the difference of the value of f_{t_0} at which $p = q$ and the value of tangent line $f_{t_0}(p_0) + (q - p_0)\frac{d}{dp}f_{t_0}(p_0)$, or simply, the difference of the curve and its tangent line.

In summary, we have:

Theorem 4.1 *Suppose $u \in C^1(J, \mathbb{R}^N)$ is a strong minimum of I, then*

$$\mathfrak{E}_L(t, u(t), \dot{u}(t), \dot{u}(t) + \xi) \geq 0, \quad \forall \xi \in \mathbb{R}^N, \quad \forall t \in J. \tag{4.1}$$

4.3 Extremal fields and strong minima

In this section, we turn our attention to the sufficient condition of a strong minimum. Given a function u, we compare u with nearby C^0 functions sharing common endpoints.

Conventionally, we call the graph $\gamma_0 = \{(t, u_0(t)) \in J \times \mathbb{R}^N \mid t \in J\}$ of u_0, a solution of the E-L equation, an extremal curve. We will embed the extremal curve γ_0 into its nearby extremal curves.

Given $J = [t_0, t_1]$, a Lagrangian $L \in C^1(J \times \mathbb{R}^N \times \mathbb{R}^N)$, the functional

$$I(u) = \int_J L(t, u(t), \dot{u}(t)) \, dt,$$

and u_0, a solution of the E-L equation of I.

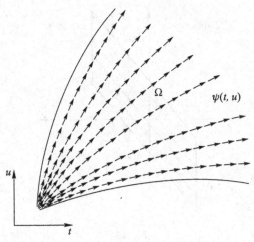

Fig. 4.2

Suppose we can extend u_0 to a larger interval $J_1 = (a, b) \supset J$ and suppose $\{(t, u(t, \alpha)) \mid t \in J_1, \alpha \in B_{\epsilon_1}(\theta) \subset \mathbb{R}^N, \epsilon_1 > 0\}$ is a family of sufficiently smooth extremal curves of I.

Definition 4.2 (A field of extremals) Let Ω be a simply connected neighborhood of $\{(t, u(t, \alpha)) \mid t \in J_1, \alpha \in B_\epsilon(\theta) \subset \mathbb{R}^N\}$ $(\epsilon \in (0, \epsilon_1))$. If $\psi \in C^1(\Omega, \mathbb{R}^N)$ is a vector field which satisfies:

1. every solution of $\dot{u}(t) = \psi(t, u(t))$ is a solution of the E-L equation of I;
2. $\det(\partial_{\alpha_i} u_j(t, \alpha)) \neq 0$;
3. $\forall (t_1, u_1) \in \Omega$, $\exists$ a unique $\alpha_1 \in B_{\epsilon_1}(\theta)$, such that $u(t_1, \alpha_1) = u_1$;
4. $u(t, 0) = u_0(t)$,

then Ω is said to be a field of extremals and ψ is said to be its directional field (flow) (see Figure 4.2).

Example 4.2 Let $L = \frac{1}{2}p^2$, then the E-L equation $\ddot{u} = 0$ has solutions $u_\lambda = mt + \lambda$. Hence, $\Omega = \{(t, mt + \lambda) \mid (t, \lambda) \in \mathbb{R}^1 \times \mathbb{R}^1\}$ and $\psi(t, u) = m$ is a field of extremals and its directional field, where m is a constant (see Figure 4.3).

Example 4.3 Let $L = \sqrt{(1 + p^2)}$, then the E-L equation $\ddot{u} = 0$ has solutions $u_\lambda = \lambda t$. Hence, $\Omega = \{(t, \lambda t) \mid (t, \lambda) \in (t_0, \infty) \times \mathbb{R}^1\}$. When $t_0 > 0$, Ω and $\psi(t, u) = \frac{u}{t}$ is a field of extremals and its directional field (see Figure 4.4).

Example 4.4 Let $L = \frac{1}{2}(p^2 - u^2)$, then the E-L equation $\ddot{u} = -u$ has solutions $u_\lambda = \sin(t + \lambda)$, $\forall \lambda \in \mathbb{R}^1$. For any open interval J, let $\Omega = \{(t, \sin(t + \lambda)) \mid (t, \lambda) \in J \times (-1, +1)\}$. Although Ω is covered by extremal

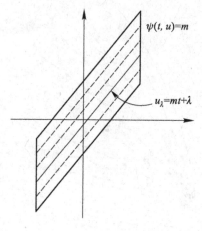

Fig. 4.3

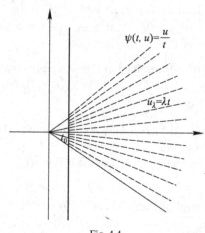

Fig. 4.4

curves, there are two distinct extremal curves passing through each point of Ω, so it is not a field of extremals.

In contrast, the family of extremal curves $u_\lambda = \lambda \sin t \ (\forall \lambda \in \mathbb{R}^1)$ generates a field of extremals and its directional field where $\Omega = \{(t, \lambda \sin t) \mid (t, \lambda) \in (\epsilon, \pi - \epsilon) \times \mathbb{R}^1\} \ (\epsilon \in (0, \frac{\pi}{2}))$ and $\psi(t, u) = u \cot t$.

Suppose $\gamma_0 = \{(t, u_0(t)) \mid t \in J\} \subset \Omega$ and ψ its directional field, we compare it with its nearby piecewise C^1 curves $\gamma = \{(t, u(t)) \mid t \in J\} \subset \Omega$ with common endpoints $(u(t_0) = u_0(t_0),\ u(t_1) = u_0(t_1))$. Since

$$I(u) = \int_\gamma L dt = \int_J L(t, u(t), \dot{u}(t)) dt$$

and ψ is a directional field of Ω, it follows that

$$I(u_0) = \int_{\gamma_0} [(L(t, u, \psi(t, u)) - \psi(t, u) L_p(t, u, \psi(t, u))) dt + L_p(t, u, \psi(t, u)) du].$$

If the integral is independent of path, then

$$\begin{aligned}
I(u_0) &= \int_J L(t, u_0(t), \dot{u}_0(t)) dt \\
&= \int_\gamma [(L(t, u, \psi(t, u)) - \psi(t, u) L_p(t, u, \psi(t, u))) dt + L_p(t, u, \psi(t, u)) du] \\
&= \int_J [(L(t, u(t), \psi(t, u(t)))) - (\dot{u}(t) - \psi(t, u(t))) L_p(t, u(t), \psi(t, u(t))))] dt.
\end{aligned}$$

Thus,

$$\begin{aligned}
I(u) - I(u_0) &= \int_J [L(t, u(t), \dot{u}(t)) - L(t, u(t), \psi(t, u(t))) \\
&\quad - (\dot{u}(t) - \psi(t, u(t))) L_p(t, u(t), \psi(t, u(t)))] dt \\
&= \int_J \mathfrak{E}(t, u(t), \psi(t, u(t)), \dot{u}(t)) dt.
\end{aligned}$$

If we further assume

$$\mathfrak{E}(t, u, \psi(t, u), p) \geq 0, \quad \forall (t, u, p) \in \Omega \times \mathbb{R}^N,$$

then

$$I(u) \geq I(u_0),$$

which means u_0 is a strong minimum.

We next address the following questions:

1. Is it possible to embed the extremal curve of u_0 into a field of extremals?

By "embed" we mean there exists an open interval $J_1 \supset J$, a continuous function $u : J_1 \times B_\epsilon \to \mathbb{R}^N$ such that $\forall \alpha = (\alpha_1, \alpha_2, \ldots, \alpha_N) \in B_\epsilon \subset \mathbb{R}^N$, $u(t, \alpha)$ is an extremal curve for which $u_0(t) = u(t, 0)|_J$, $\forall t \in J$ and $\Omega = \{t, u(t, \alpha) \mid t \in J_1, \alpha \in B_\epsilon\}$ is a field of extremals.

2. Why is the above integral independent of path?

We now answer the first question: under what condition can a given extremal curve γ_0 be embedded into a field of extremals Ω?

For simplicity, we shall only present the argument for the case $N = 1$.

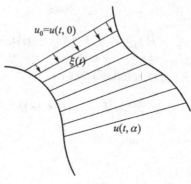

Fig. 4.5

Lemma 4.1 *If $L \in C^3$ and $\{u(t,\alpha) \in C^2(J \times (-\epsilon, \epsilon))\}$ is a family of solutions of the E-L equation, then*

$$\xi(t) = \frac{\partial}{\partial \alpha} u(t,s) \Big|_{\alpha=0}$$

is a Jacobi field along $u_0(t) = u(t,0)$ (see Figure 4.5).

Proof Denote $u_\alpha = u(t,\alpha)$ and differentiate the E-L equation:

$$\frac{d}{dt} L_p(t, u_\alpha(t), \dot{u}_\alpha(t)) = L_u(t, u_\alpha(t), \dot{u}_\alpha(t))$$

with respect to α at $\alpha = 0$, by letting $\tau = (t, u_0(t), \dot{u}_0(t))$, we obtain:

$$\frac{d}{dt}\left(L_{pp}(\tau)\dot{\xi}(t) + L_{pu}(\tau)\xi(t)\right) = L_{pu}(\tau)\dot{\xi}(t) + L_{uu}(\tau)\xi(t),$$

which means

$$J_{u_0}(\xi) = 0. \qquad \qquad \square$$

Lemma 4.2 *Assume $N = 1$, $L \in C^3$, and $u_0 \in C^2$ is a solution of its E-L equation. Suppose the strict Legendre–Hadamard condition along $u_0(t)$ holds: $L_{pp}(t, u_0(t), \dot{u}_0(t)) > 0$ for all $t \in (t_0, t_1)$. If on $(t_0, t_1]$, $(t_0, u_0(t_0))$ has no conjugate point along the extrema curve γ_0 of u_0, then γ_0 can be embedded into a field of extremals Ω, a simply connected region generated by the family of solution curves.*

Proof 1. From $L_{pp}(t, u_0(t), \dot{u}_0(t)) > 0$ for all $t \in J = [t_0, t_1]$ and the E-L equation, the solution u_0 can be extended to a larger interval $J_1 = (a, b) \supset J$.

2. $\forall \alpha \in \mathbb{R}^1$ with $|\alpha| < \epsilon_0$ sufficiently small, solving the initial value problem of the E-L equation:

$$E_L(\varphi(\cdot, \alpha)) = 0, \quad \varphi(a, \alpha) = u_0(a), \quad \varphi_t(a, \alpha) = \dot{u}_0(a) + \alpha,$$

it yields a family of solution $\varphi(t, \alpha)$, $t \in (a, b)$, and $|\alpha| < \epsilon_0$. By the uniqueness of solutions of ordinary differential equations, $\varphi(t, 0) = u_0(t)$. This fulfills requirement (4) in Definition 4.2.

3. Define $\Omega_\epsilon = \{(t, \varphi(t, \alpha)) \mid t \in (a, b), |\alpha| < \epsilon\}$, $\epsilon < \epsilon_0$. By Lemma 4.1,

$$\xi(t) = \partial_\alpha \varphi(t, 0)|_{\alpha=0}$$

is a Jacobi field along u_0 which satisfies:

$$\xi(a) = 0 \quad \text{and} \quad \dot{\xi}(a) = 1.$$

By our assumption, it has no conjugate point, and Lemma 3.3 shows that there exists $\xi(t) > 0$ for all $t \in [a, t_1]$. According to the continuous dependence of parameters, for $0 < \epsilon_1 < \epsilon_0$,

$$\partial_\alpha \varphi(t, \alpha) > 0, \quad |\alpha| < \epsilon_1.$$

This fulfills requirement (2) in Definition 4.2.

Using the Implicit Function Theorem, we can find $0 < \epsilon_2 < \epsilon_1$, $\forall\, (t, u) \in \Omega_{\epsilon_2}$, the equation

$$u = \varphi(t, \alpha)$$

has a unique continuously differentiable solution $\alpha = w(t, u) \in B_{\epsilon_2}(0)$. This fulfills requirement (3) in Definition 4.2.

4. Let $\Omega = \Omega_{\epsilon_2}$, it contains γ_0 and it is homeomorphic to a quadrilateral, whence it is simply connected. Furthermore, the directional field $\psi(t, u)$ generated by $\varphi(t, u)$ is

$$\psi(t, u) = \partial_t \varphi(t, w(t, u)).$$

It follows that ψ is defined everywhere in Ω and $\varphi(t, \alpha)$ describes a family of solution curves of the equation

$$\dot{u} = \partial_t \varphi(t, \alpha) = \partial_t \varphi(t, w(t, u)) = \psi(t, u).$$

This fulfills requirement (1) of Definition 4.2 (see Figure 4.6). $\qquad\square$

Remark 4.1 When $N > 1$, Lemma 4.2 still holds. It states: let $L \in C^3([t_0, t_1] \times \mathbb{R}^N \times \mathbb{R}^N)$ and $u_0 \in C^2([t_0, t_1] \times \mathbb{R}^N)$ be a solution of its E-L equation. Suppose for all $t \in [t_0, t_1]$, the matrix $L_{pp}(t, u_0(t), \dot{u}_0(t))$ is invertible. If on $(t_0, t_1]$, $(t_0, u_0(t_0))$ has no conjugate point along the extrema curve γ_0 of u_0, then γ_0 can be embedded into a field of extremals Ω, a simply connected region generated by the family of solution curves.

The proof resembles the proof of Lemma 4.2. We need only replace the scalar $\alpha \in \mathbb{R}^1$ in step 2 above by a vector $\alpha \in \mathbb{R}^N$. The resulting solution $\varphi(t, \alpha)$ of the E-L equation then satisfies:

$$\partial_\alpha \varphi_j(a, \alpha) = 0, \quad \partial_{\alpha_i} \partial_t \varphi_j(a, \alpha) = \delta_{ij}, \quad 1 \le i, j \le N.$$

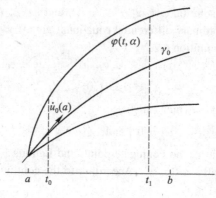

Fig. 4.6

We modify the Jacobi field in the third step above by $\{w_i(t, \alpha) = \partial_{\alpha_i} \partial_t \varphi(a, \alpha), \; i = 1, \ldots, N\}$. Since

$$w_i(a, 0) = 0, \quad \partial_{\alpha_i} w_i(a, 0) = e_i, \quad i = 1, \ldots, N$$

and on J_1, it has no conjugate point along γ_0, we have:

$$\det(\partial_\alpha \partial_t \varphi_j(t, \alpha)) \neq 0, \quad \forall \, t \in J_1.$$

The rest of the proof remains the same.

4.4 Mayer field, Hilbert's invariant integral

We now examine the second question: under what condition is the integral in Theorem 4.1 independent of path?

Let

$$\begin{cases} R_i(t, u) = L_{p_i}(t, u, \psi(t, u)), \\ H(t, u) = \langle \psi(t, u), L_p(t, u, \psi(t, u)) \rangle - L(t, u, \psi(t, u)), \end{cases}$$

where $\langle \cdot, \cdot \rangle$ denotes the standard inner product on $\mathbb{R}^N$ and the 1-form

$$\omega = \sum_{i=1}^{N} R_i \, du_i - H \, dt.$$

We show that

$$d\omega = 0.$$

Definition 4.3 A field of extremals is called a Mayer field if it satisfies the following compatibility condition:

$$\partial_{u_i} L_{p_j}(t, u(t), \psi(t, u(t))) = \partial_{u_j} L_{p_i}(t, u(t), \psi(t, u(t))), \quad \forall \, 1 \leq i, j \leq N.$$

Corollary 4.1 *For $N = 1$, every field of extremals is a Mayer field.*

Given a field of extremals and its directional field (Ω, ψ), we introduce

$$D_\psi = \partial_t + \sum_{i=1}^N \psi_i \partial_{u_i} + \sum_{i=1}^N \left(\partial_t \psi_i + \sum_{k=1}^N \psi_k \partial_{u_k} \psi_i \right) \partial_{p_i}$$

and we have:

Lemma 4.3 *Let $L \in C^2(J \times \mathbb{R}^N \times \mathbb{R}^N)$. Then (Ω, ψ) is a field of extremals and its directional field if and only if for any integral curve $(t, u(t))$, we have:*

$$D_\psi L_p(t, u(t, \alpha), \psi(t, u(t, \alpha))) = L_u(t, u(t, \alpha), \psi(t, u(t, \alpha))).$$

Proof We henceforth denote $\tilde{L} := \tilde{L}(t, \alpha) = L(t, u(t, \alpha), \psi(t, u(t, \alpha)))$ and likewise for $\tilde{L}_{u_i}$ and $\tilde{L}_{p_i}$. (Ω, ψ) is a field of extremals and its directional field if and only if $\dot{u}(t) = \psi(t, u(t))$ with the E-L equation $\tilde{L}_u = \frac{d}{dt} \tilde{L}_p$. We compute directly that

$$\tilde{L}_{u_i} = \frac{d}{dt} \tilde{L}_{p_i}$$

$$= \left(\partial_t + \sum_{j=1}^N \left[\psi^j \partial_{u_j} + \frac{d}{dt} \psi_j(t, u(t)) \partial_{p_j} \right] \right) \tilde{L}_{p_i}$$

$$= \left(\partial_t + \sum_{j=1}^N \left[\psi^j \partial_{u_j} + \left(\partial_t \psi_j + \Sigma_{k=1}^N \psi_k \partial_{u_k} \psi_j \right) \partial_{p_j} \right] \right) \tilde{L}_{p_i}$$

$$= D_\psi \tilde{L}_{p_i}.$$

Lemma 4.4 *If $L \in C^2(J \times \mathbb{R}^N \times \mathbb{R}^N)$, then Ω is a Mayer critical field if and only if $d\omega = 0$, i.e.*

$$\partial_t R_i = -\partial_{u_i} H, \quad 1 \leq i \leq N.$$

Proof By direct computation, we have:

$$\partial_t R_i = \left(\partial_t + \sum_{j=1}^N \psi_t^j \partial_{p_j} \right) \tilde{L}_{p_j}.$$

It follows from the compatibility condition that

$$-H_{u_i} = \partial_{u_i}(L - \psi L_p)(t, u, \psi(t, u))$$

$$= \left(\tilde{L}_{u_i} + \partial_{u_i} \psi^j \partial_{p_j} \tilde{L} \right) - \partial_{u_i} \sum_{j=1}^N \psi^j \tilde{L}_{p_j} - \sum_{j=1}^N \psi^j \partial_{u_j} \tilde{L}_{p_i}$$

$$= \tilde{L}_{u_i} - \sum_1^N \left[\psi^j \left(\partial_{u_j} + \sum_1^N \psi_{u_j}^k \partial_{p_k} \right) \right] \tilde{L}_{p_i}.$$

Hence,

$$\partial_t R_i + \partial_{u_i} H$$

$$= \left(\partial_t + \sum_{i=1}^{N} \psi_t^i \partial_{p_i} \right) \tilde{L}_p + \sum_{j=1}^{N} \left(\psi^j \partial_{u_j} + \sum_{k=1}^{N} \psi^k \partial_{u_k} \psi^j \partial_{p_j} \right) \tilde{L}_{p_i} - \tilde{L}_{u_i}$$

$$= D_\psi \tilde{L}_{p_i} - \tilde{L}_{u_i}.$$

$d\omega = 0$ now follows immediately from Lemma 4.3. Conversely, using $d\omega = 0$ and the above equality, the reader can derive the compatibility condition.   □

Given a Mayer field (Ω, ψ), both R and H are already defined. If γ is any curve joining the points $(t_0, u(t_0))$ and (t, u), the line intergal

$$S(t, u) = \int_{(t_0, u(t_0))}^{(t,u)} \sum_{i=1}^{N} \left(R_i(t, u) \, du_i - H(t, u) \, dt \right)$$

$$= \int_\gamma (L_p \cdot du + (L - \psi L_p) dt)$$

is independent of path. We shall call this line intergal the Hilbert invariant integral.

4.5   A sufficient condition for strong minima

The case $N = 1$ is relatively straightforward, we have the following:

Theorem 4.2 *For $N = 1$ and $L \in C^3$, suppose on $(t_0, t_1]$, $(t_0, u_0(t_0))$ has no conjugate point along γ_0, the extremal curve corresponding to the solution of the E-L equation u_0. Let (Ω, ψ) be a field of extremals with a directional field, in which u_0 is embedded. If $L_{pp}(t, u, p) > 0$, $\forall\, (t, u, p) \in \Omega \times \mathbb{R}^1$, then u_0 is a strong minimum of I.*

Proof Notice

$$\mathfrak{E}(t, u, \psi(t, u), q) = L(t, u, q) - L(t, u, \psi(t, u)) - (q - \psi(t, u)) L_p(t, u, \psi(t, u))$$

$$= L_{pp}(t, u, v) \geq 0, \quad \forall\, (t, u, q) \in \Omega \times \mathbb{R}^N,$$

where v is in between $\psi(t, u)$ and q. Our assertion follows from Lemmas 4.2 and 4.3, and the argument used in §4.3.   □

Example 4.5 Let

$$M = \{v \in C^1([0,a]) \mid v(0) = (\cosh a)^{-1}, v(a) = 1\}.$$

Suppose the E-L equation $\ddot{u} - u = 0$ corresponding to

$$I(u) = \int_0^a (\dot{u}^2 + u^2)dt$$

has a solution

$$u_0 = \frac{\cosh t}{\cosh a}.$$

Since the Jacobi equation $\ddot{\varphi} - \varphi = 0$ has no conjugate points and

$$\mathfrak{E}(t, u, p, q) = (u^2 + q^2) - (u^2 + p^2) - 2p(q - p) = (q - p)^2 \geq 0,$$

u_0 must be a strong minimum.

Example 4.6 Find the weak minima and the strong minima of the functional

$$I(u) = \int_0^1 \dot{u}^2(1 + \dot{u})^2 dt$$

with the boundary conditions $u(0) = 0$ and $u(1) = m$.

Clearly, I has minimal value 0. The solution of the E-L equation which achieves the value zero must satisfy either $\dot{u} = 0$ or $\dot{u} = -1$.

Unless $m = 0$ and $u(t) = 0$, or $m = -1$ and $u(t) = -t$, there can be no other C^1 solution. However, for Lipschitz functions, if there exists a solution, then it is necessary to have $-1 \leq m \leq 0$, in which case, they are

$$u_1 = \begin{cases} -t & 0 \leq t \leq -m, \\ m & -m \leq t \leq 1. \end{cases}$$

or

$$u_2 = \begin{cases} 0 & 0 \leq t \leq 1 + m, \\ 1 + m - t & 1 + m \leq t \leq 1. \end{cases}$$

We calculate the second derivatives of the Lagrangian:

$$L_{uu} = L_{up} = 0, \quad L_{pp} = 2(6p^2 + 6p + 1) = 12(p - p_1)(p - p_2).$$

$L_{pp} = 0$ has zeros

$$p_1 = -\frac{1}{2} - \frac{\sqrt{3}}{6} \quad \text{and} \quad p_2 = -\frac{1}{2} + \frac{\sqrt{3}}{6}.$$

When $p \notin [p_1, p_2]$, $L_{pp} > 0$. Regardless of whether we are considering the C^1 solutions or the Lip solutions u_1 and u_2, their slopes are equal to 0 or -1, which are outside the interval, hence $L_{pp}(t, u(t), \dot{u}(t)) > 0$. Their corresponding Jacobi operator

$$J_u(\varphi) = 2(6p^2 + 6p + 1)\ddot{\varphi}$$

has no conjugate points, whence u_1 and u_2 are both strict weak minima.

Lastly, we take a look at the Weierstrass excess function

$$\mathfrak{E}(t, u, p, q) = q^2(1+q)^2 - p^2(1+p)^2 - (q-p)(4p^3 + 6p^2 + 2p)$$

$$= [q(1+q) - p(1+p)]^2 + 2p(1+p)(q-p)^2$$

$$\geq 0.$$

This implies u_1 and u_2 are both strong minima as well.

Utilizing the concept of Mayer field, we have the following theorem regarding $N > 1$.

Theorem 4.3 *Suppose the E-L equation of I has a solution u_0. Suppose its corresponding extremal curve γ_0 can be embedded into a family of extremal curves such that they define a Mayer field (Ω, ψ). Furthermore, if*

$$\mathfrak{E}(t, u, \psi(t, u), p) \geq 0, \quad \forall \, (t, u, p) \in \Omega \times \mathbb{R}^N,$$

then u_0 is a strong minimum of I.

In fact, for $N > 1$, a result similar to Theorem 4.2 also holds.

Theorem 4.4 *Let $L \in C^3(J \times \mathbb{R}^N \times \mathbb{R}^N, \mathbb{R}^1)$. Suppose*

(1) *$(t_0, u_0(t_0))$ has no conjugate points along the critical curve γ_0,*
(2) *$(L_{p_i p_j}(t, u_0(t), \dot{u}_0(t))) \, \forall t \in J$ is positive definite,*
(3) *$\mathfrak{E}(t, u, \psi(t, u), p)) \geq 0, \forall \, (t, u, p) \in \Omega \times \mathbb{R}^N, p \neq \psi(t, u)$,*

then u_0 is a strong minimum of I.

4.6* The proof of Theorem 4.4 (for the case $N > 1$)

So far, we already know: if $N = 1$, then any extremal field (Ω, ψ) is a Mayer field. However, for $N > 1$, we are seeking the conditions for which an extremal curve can be embedded into a Mayer field.

For a given Lagrangian L, we denote $\tilde{L}(t, \alpha) = L(t, \varphi(t, \alpha), \dot{\varphi}(t, \alpha))$, where $\dot{\varphi}(t, \alpha) = \varphi_t(t, \alpha)$. Likewise, we denote $\tilde{L}_{u_i}$ and $\tilde{L}_{p_i}$. By direct computation, we have:

$$
d\omega = d\left(\sum_{i=1}^{N} R_i \, du_i - H \, dt \right)
$$

$$
= d\left(\sum_{i=1}^{N} \tilde{L}_{p_i} \, du^i + \left(\tilde{L} - \sum_{i=1}^{N} \dot{\varphi}^i \tilde{L}_{p_i} \right) dt \right)
$$

$$
= d\left(\tilde{L} \, dt + \sum_{i=1}^{N} \tilde{L}_{p_i} (du^i - \dot{\varphi}^i \, dt) \right)
$$

$$
= \sum_{k=1}^{N} \tilde{L}_{u_k} \, d\varphi^k \wedge dt + \sum_{k=1}^{N} \tilde{L}_{p_k} \, d\dot{\varphi}^k \wedge dt - \sum_{i=1}^{N} \tilde{L}_{p_i} \, d\dot{\varphi}^i \wedge dt
$$

$$
+ \sum_{i=1}^{N} \sum_{l=1}^{N} d\tilde{L}_{p_i} \varphi^i_{\alpha_l} \, d\alpha^l
$$

$$
= \sum_{k=1}^{N} \sum_{l=1}^{N} d\tilde{L}_{u_k} \varphi^k_{\alpha_l} d\alpha^l \wedge dt + \sum_{i=1}^{N} \sum_{l=1}^{N} \partial_t \tilde{L}_{p_i} \varphi^i_{\alpha_l} \, dt \wedge d\alpha^l
$$

$$
+ \sum_{m=1}^{N} \sum_{i=1}^{N} \sum_{l=1}^{N} \partial_{\alpha^m} \tilde{L}_{p_i} \varphi^i_{\alpha_l} \, d\alpha^m \wedge d\alpha^l
$$

$$
= \sum_{i=1}^{N} \sum_{l=1}^{N} (\tilde{L}_{u_i} - \partial_t \tilde{L}_{p_i}) \varphi^i_{\alpha_l} d\alpha^l \wedge dt + \sum_{m=1}^{N} \sum_{i=1}^{N} \sum_{l=1}^{N} \partial_{\alpha^m} \tilde{L}_{p_i} \varphi^i_{\alpha_l} \, d\alpha^m \wedge d\alpha^l.
$$

Denote

$$
[\alpha^l, \alpha^m] = \sum_{i=1}^{N} (\partial_{\alpha^l} \tilde{L}_{p_i} \varphi^i_{\alpha^m} - \partial_{\alpha^m} \tilde{L}_{p_i} \varphi^i_{\alpha^l})
$$

and we call it the Lagrange bracket. Using the Lagrange bracket, we can rewrite the above calculation as the following formula:

$$
d\omega = \sum_{i=1}^{N} \sum_{l=1}^{N} E_L(\varphi)_i \varphi^i_{\alpha^l} d\alpha^l \wedge dt + \sum_{1 \le l < m \le N} [\alpha^l, \alpha^m] \, d\alpha^l \wedge d\alpha^m. \quad (4.2)
$$

Consequently, we have:

Lemma 4.5 *Let* $L \in C^3(J \times \mathbb{R}^n \times \mathbb{R}^N)$. *Suppose* (Ω, ψ) *is an extremal field determined by a family of extremal curves* (φ, α), *then* (Ω, ψ) *is a Mayer field if and only if*

$$
E_L(\varphi(\cdot, \alpha)) = 0, \quad \forall \alpha \in \mathbb{R}^N, \quad [\alpha^l, \alpha^m] = 0, \quad 1 \le l, m \le N.
$$

Proof Suppose (Ω, ψ) is a Mayer field, then by the invariance of differential forms, the left-hand side of (4.2) is equal to zero. Since the extremal curve $\varphi(\cdot, \alpha)$ satisfies the E-L equation, the first term on the right-hand side is zero. Hence, the Lagrange bracket equals zero.

Conversely, if

$$E_L(\varphi(\cdot, \alpha)) = 0, \quad \forall \alpha \in \mathbb{R}^N, \quad [\alpha^l, \alpha^m] = 0, \quad 1 \le l, m \le N,$$

then ω is closed. Furthermore, since (Ω, ψ) is an extremal field, (Ω, ψ) is a Mayer field. $\qquad\square$

For the Lagrange bracket, we have:

Lemma 4.6 *Suppose $L \in C^3(J \times \mathbb{R}^n \times \mathbb{R}^N)$ and (Ω, ψ) is an extremal field, then*

$$\frac{\partial}{\partial t}[\alpha^l, \alpha^m] = 0, \quad \forall l, m.$$

Proof Using conditions 1–4 in the definition of extremal fields, we compute directly:

$$\frac{\partial}{\partial t}[\alpha^l, \alpha^m] = \sum_{i=1}^{N} \left[\frac{\partial \tilde{L}_{u_i}}{\partial \alpha^l} \varphi^i_{\alpha^m} + \frac{\partial \tilde{L}_{p_i}}{\partial \alpha^l} \dot{\varphi}^i_{\alpha^m} - \frac{\partial \tilde{L}_{u_i}}{\partial \alpha^m} \varphi^i_{\alpha^l} - \frac{\partial \tilde{L}_{p_i}}{\partial \alpha^m} \dot{\varphi}^i_{\alpha^l} \right].$$

After differentiating $\frac{\partial \tilde{L}_{u_i}}{\partial \alpha^l}, \frac{\partial \tilde{L}_{u_i}}{\partial \alpha^m}, \frac{\partial \tilde{L}_{p_i}}{\partial \alpha^l}, \frac{\partial \tilde{L}_{p_i}}{\partial \alpha^m}$, and using the fact that L is twice differentiable, all terms will cancel, which leaves us

$$\frac{\partial}{\partial t}[\alpha^l, \alpha^m] = 0. \qquad\square$$

In the following, we will strengthen Remark 4.1 for the higher dimensional cases, embedding a critical curve γ_0 defined on $J = [t_0, t_1]$ into a Mayer field (Ω, ψ).

Theorem 4.5 *Suppose the extremal curve γ_0 has no conjugate points on J. Suppose also $(L_{p_i p_j}(t, u(t), \dot{u}(t))$ is invertible for all $t \in J$, then γ_0 can be embedded into a family of extremal curves such that this family of curves defines a Mayer field (Ω, ψ).*

Proof Based on Remark 4.1, we obtain a field of extremals (Ω, ψ), we now show that this is a Mayer field. In fact, using $\partial_{\alpha_i} \varphi_j(a, \alpha) = 0, \forall i, j$, we can deduce that $[\alpha^l, \alpha^m](a, \alpha) = 0, \forall l, m \in [1, N]$. By Lemma 4.6, $\frac{d}{dt}[\alpha^l, \alpha^m](t, \alpha) = 0$, hence $[\alpha^l, \alpha^m](t, \alpha) = 0$. This means (Ω, ψ) is a Mayer field. $\qquad\square$

Exercises

1. Use the Weierstrass excess function to verify that u_0 is not a strong minimum of $I(u) = \int_J (\dot{u}^2 + \dot{u}^3)\, dt$.

2. Verify that for $I(u) = \int_0^2 \sqrt{u(1 + \dot{u}^2)}$ and $M = \{u \in C^1([0,2]) \mid u(0) = 2, u(2) = 5\}$, there is a two-parameter family of solutions to the E-L equation

$$u(t, \alpha, \beta) = \alpha + \frac{1}{\alpha}\left(\frac{t + \beta}{2}\right)^2.$$

When $(\alpha, \beta) = (1, 2)$, $u(t, 1, 2) \in M$.

Use $(u(t, \alpha, \beta))$ to determine two independent Jacobi fields.

3. Suppose $I(u) = \int_a^b \sqrt{1 + \dot{u}}\, dt$ is defined on $C_0^1(0, b)$, describe the field of extremals $(\psi(t, u))$ containing $u_0 = 0$. Also, verify that it is a strong minimum.

4. Let $I(u) = \int_1^2 (\dot{u}^2 + t^2 \dot{u}^2)\, dt$ and $M = \{v \in C^1([1, 2]) \mid v(1) = 1, v(2) = 2\}$. Verify that

$$u_0(t) = -\frac{2}{t} + 3$$

is a strong minimum.

5. Suppose $u \in C^1(J \times \mathbb{R}^N, \mathbb{R}^N)$ satisfies $E_l(u(t, \alpha)) = 0, \forall t \in J, \forall \alpha \in \mathbb{R}^N$, $u(0, \alpha) = \theta$, $\partial_t u(0, \alpha) = \alpha$, and $u_0 = u(\cdot, 0)$. Prove that

   (1) $\partial_{\alpha_i} u(t, \alpha)|_{\alpha=0} (i = 1, \ldots, N)$ are Jacobi fields.
   (2) If $(t, u_0(t))$ has no conjugate points on J, then $\det(\partial_{\alpha_i} u_j(t, \alpha)) \neq 0$, $\forall t \in J$.

6. Suppose $I(u) = \int_a^b [\frac{1}{2}\dot{u}^2 - V(u)]\, dt$.

   (1) When $V(u) = \cos u$, write down the Jacobi operator and the Jacobi field along the E-L equation solution $u = 0$.
   (2) When $V \in C^2(\mathbb{R}^1)$, $V''(u) \leq 0$, show that u_0 is a strict weak minimum of I.

Lecture 5

The Hamilton–Jacobi theory

5.1 Eikonal and the Carathéodory system of equations

Given a Lagrangian $L : J \times \mathbb{R}^N \times \mathbb{R}^N \longrightarrow \mathbb{R}^1$ and its corresponding functional

$$I(u) = \int_J L(t, u(t), \dot{u}(t)) dt,$$

we know from the previous lecture that for a field of extremals and its directional field (Ω, ψ), there exists a 1-form

$$\omega = L_p(t, u, \psi(t, u)) du - (\langle \psi(t, u), L_p(t, u, \psi(t, u)) \rangle$$

$$- L(t, u, \psi(t, u))) dt,$$

where $\langle \cdot, \cdot \rangle$ denotes the standard inner product on $\mathbb{R}^N$.

In order for (Ω, ψ) to be a Mayer field, it is necessary and sufficient to require ω to be a closed form.

As a consequence, by the Mayer field, ω defines a single-valued function g:

$$g(t, u) - g(t_0, u_0) = \int_\gamma \omega, \tag{5.1}$$

where γ is any curve connecting (t_0, u_0) and (t, u). We call this single-valued function g an eikonal.

Remark 5.1 The eikonal has the following physical meaning: consider the line integral of ω along an extremal curve $\gamma = (t, u(t))$, $\dot{u}(t) = \psi(t, u(t))$:

61

$$g(t_2, u(t_2)) - g(t_1, u(t_1)) = \int_\gamma \omega$$

$$= \int_J ((\langle L_p(t, u(t), \psi(t, u(t))), du \rangle$$
$$- [\langle \psi(t, u(t)), L_p(t, u(t), \psi(t, u(t))) \rangle$$
$$- L(t, u(t), \psi(t, u(t)))]) dt$$
$$= \int_J \langle L_p(t, u(t), \dot{u}(t)), \dot{u}(t) \rangle$$
$$- [\langle \dot{u}(t), L_p(t, u(t), \dot{u}(t)) \rangle - L(t, u(t), \dot{u}(t))] dt$$
$$= \int_J L(t, u(t), \dot{u}(t)) dt.$$

This shows that the difference of an eikonal g between any two points along a given extremal curve is equal to the line integral of the Lagrangian along that curve.

In optics, the Lagrangian represents the distance of a ray of light traveled in an instance divided by its speed. The line integral is therefore equal to the time elapsed when the light ray travels from point one to point two. Consequently, the wavefront of a light ray from a one point source (or the phase front of waves) can be expressed via the level sets $g(t, u) = $ const.

The definition shows directly that on the field of extremals (Ω, ψ), the eikonal g satisfies the following Carathéodory system of equations:

$$\begin{cases} \nabla_u g(t, u) = L_p(t, u, \psi(t, u)) \\ \partial_t g(t, u) = L(t, u, \psi(t, u)) - \langle \psi(t, u), L_p(t, u, \psi(t, u)) \rangle. \end{cases} \tag{5.2}$$

5.2 The Legendre transformation

Suppose the derivative function f' of $f \in C^2(\mathbb{R}^1, \mathbb{R}^1)$ has an inverse function ψ. Denote $x = \psi(\xi)$. We call

$$f^*(\xi) = \xi x - f(x) = \xi \psi(\xi) - f \circ \psi(\xi)$$

the Legendre transform of f.

The Legendre transformation can be extended to multivariate functions. Let $f \in C^2(\mathbb{R}^N, \mathbb{R}^1)$. Suppose the gradient $\xi = \nabla f(x)$ has inverse mapping ψ. Denote $x = \psi(\xi)$. We call

$$f^*(\xi) = \langle \xi, x \rangle - f(x) = \langle \xi, \psi(\xi) \rangle - f \circ \psi(\xi)$$

the Legendre transform of f.

We describe the geometric meaning of the Legendre transformation as follows: denote G the graph of f $\{(x, y) \in \mathbb{R}^N \times \mathbb{R}^1 \mid y = f(x)\}$, the tangent hyperplane at the point $P = (x, y)$ is $\{(\alpha, \beta) \mid \beta - f(x) = \langle \nabla f(x), \alpha - x \rangle\}$. So any point $Q = (\alpha, \beta)$ on the hyperplane must satisfy:

$$\beta - \langle \nabla f(x), \alpha \rangle = f(x) - \langle \nabla f(x), x \rangle,$$

i.e.

$$\beta - \langle \xi, \alpha \rangle = -f^*(\xi).$$

$-f^*(\xi)$ is the intercept of the hyperplane on the β-axis (see Figure 5.1).

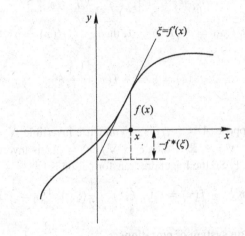

Fig. 5.1

The Legendre transformation has the following properties:
(1) If $f \in C^s$, $s \geq 2$, then $f^* \in C^s$.

Proof In fact, since $\psi \in C^{s-1}$, $f^* \in C^{s-1}$. But by definition,

$$f^*(\xi) = \langle \xi, \psi(\xi) \rangle - f \circ \psi(\xi),$$

and since $\xi = \nabla f(\psi(\xi))$, we have:

$$\nabla f^*(\xi) = \psi(\xi) + \langle \xi, \nabla \psi(\xi) \rangle - \langle \nabla f(\psi(\xi)), \nabla \psi(\xi) \rangle = \psi(\xi),$$

whence $f^* \in C^s$. □

(2) $f^{**} = f$, i.e. the Legendre transformation is reflexive.
In fact, by (1), $x = \psi(\xi) = \nabla f^*(\xi)$,

$$f^{**}(x) = \langle \xi, x \rangle - f^*(\xi) = f(x).$$

In order to emphasize this symmetry, we write it as

$$f(x) + f^*(\xi) = \langle \xi, x \rangle, \quad \xi = \nabla f(x), \quad x = \nabla f^*(\xi).$$

(3)

$$\frac{\partial^2}{\partial \xi_i \partial \xi_j} f^*(\xi)\Big|_{\xi=\nabla f(x)} = \left(\frac{\partial^2 f}{\partial x_i \partial x_j}(x)\right)^{-1}.$$

This is because

$$x = \psi(\nabla f(x)) = (\nabla f^*)((\nabla f)(x)),$$

differentiating on both sides, we have the identity matrix

$$I = \frac{\partial^2}{\partial \xi_i \partial \xi_j} f^*(\xi)\Big|_{\xi=\nabla f(x)} \cdot \frac{\partial^2}{\partial x_i \partial x_j} f(x).$$

Example 5.1 Let $f(x) = x^p/p$, $p > 1$, then $\xi = f'(x) = x^{p-1}$, so $x = \xi^{\frac{1}{p-1}}$. By definition,

$$f^*(\xi) = \xi \cdot x - f(x) = \frac{1}{p'} \xi^{p'},$$

where $\frac{1}{p'} + \frac{1}{p} = 1$.

Example 5.2 Suppose $A = (a_{ij})$ is a symmetric invertible $N \times N$ matrix. Let $f(x) = \frac{1}{2}\langle Ax, x\rangle$, $\forall\, x \in \mathbb{R}^N$, then $\xi = \nabla f(x) = Ax$ is invertible and we can solve for $x = A^{-1}\xi$. So the Legendre transformation is

$$f^*(\xi) = \langle x, \xi\rangle - f(x) = \langle A^{-1}\xi, \xi\rangle - \frac{1}{2}\langle \xi, A^{-1}\xi\rangle = \frac{1}{2}\langle A^{-1}\xi, \xi\rangle.$$

5.3   The Hamilton system of equations

Given a Lagrangian $L : \mathbb{R}^1 \times \mathbb{R}^N \times \mathbb{R}^N \xrightarrow{C^2} \mathbb{R}^1$.
Suppose $\det\left(L_{p_i p_j}(t, u, p)\right) \neq 0$. Let

$$\xi_i = L_{p_i}(t, u, p), \qquad 1 \le i \le N.$$

Using the Implicit Function Theorem, we can solve the system of equations locally

$$p_i = \varphi_i(t, u, \xi), \ \xi = (\xi_1, \xi_2, \ldots, \xi_N), \qquad 1 \le i \le N.$$

Fix (t, u), as a function of p, we apply the Legendre transformation on L and let

$$H(t, u, \xi) = L^*(t, u, \xi) = \sum_1^N \xi_i p_i - L(t, u, p)\big|_{p=\varphi(t,u,\xi)}. \qquad (5.3)$$

We call H the Hamiltonian.

Since the Legendre transformation is reflexive, if H is the Legendre transform of L, L is vice versa the Legendre transform of H.

The E-L equation corresponding to the Lagrangian L is a second order differential equation

$$\frac{d}{dt}L_p(t, u(t), \dot{u}(t)) - L_u(t, u(t), \dot{u}(t)) = 0,$$

which in turn can be written as a system of first order equations:

$$\begin{cases} \dot{u}(t) = p(t), \\ \dfrac{d}{dt}L_p(t, u(t), p(t)) - L_u(t, u, p(t)) = 0, \end{cases}$$

whose solution is $(u(t), p(t))$.

If we further assume $L \in C^3$, then by differentiating both sides of (5.3), we have:

$$\begin{aligned} H_t dt + \langle H_u, du \rangle + \langle H_\xi, d\xi \rangle \\ = \langle \xi, dp \rangle + \langle p, d\xi \rangle - L_t dt - \langle L_u, du \rangle - \langle L_p, dp \rangle \\ = -L_t dt - \langle L_u, du \rangle + \langle p, d\xi \rangle. \end{aligned}$$

Hence, the following relations hold:

$$H_\xi(t, u, \xi) = p, \qquad L_p(t, u, p) = \xi,$$
$$H_u(t, u, \xi) + L_u(t, u, p) = 0, \quad H_t(t, u, \xi) + L_t(t, u, p) = 0.$$

For $(u(t), p(t))$, a solution of the E-L equation, let $\xi(t) = L_p(t, u(t), p(t))$, then

$$\begin{aligned} \dot{\xi}(t) &= \frac{d}{dt}L_p(t, u(t), \dot{u}(t)) \\ &= L_u(t, u(t), \dot{u}(t)) \\ &= L_u(t, u(t), p(t)) \\ &= -H_u(t, u(t), \xi(t)) \end{aligned}$$

and

$$\dot{u}(t) = p(t) = H_\xi(t, u(t), \xi(t)).$$

Thus, $(u(t), \xi(t))$ satisfies the Hamilton system of equations (we shall abbreviate it by H-S):

$$\boxed{\dot{\xi}(t) = -H_u(t, u(t), \xi(t)), \quad \dot{u}(t) = H_\xi(t, u(t), \xi(t)).} \quad (5.4)$$

Conversely, given an H-S solution $(u(t), \xi(t))$, by letting $p(t) = \dot{u}(t)$, and by $\xi(t) = L_p(t, u(t), p(t))$, we can deduce $(u(t), p(t))$ is a solution of the E-L equation:

$$\frac{d}{dt}L_p(t, u(t), \dot{u}(t)) = \dot{\xi}(t) = -H_u(t, u(t), \xi(t)) = L_u(t, u(t), p(t)).$$

From this, we can establish the following one-to-one correspondence between the E-L equation and H-S:

$$\text{(E-L)} \quad \longleftrightarrow \quad \text{(H-S)}$$
$$(u(t), p(t)) \longleftrightarrow (u(t), \xi(t)) = (u(t), L_p(t, u(t), p(t))).$$

The H-S is also the E-L equation of the functional:

$$F(u, \xi) = \int_J [\langle \dot{u}(t), \xi(t) \rangle - H(t, u(t), \xi(t))] dt.$$

Its corresponding 1-form is

$$\alpha = \xi du - H dt,$$

which we call the Poincaré–Cartan invariant.

Since H is the Legendre transform of L, L is also the Legendre transform of H. This implies:

$$L(t, u(t), \dot{u}(t)) = \langle \dot{u}(t), \xi(t) \rangle - H(t, u(t), \xi(t)). \tag{5.5}$$

From this, we see that the integrands in the functional F and the functional I are in fact the same function expressed via different variables, while the Poincaré–Cartan invariant is indeed the differential version of the Hilbert invariant integral from the previous lecture.

It is worth noting that the functional $F(u, \xi)$ corresponding to the H-S is not bounded below, hence it is impossible to possess any minimal value. The solutions of the H-S are merely "critical points" of the functional F.

Example 5.3 For a given collection of particles in classical mechanics (cf. Example 2.1 in Lecture 2), let $q = (q_1, \ldots, q_N)$, $p = (p_1, \ldots, p_N)$, $T(p) = \frac{1}{2} \sum_1^N a_{ij} p_i p_j$, and $V = V(q_1, \ldots, q_N)$, its Lagrangian is

$$L(t, q, p) = T(p) - V(q),$$

whose corresponding E-L equation is

$$\frac{d}{dt} \sum_1^N a_{ij} \dot{q}_j = -\partial_{q_i} V(q), \quad i = 1, \ldots, N.$$

Here the Hamiltonian

$$H(q, \xi) = \frac{1}{2} \sum_1^N a^{ij} \xi_i \xi_j + V(q)$$

is the energy of the collection of particles, where (a^{ij}) is the inverse matrix of (a_{ij}). The corresponding H-S is

$$\begin{cases} \dot{\xi}_i = -\partial_{q_i} V(q), \\ \dot{q}_i = \sum_{j=1}^N a^{ij} \xi_j, \quad i = 1, 2, \cdots, N. \end{cases}$$

When the Hamiltonian H is independent of t, $\forall c \in \mathbb{R}^1$, let $H^{-1}(c) = \{(u, \xi) \in \mathbb{R}^N \times \mathbb{R}^N \mid H(u, \xi) = c\}$ be a level set of the Hamiltonian.

Theorem 5.1 *The solution curve $\{(t, u(t), \xi(t) \mid \forall t\}$ of a Hamiltonian system remains on the same level set.*

Proof Suppose $(u(t), \xi(t))$ is a solution of

$$\begin{cases} \dot{\xi}(t) = -H_u(u(t), \xi(t)), \\ \dot{u}(t) = H_\xi(u(t), \xi(t)), \end{cases}$$

then using

$$\frac{d}{dt} H(u(t), \xi(t)) = \langle H_u(u(t), \xi(t)), \dot{u}(t) \rangle + \langle H_\xi(u(t), \xi(t)), \dot{\xi}(t) \rangle = 0,$$

it is immediate for all t, $H(u(t), \xi(t)) = \text{const.}$ □

Applying this theorem to a collection of moving particles, it asserts: the total energy of an isolated system remains constant (the law of conservation of energy).

5.4 The Hamilton–Jacobi equation

Given a Hamiltonian $H = H(t, u, \xi)$, we call the first order partial differential equation

$$\boxed{\partial_t S(t, u) + H(t, u, \nabla_u S(t, u)) = 0 \qquad (5.6)}$$

the Hamilton–Jacobi equation (we shall abbreviate it by H-J equation), where $S = S(t, u)$ is a function of $N + 1$ variables.

The H-J equation is a fundamental equation in both classical mechanics and quantum mechanics.

We know for a Mayer field (Ω, ψ), its eikonal g satisfies the Carathéodory system of equations (5.2):

$$\begin{cases} \nabla_u g(t, u) = L_p(t, u, \psi(t, u)) \\ \partial_t g(t, u) = L(t, u, \psi(t, u)) - \langle \psi(t, u), L_p(t, u, \psi(t, u)) \rangle. \end{cases}$$

Substituting the Legendre transform $\xi = L_p(t, u, \psi(t, u)) = \nabla_u g(t, u)$ into the Carathéodory system of equations, we see that g satisfies the first order partial differential equation:

$$\partial_t g(t, u) + H(t, u, \nabla_u g(t, u)) = 0, \qquad (5.6')$$

where H is the Legendre transform of the Lagrangian L.

Based on the relationship of the Legendre transforms of the Lagrangian and the Hamiltonian, for a given Hamiltonian H, we can write L and transforming a solution $(u(t), \xi(t))$ of the H-S:

$$\begin{cases} \dot{\xi}_i(t) = -H_{u_i}(u(t), \xi(t)), \\ \dot{u}_i(t) = H_{\xi_i}(u(t), \xi(t)), \qquad 1 \leq i \leq n \end{cases}$$

into a solution $(u(t), p(t))$ of the E-L equation. After integrating, we can express the eikonal $g(t, u)$.

It is evident, if we have a solution $(u(t), \xi(t))$ of the Hamiltonian system, by letting

$$p(t) = H_\xi(t, u(t), \xi(t)),$$

we can then obtain the solution $(u(t), p(t))$ of the E-L equation.

Moreover, using (5.5), we can obtain the Lagrangian $L(t, u(t), p(t))$, whence,

$$g(t, u(t)) - g(t_0, u(t_0)) = \int_{t_0}^t L(t, u(t), p(t))dt$$

is the solution of the eikonal equation.

That being said, we can derive solutions of the H-J equation from the solutions of the Hamiltonian system by choosing arbitrary initial values. However, it should be noted that the Hamiltonian system is a system of ordinary differential equations, whereas the H-J equation is a first order partial differential equation.

Example 5.4 (Propagation of light through a medium) Denote the density of a medium at the point $(t, u) \in \mathbb{R}^1 \times \mathbb{R}^n$ by $\rho(t, u)$. Using the speed of light in vacuum as a unit, suppose the light speed at the point is $\frac{1}{\rho(t,u)}$, then the corresponding Lagrangian is

$$L(t, u, p) = \rho(t, u)\sqrt{1 + p^2},$$

here

$$\xi = L_p = \rho \cdot \frac{p}{\sqrt{1 + p^2}},$$

$$p = \frac{\xi}{\sqrt{\rho^2 - \xi^2}},$$

$$H(t, u, \xi) = p\xi - L(t, u, p) = -\sqrt{\rho^2 - \xi^2}.$$

The eikonal g satisfies the eikonal equation

$$g_t = \sqrt{\rho^2 - |\nabla_u g|^2}.$$

This is a H-J equation. Sometimes, we also write it as

$$g_t^2 + |\nabla_u g|^2 = \rho^2$$

and

$$g(t, u) = \text{const.}$$

is the wavefront of light.

The corresponding directional field is

$$\psi(t, u) = H_\xi(t, u, \nabla_u g) = \frac{\nabla_u g}{\sqrt{\rho^2 - |\nabla_u g|^2}} = \frac{\nabla_u g}{\partial_t g}.$$

Since

$$\dot{u} = \psi(t, u), \qquad (\dot{t}, \dot{u}) = (1, \dot{u}) = (\partial_t g)^{-1}(\partial_t g, \nabla_u g),$$

its integral curve therefore follows the normal direction of the wavefront $g(t, u) =$ const. in the (t, u)-space. In other words, "a ray of light travels perpendicularly to the wavefronts."

5.5* Jacobi's Theorem

On the other hand, we can also derive the solutions of H-S from the solutions of the H-J equation.

Definition 5.1 Let $g = g(t, u_1, \ldots, u_N; \lambda_1, \cdots, \lambda_N)$ be a family of solutions of the H-J equation depending on the N independent parameters $(\lambda_1, \ldots, \lambda_N) \in \Lambda$ (here $\Lambda \subset \mathbb{R}^N$ is a region). If $\det(g_{u_i \lambda_j}) \neq 0$, then it is called a complete integral.

Theorem 5.2 (Jacobi) *Let the C^2 function $g(t, u_1, \ldots, u_N; \lambda_1, \ldots, \lambda_N)$ be a complete integral of the H-J equation. Given $2N$ parameters $(\alpha, \beta) = (\alpha_1, \ldots, \alpha_N, \beta_1, \ldots, \beta_N)$, suppose the function*

$$\begin{cases} u = U(t, \alpha, \beta), \\ p = P(t, \alpha, \beta), \end{cases}$$

satisfies

$$\begin{cases} g_{\alpha_i}(t, U(t, \alpha, \beta), \alpha) = -\beta_i \\ P_i(t, \alpha, \beta) = g_{u_i}(t, U(t, \alpha, \beta), \alpha) \quad i = 1, 2, \ldots, N, \end{cases} \tag{5.7}$$

then they form a family of solutions of the Hamiltonian system.

Proof 1. First, differentiating the H-J equation (5.6′) with respect to α_i, we have:

$$g_{t, \alpha_i} + \sum_{k=1}^{N} H_{\xi_k}(t, u, \nabla_u g) g_{\alpha_i, u_k} = 0, \qquad i = 1, 2, \ldots, N.$$

Substituting $u = U(t, \alpha, \beta)$ and we have:

$$g_{t, \alpha_i}(t, U(t, \alpha, \beta), \alpha)$$

$$+ \sum_{k=1}^{N} H_{\xi_k}(t, U(t, \alpha, \beta), P(t, \alpha, \beta)) g_{\alpha_i, u_k}(t, U(t, \alpha, \beta), \alpha) = 0. \tag{5.8}$$

Differentiating the first equation in (5.7) with respect to t, we have:

$$g_{t,\alpha_i}(t, U(t,\alpha,\beta), \alpha) + \sum_{k=1}^{N} g_{\alpha_i,u_k}(t, U(t,\alpha,\beta), \alpha)\dot{U}_k(t,\alpha,\beta) = 0. \qquad (5.9)$$

Subtracting (5.9) by (5.8), it yields:

$$\sum_{k=1}^{N} [\dot{U}_k(t,\alpha,\beta) - H_{\xi_k}(t, U(t,\alpha,\beta), P(t,\alpha,\beta))]g_{\alpha_i,u_k}(t, U(t,\alpha,\beta), \alpha) = 0.$$

Since the matrix (g_{α_i,u_k}) is invertible, it follows that

$$\dot{U}_k(t,\alpha,\beta) = H_{\xi_k}(t, U(t,\alpha,\beta), P(t,\alpha,\beta)),$$

which constitutes a set of equations of the H-S.

2. Next, by differentiating the H-J equation (5.6′) with respect to u_i, we get

$$g_{t,u_i}(t, u, \alpha) + H_{u_i}(t, u, \nabla_u g(t, u, \alpha))$$

$$+ \sum_{k=1}^{N} H_{\xi_k}(t, u, \nabla_u g(t, u, \alpha))g_{u_i,u_k}(t, u, \alpha) = 0.$$

Substituting $u = U(t,\alpha,\beta)$ and $P(t,\alpha,\beta) = \nabla_u g(t, U(t,\alpha,\beta), \alpha)$ into the above equation, we have

$$-H_{u_i}(t, U, P) = g_{t,u_i}(t, U, \alpha) + \sum_{1}^{N} g_{u_i u_k}(t, U, \alpha)\dot{U}_k. \qquad (5.10)$$

Differentiating the second equation in (5.7) with respect to t, we then have

$$\dot{P}_i(t,\alpha,\beta) = g_{u_i,t}(t, U, \alpha) + \sum_{k=1}^{N} g_{u_i,u_k}(t, U, \alpha)\dot{U}_k(t,\alpha,\beta), \qquad (5.11)$$

i.e.

$$\dot{P}_i(t,\alpha,\beta) = -H_{u_i}(t, U(t,\alpha,\beta), P(t,\alpha,\beta)).$$

This yields the other set of equations of the H-S. $\qquad\qquad\qquad\square$

Remark 5.2 The significance of the Jacobi's Theorem is that one can express the general solutions of the H-S via the solutions of the H-J equation. The approach is to solve the system of N implicit function equations:

$$g_{\alpha_i}(t, u, \alpha) = -\beta_i, \qquad i = 1,\ldots,N$$

to obtain

$$u = U(t,\alpha,\beta) \qquad (5.12)$$

and then substituting it back into the eikonal g to get

$$p = P(t,\alpha,\beta) = \partial_u g(t, U(t,\alpha,\beta), \alpha). \qquad (5.13)$$

The set (u, p) is precisely the desired general solutions of the H-S.

Remark 5.3 Complete integrals and general solutions have different meanings. This is because the H-J equation is a first order partial differential equation, and according to the uniqueness of solution to a Cauchy problem, its general solution should contain an arbitrary function $\varphi(u)$, not just $2N$ independent parameters.

However, the general solutions of the H-S can be determined by a complete integral g of the H-J equation. In other words, for a given initial value (u_0, ξ_0), to solve the initial value problem of the H-S

$$\dot{u} = H_\xi(t, u, \xi), \ \dot{\xi} = -H_u(t, u, \xi), \ u(0) = u_0, \ \xi(0) = \xi_0,$$

we may proceed as follows:

Once (u_0, ξ_0) is given, since $\det(g_{u_i \alpha_j}) \neq 0$, we can first apply the Implicit Function Theorem to the second equation of the system

$$\begin{cases} \nabla_\alpha g(0, u_0, \alpha) = -\beta, \\ \xi_0 = \nabla_u g(0, u_0, \alpha) \end{cases}$$

to solve for

$$\alpha_0 = \alpha(u_0, \xi_0).$$

Setting

$$\beta_0 = -\nabla_\alpha g(0, u_0, \alpha_0)$$

and plugging them into (5.12) and (5.13), we will end up with the solution of the H-S with initial value (u_0, ξ_0).

Example 5.5 (Harmonic Oscillators) Given the Lagrangian $L = \frac{1}{2}(mp^2 - ku^2)$, where m and k are positive constants. Then the Hamiltonian is

$$H = \frac{1}{2}\left(\frac{p^2}{m} + ku^2\right).$$

The corresponding H-S

$$\begin{cases} \dot{u} = \dfrac{p}{m}, \\ \dot{p} = -ku. \end{cases}$$

has solution

$$\begin{cases} u = C \sin \sqrt{\dfrac{k}{m}}(t + t_0), \\ p = C\sqrt{mk} \cos \sqrt{\dfrac{k}{m}}(t + t_0), \end{cases}$$

where t_0 and C are arbitrary constants.

We now use the H-J equation

$$g_t + H(t, u, g_u) = 0$$

and the Jacobi's Theorem to express the H-S solution.

Consider the special eikonal $g(t, u, \alpha) = \varphi(u) - \alpha t$, where α is a parameter and φ is a function yet to be determined. So the H-J equation is

$$\frac{1}{2}\left(\frac{\varphi'^2(u)}{m} + ku^2\right) = \alpha,$$

i.e.

$$\varphi'(u) = \sqrt{m(2\alpha - ku^2)}.$$

Solving this, we have:

$$g = g(t, u, \alpha) = \int_0^u \sqrt{m(2\alpha - ku^2)}du - \alpha t.$$

We now solve the equation

$$\beta = -g_\alpha = t - \sqrt{\frac{m}{k}}\arcsin\left(\sqrt{\frac{k}{2\alpha}}u\right)$$

to get

$$u = \sqrt{\frac{2\alpha}{k}}\sin\left(\sqrt{\frac{k}{m}}(t - \beta)\right).$$

Substituting it back into g, it follows that

$$p = g_u = \varphi'(u) = \sqrt{2\alpha m}\cos\left(\sqrt{\frac{k}{m}}(t - \beta)\right).$$

This gives the solution of the Hamiltonian system of a harmonic oscillator involving the two parameters α, β.

Exercises

1. For each of the given Lagrangian, find the Hamiltonian and solve the Hamiltonian system.

   (1) $L = (p + ku)^2$, $k \neq 0$.
   (2) $L = e^{-u}\sqrt{1 + p^2}$.

   Write the H-J equation and find its complete solutions.

2. Suppose $\forall\,(t, u)$, $L(t, u, p)$ is convex in p. Prove that

   $$H(t, u, \xi) = \sup_{p \in \mathbb{R}^N}\{\langle p, \xi\rangle - L(t, u, p)\}.$$

Lecture 6

Variational problems involving multivariate integrals

Previously, we have discussed variational problems involving single integrals. However, when the unknown is a multivariate (or vector-valued) function, we are confronted with variational problems involving multivariate integrals. In this lecture, we extend the theory of calculus of variations from a single integral setting to a multivariate integral setting, including the E-L equation, the criteria for weak and strong minima, Jacobi fields, and the Weierstrass excess function, etc.

Let $\Omega \subset \mathbb{R}^n$ be a bounded region with $\partial\Omega \in C^1$. Given a Lagrangian $L = L(x, u, p) \in C^2(\bar{\Omega} \times \mathbb{R}^N \times \mathbb{R}^{nN})$ and a boundary function $\Phi \in C^1(\partial\Omega, \mathbb{R}^N)$, we want to minimize the functional

$$I(u) = \int_\Omega L(x, u(x), \nabla u(x))dx,$$

under the boundary condition $u \in M := \{v \in C^1(\bar{\Omega}, \mathbb{R}^N) \mid v|_{\partial\Omega} = \Phi|_{\partial\Omega}\}$.

For simplicity, we introduce the following notation:

$$x = (x_\alpha)_1^n = (x_1, \ldots, x_n),$$
$$u = (u^i)_1^N = (u^1, \ldots, u^N),$$
$$p = (p_\alpha^i)_{1 \le i \le N, 1 \le \alpha \le n},$$
$$p^i = (p_1^i, \ldots, p_n^i),\ 1 \le i \le N,$$
$$\partial_\alpha u = \frac{\partial u}{\partial x_\alpha},\ \alpha = 1, \ldots, n,$$
$$\nabla u = \left(\frac{\partial u^i}{\partial x_\alpha}\right) = (u_\alpha^i).$$

6.1 Derivation of the Euler–Lagrange equation

$u_0 \in M$ is said to be a minimum of I on M, if

$$I(u) \ge I(u_0), \quad \forall u \in U,$$

where $U \subset M$ is a neighborhood of u_0. When using the C^1-topology, u_0 is called a weak minimum, when using the C^0-topology, u_0 is called a strong minimum.

Similar to the single variable setting, assuming such u_0 exists, we seek a necessary condition for which it must satisfy, namely the E-L equation.

For brevity, we denote $\tau = (x, u_0(x), \nabla u_0(x))$.

Theorem 6.1 *Suppose $L \in C^2$, $u_0 \in C^2$, and u_0 is a minimum of I on M, then it satisfies the following E-L equation*

$$\sum_{\alpha=1}^{n} \frac{\partial L_{p_\alpha^i}(x, u_0(x), \nabla u_0(x))}{\partial x_\alpha} - L_{u^i}(x, u_0(x), \nabla u_0(x)) = 0, \quad 1 \leq i \leq N.$$

Proof $\forall \, \varphi \in C_0^1(\bar{\Omega}, \mathbb{R}^N)$, consider the 1-variable function $s \mapsto g(s) := I(u_0 + s\varphi)$. Since 0 is a minimum of g, we deduce that

$$0 = g'(0) = \sum_{i=1}^{N} \int_{\Omega} \left[L_{u^i}(\tau)\varphi^i(x) + \sum_{\alpha=1}^{n} L_{p_\alpha^i}(\tau)\partial_{x_\alpha}\varphi^i(x) \right] dx$$

$$= \sum_{i=1}^{N} \int_{\Omega} \left[L_{u^i}(\tau) - \sum_{\alpha=1}^{n} L_{p_\alpha^i}(\tau) \right] \varphi^i(x) dx.$$

The assertion follows from a result (Lemma 6.1) similar to the Du Bois Reymond's lemma. $\qquad\square$

We will prove a generalized version of the Du Bois Reymond's lemma in higher dimensions. But first, we introduce the 1-variable bump function:

$$\psi(t) = \begin{cases} \exp\left(\dfrac{-1}{1 - |t|^2}\right) & |t| < 1 \\ 0 & |t| > 1. \end{cases}$$

For the multivariable $x = (x_1, \ldots, x_n)$, let

$$\varphi(x) = c_n^{-1} \psi(|x|),$$

where $|x| = (x_1^2 + \cdots + x_n^2)^{1/2}$ and $c_n = \int_{\mathbb{R}^n} \psi(|x|) dx$.

For $\epsilon > 0$ sufficiently small, we further define

$$\varphi_\epsilon(x) = \epsilon^{-n} \varphi\left(\frac{x}{\epsilon}\right).$$

Given a region $\Omega \subset \mathbb{R}^N$, for any $\delta > 0$ sufficiently small, we denote $\Omega_\delta = \{x \in \Omega \mid d(x, \partial\Omega) \geq \delta, \, |x| \leq 1/\delta\}$.

Suppose $u \in L^1_{\text{loc}}(\Omega)$, let

$$u_\delta(x) = \int_{\Omega} u(y)\varphi_\delta(x - y) dy, \quad \forall \, x \in \Omega_\delta.$$

Using the change of variable $z = (y - x)/\delta$, we obtain:

$$u_\delta(x) = \int_{B_1(\theta)} u(x + \delta z)\varphi(z)dz, \quad \forall\, x \in \Omega_\delta,$$

where $B_1(\theta)$ is the unit ball in $\mathbb{R}^n$ centered at the origin. Hence, $\forall\, \delta_0 > 0$,

$$\int_{\Omega_{\delta_0}} |u(x) - u_\delta(x)|dx \le \int_{B_1(\theta)} \varphi(z)dz \int_{\Omega_{\delta_0}} |u(x) - u(x + \delta z)|dx \to 0, \text{ as } \delta \to 0.$$

In fact, if we use a continuous function v defined on $\bar{\Omega}_\delta$ instead of u, the above limit clearly still holds. Furthermore, since $C(\bar{\Omega}_{\delta_0})$ is dense in $L^1(\Omega_{\delta_0})$, replacing it by $u \in L^1_{\text{loc}}(\Omega)$, the limit remains zero.

In summary, we have:

Lemma 6.1 *Suppose* $u \in L^1_{\text{loc}}(\Omega)$, *then* $\forall\, \delta_0 > 0$,

$$\int_{\Omega_{\delta_0}} |u(x) - u_\delta(x)|dx \to 0 \quad \text{as } \delta \to 0. \tag{6.1}$$

Corollary 6.1 *Suppose* $u \in L^1_{\text{loc}}(\Omega)$ *and*

$$\int_\Omega u(x)\varphi(x)dx = 0, \quad \forall\, \varphi \in C_0^\infty(\Omega),$$

then $u(x) = 0$ *a.e. for* $x \in \Omega$.

Proof Note that $\forall\, \delta_0 > 0$, since

$$\varphi_\delta(x - y) \in C_0^\infty(\Omega), \quad \forall\, x \in \Omega_{\delta_0}, \quad \forall\, \delta < \delta_0$$

and by assumption, one has

$$u_\delta(x) = \int_\Omega u(y)\varphi_\delta(x - y)dy = 0, \quad \forall\, x \in \Omega_\delta, \quad \forall\, \delta < \delta_0.$$

It follows from Lemma 6.1 that $u_\delta(x) \to u(x)$ a.e. for $x \in \Omega_{\delta_0}$, namely $u(x) = 0$ a.e. for $x \in \Omega_{\delta_0}$. Since $\delta_0 > 0$ is arbitrary, it is immediate $u(x) = 0$ a.e. for $x \in \Omega$. $\qquad\square$

Just like in the 1-variable setting, we call $E_L : u \mapsto v = (v_1, \ldots, v_N)$, where

$$v_i = \sum_{\alpha=1}^n \frac{\partial L_{p_\alpha^i}(x, u(x), \nabla u(x))}{\partial x_\alpha} - L_{u^i}(x, u(x), \nabla u(x)), \qquad i = 1, \ldots, N$$

the Euler–Lagrange operator of L.

Remark 6.1 Without the hypothesis $u \in C^2$, one can still define the E-L operator by interpreting the term $\frac{\partial}{\partial x_\alpha}$ in front of $L_{p_\alpha^i}(x, u(x), \nabla u(x))$ as derivatives of a distribution, so the E-L equation holds in the sense of distributions.

Similar to Remarks 2.2 and 2.3 in Lecture 2, when dealing with variational problems involving multivariate integrals, we can change the domain M of a functional from $C^1(\bar{\Omega}, \mathbb{R}^N)$ to $\text{Lip}(\bar{\Omega}, \mathbb{R}^N)$, or more specially, to $PWC^1(\bar{\Omega}, \mathbb{R}^N)$, the class of piecewise C^1 functions. $u \in PWC^1(\bar{\Omega}, \mathbb{R}^N)$ if there exists a finite collection of $n-1$ dimensional piecewise C^1-hypersurfaces $\{S_1, \ldots, S_k\}$ such that the continuous function $u \in C^1(\bar{\Omega} \setminus \bigcup_{j=1}^{k} S_j, \mathbb{R}^N)$ and u has normal derivatives on both sides of S_j.

Example 6.1 (Dirichlet integrals) Assume $N = 1$, $L(p) = |p|^2/2$, for the functional

$$D(u) = \frac{1}{2} \int_\Omega |\nabla u(x)|^2 dx,$$

its E-L equation is

$$\Delta u = \sum_{\alpha=1}^{n} \frac{\partial^2 u}{\partial x_\alpha^2} = 0, \quad \forall x \in \Omega.$$

This is a harmonic equation, also called the Laplace equation.

We can also consider a more generalized variational problem, such as

$$L(x, u, p) = \frac{1}{2}|p|^2 - \frac{a(x)}{\alpha+1}|u|^{\alpha+1}, \quad \alpha > 0,$$

where $a \in C(\bar{\Omega})$. Its E-L equation is

$$-\Delta u(x) = a(x)|u|^{\alpha-1}u, \quad \forall x \in \Omega.$$

Example 6.2 (Wave equations) Denote $\mathbb{R}^1 \times \mathbb{R}^3$ the time-space continuum, a given point has coordinate (t, x), where t represents time and $x = (x_1, x_2, x_3)$ represents its spacial location. We use $u = u(t, x_1, x_2, x_3)$ to represent the propagation of an elastic wave in the region $\Omega \subset \mathbb{R}^1 \times \mathbb{R}^3$.

The kinetic energy of the elastic wave is

$$T = \frac{1}{2} \int_\Omega |\partial_t u(t, x)|^2 dt dx$$

and the potential energy is

$$U = \frac{1}{2} \int_\Omega |\nabla_x u(t, x)|^2 dt dx.$$

The Lagrangian is

$$L(u) = T(u) - U(u).$$

According to the principle of stable action, the propagation of the elastic wave is a solution of the E-L equation

$$\Box u = \partial_t^2 u - \Delta u = 0$$

of L. This is known as the d'Alembert equation.

Likewise, if there are internal forces and or external forces involved, then there are some added terms to the potential energy. For instance,

$$U = \frac{1}{2} \int_\Omega (|\nabla_x u(t, x)|^2 + M^2 |u(t, x)|^2) dt dx,$$

where $M > 0$ is a constant. In this case, the corresponding E-L equation is a Klein–Gordan equation:

$$\Box u - M^2 u = \partial_t^2 u - \triangle u + M^2 u = 0.$$

Another example is

$$U = \int_\Omega \left(\frac{1}{2} |\nabla_x u(t, x)|^2 + \frac{1}{4} |u(t, x)|^4 \right) dt dx,$$

then the corresponding E-L equation is a nonlinear wave equation:

$$\Box u + u^3 = \partial_t^2 u - \triangle u + u^3 = 0.$$

Example 6.3 (Minimal surfaces) Let $\Omega \subset \mathbb{R}^n$. Given $u \in C^1(\bar{\Omega})$, its graph $\{(x, u(x)) \mid x \in \bar{\Omega}\}$ is a hypersurface, whose area is given by

$$A(u) = \int_\Omega \sqrt{1 + |\nabla u(x)|^2} dx.$$

Under the boundary condition $u|_{\partial\Omega} = \Phi$, the hypersurface minimizing the area satisfies the E-L equation

$$\text{div} \left(\frac{\nabla u(x)}{\sqrt{1 + |\nabla u(x)|^2}} \right) = 0, \quad \forall x \in \Omega. \tag{6.2}$$

Notice the mean curvature of the hypersurface is

$$H = \frac{1}{n} \text{div} \left(\frac{\nabla u(x)}{\sqrt{1 + |\nabla u(x)|^2}} \right).$$

Suppose we are given the mean curvature function $H(x)$, $x \in \Omega$, then u satisfies the mean curvature equation

$$\text{div} \left(\frac{\nabla u(x)}{\sqrt{1 + |\nabla u(x)|^2}} \right) = nH(x), \quad \forall x \in \Omega. \tag{6.3}$$

It is worth noting, (6.3) is the E-L equation of the functional

$$I(u) = \int_\Omega \left(\sqrt{1 + |\nabla u(x)|^2} + H(x) u(x) \right) dx.$$

In particular, when $n = 2$, the above equation reduces to

$$(1 + u_y^2) u_{xx} - 2 u_x u_y u_{xy} + (1 + u_x^2) u_{yy} = 2H(x)(1 + u_x^2 + u_y^2)^{\frac{3}{2}}. \tag{6.4}$$

Comparing (6.2) and (6.3), the mean curvature with mean curvature zero equation coincides with the minimal surface equation. Therefore, the zero mean curvature surfaces are usually called minimal surfaces.

We now find a special solution $u(x, y) = f(x) + g(y)$ of the minimal surface equation. Substituting this into (6.3) with $H = 0$, it yields

$$(1 + g'(y)^2)f''(x) + (1 + f'(x)^2)g''(y) = 0,$$

i.e.

$$\frac{f''(x)}{1 + f'(x)^2} = c = -\frac{g''(y)}{1 + g'(y)^2}.$$

From which, we find $\arctan f'(x) = cx$ or equivalently

$$f(x) = -\frac{1}{c} \ln |\cos cx|.$$

Likewise,

$$g(y) = \frac{1}{c} \ln |\cos cy|.$$

Thus,

$$u(x, y) = \frac{1}{c} \ln \left| \frac{\cos cy}{\cos cx} \right|.$$

The minimal surface defined by u is known as the Scherk surface.

Example 6.4 (Maxwell's equations) Consider a point with space-time coordinate $(x^0, x^1, x^2, x^3) \in \mathbb{R}^1 \times \mathbb{R}^3$, where $x = (x^1, x^2, x^3)$ denotes its spacial coordinate and $x^0 = ct$, t is time and c is the speed of light.

In an electromagnetic field, the electric charge ρ and the electric current j are both functions of space-time. Let $E = (E_1, E_2, E_3)$ and $B = (B_1, B_2, B_3)$ denote the electric field and the magnetic field respectively, which are also functions of space-time.

Maxwell's equations can be written as

$$\begin{cases} -\frac{1}{c}\frac{\partial B}{\partial t} = \nabla \times E & \text{(Faraday's law of induction)} \\ \nabla \cdot B = 0 & \text{(Gauss's law for magnetism)} \\ \nabla \times B = \frac{1}{c}\frac{\partial E}{\partial t} + \frac{4\pi}{c}j & \text{(Ampere's circuital law)} \\ \nabla \cdot E = 4\pi\rho. & \text{(Gauss's law for electric charge)} \end{cases}$$

Since $\nabla \cdot B = 0$, there exists a magnetic potential $A = (A_1, A_2, A_3)$ such that

$$\nabla \times A = B.$$

From Faraday's law of induction,

$$\nabla \times \left(E + \frac{\partial A}{\partial x^0} \right) = \nabla \times E + \frac{1}{c}\frac{\partial B}{\partial t} = 0,$$

we can deduce that there exists an electric potential A_0 such that

$$E + \frac{\partial A}{\partial x^0} = \nabla A_0.$$

We call $A = (A_0, A_1, A_2, A_3)$ an electromagnetic potential and let

$$F_{ij} = \frac{\partial A_j}{\partial x^i} - \frac{\partial A_i}{\partial x^j},$$

it then follows that

$$(F_{ij}) = \begin{pmatrix} 0 & -E_1 & -E_2 & -E_3 \\ E_1 & 0 & B_3 & -B_2 \\ E_2 & -B_3 & 0 & B_1 \\ E_3 & B_2 & -B_1 & 0 \end{pmatrix}.$$

We now define the Lagrangian

$$L = -\frac{1}{c}\left(\frac{1}{c}\sum_{i=0}^{3} j_i A_i + \frac{1}{16\pi}\sum_{i,k=0}^{3} F_{ik}^2\right)$$

and $J = (j_0, j_1, j_2, j_3)$, where $j_0 = c\rho$ and $j = (j_1, j_2, j_3)$.

The corresponding functional

$$I(A) = \int_{\mathbb{R}^1 \times \mathbb{R}^3} L d^4 x.$$

From this, we can deduce the E-L equation

$$\frac{\partial L}{\partial A_i} - \frac{\partial}{\partial x^j}\left(\frac{\partial L}{\partial p_i^j}\right) = 0,$$

where

$$p_i^j = \frac{\partial A_i}{\partial x^j}.$$

Since

$$\frac{\partial L}{\partial A_i} = -\frac{1}{c^2} j_i, \qquad \frac{\partial L}{\partial p_i^j} = \frac{1}{4c\pi} F_{ij},$$

we have

$$\frac{\partial F_{ij}}{\partial x^j} = -\frac{4\pi}{c} j_i \qquad i = 0, 1, 2, 3.$$

This is precisely the Ampere's circuital law ($i = 1, 2, 3$) and Gauss's law for electric charges ($i = 0$).

Remark 6.2 (A_0, A_1, A_2, A_3) is not uniquely determined by E and H. In fact, for any function $f \in C^1$, we use

$$A_j'(x) = A_j(x) + \frac{\partial f}{\partial x_j}, \qquad j = 0, 1, 2, 3$$

to replace $(A_j)_0^3$, then the corresponding E, H remain unchanged. The above transformation is called a gauge transformation. If, in addition, we impose the Lorentz condition

$$-\frac{\partial A_0}{\partial x_0} + \sum_1^3 \frac{\partial A_j}{\partial x_j} = 0,$$

it will resolve the non-uniqueness issue.

6.2 Boundary conditions

Similar to single integral's variational problems, when dealing with variational problems involving multivariate integrals, depending on the requirement of the functional on the boundary of its domain, the resulting E-L equation also has to meet certain boundary conditions. Previously, we have discussed the scenario where a given domain M comes with a prescribed boundary function:

$$M = \{u \in C^1(\bar{\Omega}, \mathbb{R}^N) \mid u|_{\partial\Omega} = \Phi\}.$$

As a consequence, the resulting u^* has to satisfy not only the E-L equation

$$E_L(u) = 0,$$

but also the Dirichlet boundary condition:

$$u|_{\partial\Omega} = \Phi.$$

Suppose we change the domain to be $M = C^1(\bar{\Omega}, \mathbb{R}^N)$, in other words, on the boundary $\partial\Omega$ of Ω, we impose no condition on the functional whatsoever. Then via an argument similar to that used in the single integral setting, we see that the C^2 extremal function u^* of the functional I again satisfies the E-L differential equation

$$E_L(u^*) = 0.$$

Furthermore, using integration by parts, we have:

$$\delta I(u^*, \varphi) = \int_\Omega \left(\sum_{i=1}^n \left[L_{u^i}(x, u^*(x), \dot{u}^*(x))\varphi^i(x) \right. \right.$$

$$\left. \left. + \sum_{\alpha=1}^n L_{p_\alpha^i}(x, u^*(x), \nabla u^*(x))\partial_\alpha \varphi^i(x) \right] \right) dx$$

$$= \int_\Omega E_L(u^*)\dot\varphi + \int_{\partial\Omega} \sum_{\alpha=1}^n \sum_{i=1}^N \nu_\alpha(x) L_{p_\alpha^i}(x, u^*(x), \nabla u^*(x))\partial_\alpha \varphi^i(x) dH^{n-1}(x),$$

where $dH^{n-1}(x)$ denotes the area element of $\partial\Omega$ and $\nu(x) = (\nu_1(x), \nu_2(x), \ldots, \nu_n(x))$ denotes the unit outward normal vector of $\partial\Omega$. This gives rise to the Neumann boundary condition:

$$\sum_{i=1}^N \nu_\alpha L_{p_\alpha^i}(x, u^*(x), \nabla u^*(x))|_{\partial\Omega} = 0, \quad i = 1, \ldots, N. \tag{6.5}$$

6.3 Second order variations

Given $\Omega \subset \mathbb{R}^n$ and a Lagrangian $L \in C^2(\bar{\Omega} \times \mathbb{R}^N \times \mathbb{R}^{nN})$, we define the functional $I(u) = \int_\Omega L(x, u(x), \nabla(x))dx$. Suppose u_0 is a solution of the E-L equation, we now study the second order variation of the functional I.

$\forall \varphi \in C_0^\infty(\Omega, \mathbb{R}^N)$, we continue to use the previous 1-variable function $g(s) = I(u_0 + s\varphi)$ and we have:

$$\delta^2 I(u_0, \varphi) = g''(0) = \frac{d^2}{ds^2} I(u_0 + s\varphi)|_{s=0}$$

$$= \sum_{i,j}^N \int_\Omega \left[L_{u^i u^j}(\tau)\varphi^i(x)\varphi^j(x) + 2\sum_{\alpha=1}^n L_{u^i p_\alpha^j}(\tau)\varphi^i(x)\partial_\alpha\varphi^j(x) \right.$$

$$\left. + \sum_{\alpha,\beta=1}^n L_{p_\alpha^i p_\beta^j}(\tau)\partial_\alpha\varphi^i(x)\partial_\beta\varphi^j(x) \right] dx,$$

where $\tau = (x, u_0(x), \nabla u_0(x))$.

For simplicity, we introduce the following notation.

$$A = (L_{p_\alpha^j p_\beta^k}(x, u, p)),$$
$$B = (L_{p_\beta^j u^k}(x, u, p)),$$
$$C = (L_{u^j u^k}(x, u, p)),$$

and

$$A_{u_0} = (a_{\alpha\beta}^{jk}) = (L_{p_\alpha^j p_\beta^k}(\tau)),$$
$$B_{u_0} = (b_\beta^{jk}) = (L_{p_\beta^j u^k}(\tau)),$$
$$C_{u_0} = (c^{jk}) = (L_{u^j u^k}(\tau)).$$

Furthermore, we denote

$$Q_{u_0}(\varphi) = \delta^2 I(u_0) = \int_\Omega [A_{u_0}(\nabla\varphi, \nabla\varphi) + 2B_{u_0}(\nabla\varphi, \varphi) + C_{u_0}(\varphi, \varphi)]dx.$$

If u_0 is a weak minimum, then it is necessary to have

$$Q_{u_0}(\varphi) \geq 0, \qquad \forall \varphi \in C_0^1(\Omega, \mathbb{R}^N), \tag{6.6}$$

where $C_0^1(\Omega, \mathbb{R}^N)$ is the closure of $C_0^\infty(\Omega, \mathbb{R}^N)$ in $C^1(\bar{\Omega}, \mathbb{R}^N)$.

Conversely, suppose $u_0 \in C^1(\bar{\Omega}, \mathbb{R}^N)$ satisfies the E-L equation and if there exists $\lambda > 0$ such that

$$Q_{u_0}(\varphi) \geq \lambda \int_\Omega [|\nabla\varphi|^2 + |\varphi|^2]dx, \quad \forall \varphi \in C_0^1(\Omega, \mathbb{R}^N), \tag{6.7}$$

then u_0 is a strict minimum of I. The proof is identical to the case $n = 1$ (we refer to Theorem 3.1).

Analogous to the $n = 1$ case, we have a similar Legendre–Hadamard condition.

$\forall x_0 \in \Omega, \forall \mu > 0$, let $v \in C_0^\infty(B_1(\theta), \mathbb{R}^N)$. For μ sufficiently small, let

$$\varphi(x) = \mu v\left(\frac{x - x_0}{\mu}\right).$$

Substituting it into (6.6), it yields

$$Q_{u_0}(\varphi) = \mu^n \int_{B_1(\theta)} A_{u_0}(x_0 + \mu y)\nabla v(y)\nabla v(y)$$
$$+ 2\mu B_{u_0}(x_0 + \mu y)v\nabla v(y) + \mu^2 C_{u_0}(x_0 + \mu y)v(y)v(y)dy \geq 0.$$

Letting $\mu \to 0$, we see that

$$\sum_{j,k=1}^{N} \sum_{\alpha,\beta=1}^{n} a_{\alpha\beta}^{jk}(x_0) \int_{B_1(\theta)} \partial_\alpha v^j(y)\partial_\beta v^k(y)dy \geq 0.$$

Now for any $\rho \in C_0^\infty(B_1(\theta), \mathbb{R}^1)$ satisfying $\int_{B_1(0)} \rho^2(y)dy = 1, \forall \xi \in \mathbb{R}^N$, and $\forall \eta \in \mathbb{R}^n$, we define

$$v(y) = \xi \cos(t\eta \cdot y)\rho(y)$$

and

$$v(y) = \xi \sin(t\eta \cdot y)\rho(y).$$

Substituting these into the above inequality respectively and then adding them, furthermore, by letting $t \to \infty$, we have:

$$\sum_{j,k=1}^{N} \sum_{\alpha,\beta=1}^{n} a_{\alpha\beta}^{jk}(x_0)\xi^j\xi^k\eta_\alpha\eta_\beta + O(t^{-1}) \geq 0,$$

i.e.

$$\sum_{j,k=1}^{N} \sum_{\alpha,\beta=1}^{n} a_{\alpha\beta}^{jk}(x_0)\xi^j\xi^k\eta_\alpha\eta_\beta \geq 0.$$

This is the Legendre–Hadamard condition

$$\boxed{\begin{array}{c} \displaystyle\sum_{j,k=1}^{N} \sum_{\alpha,\beta=1}^{n} L_{p_\alpha^j p_\beta^k}(x, u_0(x), \nabla u_0(x))\xi^j\xi^k\eta_\alpha\eta_\beta \geq 0, \\[3mm] \forall (x, \xi, \eta) \in \Omega \times \mathbb{R}^N \times \mathbb{R}^n. \qquad (6.8) \end{array}}$$

If we adopt the rank-1 matrix notation

$$\pi = (\pi_\alpha^i) = (\xi^i \eta_\alpha),$$

then (6.8) can be written in the following equivalent form:

$$\sum_{j,k=1}^{N} \sum_{\alpha,\beta=1}^{n} L_{p_\alpha^j p_\beta^k}(x, u_0(x), \nabla u_0(x)) \pi_\alpha^j \pi_\beta^k \geq 0, \quad \forall \pi, \quad \text{rank}(\pi) = 1.$$

If $\exists \lambda > 0$ such that $\forall x \in \Omega$, $\forall \xi \in \mathbb{R}^N$, and $\forall \eta \in \mathbb{R}^n$,

$$\sum_{j,k=1}^{N} \sum_{\alpha,\beta=1}^{n} L_{p_\alpha^j p_\beta^k}(x, u_0(x), \nabla u_0(x)) \xi^j \xi^k \eta_\alpha \eta_\beta \geq \lambda |\xi|^2 |\eta|^2, \qquad (6.9)$$

then we call it the strict Legendre–Hadamard condition.

(6.9) also has the equivalent form of

$$\sum_{j,k=1}^{N} \sum_{\alpha,\beta=1}^{n} L_{p_\alpha^j p_\beta^k}(x, u_0(x), \nabla u_0(x)) \pi_\alpha^j \pi_\beta^k \geq \lambda \|\pi\|^2, \quad \forall \pi, \quad \text{rank}(\pi) = 1.$$

Note the norm of the matrix $\pi = (\pi_\alpha^j)$ is given by

$$\|\pi\| = \left(\sum_{\alpha=1}^{n} \sum_{j=1}^{N} (\pi_\alpha^j)^2 \right)^{\frac{1}{2}}.$$

In addition, there is a stronger condition for multi-integral variational problems:

$$\sum_{j,k=1}^{N} \sum_{\alpha,\beta=1}^{n} L_{p_\alpha^j p_\beta^k}(x, u_0(x), \nabla u_0(x)) \pi_\alpha^j \pi_\beta^k \geq \lambda \|\pi\|^2, \ \forall x \in \Omega \ \ \forall \pi \in \mathbb{R}^{n \times N},$$

known as the strong elliptical condition.

However, for $N = 1$ or $n = 1$, the strict Legendre–Hadamard condition and the strong elliptical condition agree with each other.

In summary, we have the following.

Theorem 6.2 *Let $L \in C^2$ and suppose $u_0 \in M$ is a weak minimum of I, then (6.8) holds. Conversely, if $u_0 \in M$ satisfies the E-L equation and if there exists $\lambda > 0$ such that*

$$\delta^2 I(u_0, \varphi) \geq \lambda \int_\Omega \{|\varphi(x)|^2 + |\nabla \varphi(x)|^2\} dx, \quad \forall \varphi \in C_0^1(\Omega, \mathbb{R}^N), \qquad (6.10)$$

then u_0 is a strict minimum of I.

Similarly, (6.10) implies the strict Legendre–Hadamard condition (6.9), but (6.9) is not a sufficient condition for u_0 to be a weak minimum.

6.4 Jacobi fields

For multi-integral variational problems, we also have the concept of a Jacobi field.

Let $L \in C^3$ and suppose $u_0 \in M$ is a weak minimum of I, then

$$\delta^2 I(u_0, \varphi) \geq 0, \qquad \forall \varphi \in C_0^1(\Omega, \mathbb{R}^N),$$

i.e. $Q_{u_0}(\varphi) \geq 0$.

The E-L equation of the functional Q_{u_0} is a system of homogeneous second order partial differential equations:

$$\sum_{k=1}^{N} \left[\sum_{\alpha=1}^{n} \partial_\alpha \left(a_{\alpha\beta}^{jk} \partial_\beta \varphi^k + b_\alpha^{jk} \varphi^k \right) - \left(\sum_{\beta=1}^{n} b_\beta^{jk} \partial_\beta \varphi^k + c^{jk} \varphi^k \right) \right] = 0,$$
$$j = 1, 2, \ldots, N.$$

As for $n = 1$, we call this the Jacobi equation and call

$$J_{u_0} : \varphi \mapsto (\psi^1, \ldots, \psi^N)$$

the Jacobi operator along u_0, where

$$\psi^j = \sum_{k=1}^{N} \left[\sum_{\alpha=1}^{n} \partial_\alpha \left(\sum_{\beta=1}^{n} a_{\alpha\beta}^{jk} \partial_\beta \varphi^k + b_\alpha^{jk} \varphi^k \right) - \left(\sum_{\beta=1}^{n} b_\beta^{jk} \partial_\beta \varphi^k + c^{jk} \varphi^k \right) \right],$$
$$j = 1, \ldots, N.$$

Any C^2 solution of the Jacobi equation is called a Jacobi field along u_0.

For a differential operator satisfying the strict Legendre–Hadamard condition, we have the following inequality:

Lemma 6.2 (Gårding's inequality) *Suppose $(a_{\alpha\beta}^{jk}(x))$ are uniformly continuous functions defined on $\bar{\Omega} \subset \mathbb{R}^n$ and there exists $\sigma > 0$ such that*

$$\sum_{\alpha,\beta=1}^{n} \sum_{j,k=1}^{N} a_{\alpha\beta}^{jk} \xi^j \xi^k \eta_\alpha \eta_\beta \geq \sigma |\xi|^2 |\eta|^2, \quad \forall x \in \Omega,$$

then there exist $\alpha > 0$ and $C_0 > 0$ such that for all $\varphi \in C_0^1(\Omega, \mathbb{R}^N)$,

$$\int_\Omega \sum_{\alpha,\beta=1}^{n} \sum_{j,k=1}^{N} a_{\alpha\beta}^{jk}(x) \partial_\alpha \varphi^j \partial_\beta \varphi^k \, dx \geq \alpha \int_\Omega |\nabla \varphi(x)|^2 dx - C_0 \int_\Omega |\varphi(x)|^2 dx.$$

Proof The inequality clearly holds for $N = 1$, in which case, $C_0 = 0$. It follows readily from the assumption

$$\sum_{\alpha,\beta=1}^{n} a_{\alpha\beta} \partial_\alpha u(x) \partial_\beta u(x) \geq \sigma |\nabla \varphi(x)|^2$$

and integrating on both sides.

For $N > 1$, if $(a_{\alpha\beta}^{jk}(x))$ are constants, we proceed by using the Fourier transform. Let φ vanish outside Ω, then it is defined on the entire $\mathbb{R}^n$. Let

$$\hat{\varphi}(\xi) = \int_{\mathbb{R}^n} \varphi(x) \exp\left[-2\pi i \langle \xi, x \rangle_{\mathbb{R}^n}\right] dx,$$

then

$$-2\pi i \xi_\alpha \hat{\varphi}(\xi) = \int_{\mathbb{R}^n} \partial_\alpha \varphi(x) \exp\left[-2\pi i \langle \xi, x \rangle_{\mathbb{R}^n}\right] dx.$$

According to Parsaval's equality,

$$\sum_{j,k=1}^{N} \sum_{\alpha,\beta=1}^{n} \int_{\mathbb{R}^n} a_{\alpha\beta}^{jk} \partial_\alpha \varphi^j(x) \partial_\beta \varphi^k(x) dx$$

$$= 4\pi^2 \sum_{j,k=1}^{N} \sum_{\alpha,\beta=1}^{n} \int_{\mathbb{R}^n} a_{\alpha\beta}^{jk} \xi_\alpha \xi_\beta \hat{\varphi}^j(\xi) \hat{\varphi}^k(\xi) d\xi$$

$$\geq 4\pi^2 \sigma \int_{\mathbb{R}^n} |\xi|^2 |\hat{\varphi}(\xi)|^2$$

$$= \sigma \int_\Omega |\nabla \varphi(x)|^2 dx.$$

When $(a_{\alpha\beta}^{jk}(x))$ are non-constants, we can employ the argument of partition of unity, treating the coefficients as if they are constants in each small neighborhood, and then piece them together using the estimate given above. The remainder can be combined into the second integral on the right-hand side be means of the Schwarz's inequality. Since the details of this proof is rather complicated and beyond the scope of this course, we shall omit it and refer the interested readers to K. Yosida, *Functional Analysis*, pp. 175–177.

Lemma 6.3 *Let $L \in C^2$ satisfy the strict Legendre–Hadamard condition, namely $\exists \, \sigma > 0$ such that*

$$L_{p_\alpha^i p_\beta^j}(x, u, p) \xi^i \xi^j \eta_\alpha \eta_\beta \geq \sigma |\xi|^2 |\eta|^2, \quad \forall \, (x, u, p) \in \Omega \times \mathbb{R}^N \times \mathbb{R}^{nN}.$$

Suppose u_0 is a solution of the E-L equation, and there exists $\mu > 0$ such that

$$Q_{u_0}(\varphi) \geq \mu \int_\Omega |\varphi|^2 dx \quad \forall \, \varphi \in C_0^1(\Omega, \mathbb{R}^N),$$

then there exists $\lambda > 0$ such that

$$Q_{u_0}(\varphi) \geq \lambda \int_\Omega (|\nabla \varphi|^2 + |\varphi|^2) dx \quad \forall \, \varphi \in C_0^1(\Omega, \mathbb{R}^N),$$

whence u_0 is a strict minimum of I.

Proof Since L satisfies the strict Legendre–Hadamard condition, according to Gårding's inequality, there exist $\alpha > 0$ and $C_0 > 0$ such that

$$\int_\Omega A_{u_0}(\nabla\varphi \cdot \nabla\varphi)dx \geq \int_\Omega [\alpha|\nabla\varphi|^2 - C_0|\varphi|^2]dx.$$

From

$$Q_{u_0}(\varphi) = \int_\Omega A_{u_0}(\nabla\varphi \cdot \nabla\varphi) + 2B_{u_0}(\nabla\varphi \cdot \varphi) + C_{u_0}(\varphi \cdot \varphi),$$

we deduce that there exist positive constants C_1 and C_2 such that

$$\alpha \int_\Omega |\nabla\varphi|^2 dx$$

$$\leq Q_{u_0}(\varphi) + C_1\left[\left(\int_\Omega |\nabla\varphi|^2 dx\right)^{\frac{1}{2}}\left(\int_\Omega |\varphi|^2 dx\right)^{\frac{1}{2}} + \int_\Omega |\varphi|^2 dx\right] + C_2\int_\Omega |\varphi|^2 dx$$

$$\leq \frac{\alpha}{2}\int_\Omega |\nabla\varphi|^2 dt + Q_{u_0}(\varphi) + C_2\int_\Omega |\varphi|^2 dx.$$

Moreover, using the fact

$$\int_\Omega |\varphi|^2 dx \leq \mu^{-1}Q_{u_0}(\varphi),$$

we then have

$$\int_\Omega |\nabla\varphi(x)|^2 dx \leq \frac{2}{\alpha}(1 + C_2\mu^{-1})Q_{u_0}(\varphi).$$

Combining the two inequalities, it follows that there exists $\lambda > 0$ such that

$$Q_{u_0}(\varphi) \geq \lambda \int_\Omega (|\nabla\varphi|^2 + |\varphi|^2)dx.$$

By Theorem 6.2, u_0 is a strict minimum of I. □

In addition, we provide a different criterion for strict minimum from the "eigenvalue" point of view.

Suppose $u_0 \in C^1(\bar\Omega, \mathbb{R}^N)$ is a solution of the E-L equation. We call

$$\boxed{\lambda_1 = \inf\left\{Q_{u_0}(\varphi) \,|\, \varphi \in C_0^1(\Omega, \mathbb{R}^N), \int_\Omega |\varphi(x)|^2 dx = 1\right\}}$$

the first eigenvalue of the Jacobi operator (details are given in Lecture 12).

Theorem 6.3 *Let $L \in C^2$ satisfy the strict Legendre–Hadamard condition. Suppose $u_0 \in M$ is a weak minimum of I, then $\lambda_1 \geq 0$. Furthermore, if $\lambda_1 > 0$, then u_0 is a strict weak minimum of I.*

Proof Suppose $\lambda_1 < 0$, then $\exists\ \varphi_0 \in C_0^1(\Omega, \mathbb{R}^N) \backslash \{\theta\}$ such that

$$Q_{u_0}(\varphi_0) < \frac{\lambda_1}{2} \int_\Omega |\varphi_0(x)|^2 dx < 0,$$

which contradicts (6.6). Thus, $\lambda_1 \geq 0$.

Suppose $\lambda_1 > 0$, then

$$Q_{u_0}(\varphi) \geq \frac{\lambda_1}{2} \int_\Omega |\varphi(x)|^2, \quad \forall \varphi \in C_0^1(\Omega, \mathbb{R}^N).$$

u_0 is a strict weak minimum from Lemma 6.3. $\qquad \square$

Remark 6.3 (Strong minimum) For multi-integral variational problems, we can also define the Weierstrass excess function $\mathfrak{E}_L \in C^1(\Omega \times \mathbb{R}^N \times \mathbb{R}^{nN} \times \mathbb{R}^{nN}, \mathbb{R}^1)$ and use this to state the necessary condition for a strong minimum. We define

$$\mathfrak{E}_L(x, u, p, q) = L(x, u, q) - L(x, u, p) - \sum_{\alpha=1}^{n}\sum_{\beta=1}^{N}(q_\alpha^i - p_\alpha^i)L_{p_\alpha^i}(x, u, p).$$

We have the following.

Theorem 6.4 *Suppose* $u_0 \in C^1(\bar{\Omega}, \mathbb{R}^N)$ *is a strong minimum of* I, *then*

$$\mathfrak{E}_L(x, u_0(x), \nabla u_0(x), \nabla u(x) + \pi) \geq 0, \forall x \in \Omega, \forall \pi = (\pi_\alpha^i), \text{rank}(\pi) = 1.$$

The idea of the proof follows closely to that of a single variable, however, it is more complicated, so we shall omit it here.

When $n > 1$ but $N = 1$, there is also a similar sufficient condition for strong minimum (Lichtenstein theorem), we refer to [GH], p. 390.

Exercises

1. Let $I(u) = \int_\Omega \sqrt{1 + u_x^2 + u_y^2}\, dxdy$, $\forall \varphi \in C_0^\infty(\Omega)$, find the first variation $\delta I(u, \varphi)$ and the second variation $\delta^2 I(u, \varphi)$.

2. Suppose $g = (g_{\alpha\beta}(x))_{1 \leq \alpha, \beta \leq n}$ is a continuous positive definite marix defined on a closed and bounded region Ω. Denote $\det(g)$ its determinant and $(g^{\alpha\beta}(x))$ its inverse matrix. Let

$$I(u) = \frac{1}{2}\int_\Omega \sum_{\alpha,\beta=1}^{n} g^{\alpha\beta}(x)\partial_\alpha u(x)\partial_\beta(x)\det(g)\, dx_1 \cdots dx_n.$$

   (1) find its E-L equation.
   (2) Assume $\psi \in C^1(\partial\Omega, \mathbb{R}^1)$, $u_0 \in M := \{v \in C^1(\Omega, \mathbb{R}^1) \mid v|_{\partial\Omega} = \psi\}$ is a critical point of I. Prove that u_0 is a weak minimum.
   (3) Find J_{u_0}.

Lecture Notes on Calculus of Variations

3. Find the E-L equation of the functional

$$I(u) = \int_\Omega (|\nabla u|^p - |u|^q)dx, \qquad 1 \le p, q < \infty.$$

Suppose u_0 is a critical point, find J_{u_0}.

4. Suppose $u \in C^1(\mathbb{R}^4)$ is a solution of $\Box u = 0$. Is u a critical point or minimum of the functional

$$I(u) = \int_{\mathbb{R}^4} [(\partial_t u)^2 - |\nabla_x u|^2]\, dt dx_1 dx_2 dx_3?$$

5. Suppose $F \in C^2(\mathbb{R}^1, \mathbb{R}^1)$ satisfy $|F''(t)| < \lambda_1$, which is the first eigenvalue of the Laplace operator

$$I(u) = \int_\Omega \left[\frac{1}{2}|\nabla u|^2 + F(u) \right] dx$$

with zero boundary condition. Suppose $u_0 \in C_0^1(\Omega)$ is a critical point of I, show that it is a minimum.

6. Let $\Omega \subset \mathbb{R}^2$ be a plane region, ν be a constant, $f \in L^1(\Omega)$, $M = \{w \in C^2(\Omega, \mathbb{R}^1) \mid w|_\Omega = \partial_n w|_\Omega = 0\}$, where ∂_n denotes the normal derivative. Write the E-L equation of the functional:

$$I(w) = \int_\Omega [(w_{xx} + w_{yy})^2 - 2\nu(w_{xx}w_{yy} - w_{xy}^2) + fw]\, dx dy.$$

Lecture 7

Constrained variational problems

Finding extremal values of functions includes both unconstrained and constrained problems. The extreme value problems of functionals also include both unconstrained and constrained problems. However, the constraints can be more colorful in variational problems.

7.1 The isoperimetric problem

The so-called isoperimetric problem states: for a given target functional $(\mathcal{M}, I)$, a constraint functional $(\mathcal{M}, N)$, and a prescribed constant c, find the necessary and sufficient condition for I to attain its minimum under the constraint $N(u) = c$, namely,

$$\min\{I(u) \mid u \in \mathcal{M}, \ N(u) = c\}.$$

The following example is the original source of isoperimetric problems.

Example 7.1 (The isoperimetric problem) Find a closed plane curve of a given perimeter which encloses the greatest area.

We parametrize a closed plane curve as follows:

$$\begin{cases} x = x(\theta), \\ y = y(\theta), \end{cases}$$

where $0 \le \theta \le 2\pi$ is the parameter. The enclosed area is then given by

$$A = \frac{1}{2} \int_0^{2\pi} (xy' - yx')d\theta$$

and its arclength

$$L = \int_0^{2\pi} \sqrt{x'^2 + y'^2}d\theta.$$

Since the perimeter l is given, our goal is to find a curve $(x(\theta), y(\theta))$ such that functional A is a maximum under the constraint $L = l$.

We first recall the method used on constrained optimization problems in mathematical analysis. Assume $f, g \in C^1(\Omega, \mathbb{R}^1)$, where $\Omega \subset \mathbb{R}^n$ is an open subset. Assume $g^{-1}(1) \neq \emptyset$. If $x_0 \in \Omega$ is such that the function f attains its minimum under the constraint $g(x_0) = 1$ with $\nabla g(x_0) \neq 0$:

$$f(x_0) = \min_{g^{-1}(1)} f(x),$$

then we can apply the Lagrange multiplier to turn this constrained minimization problem into an unconstrained minimization problem.

To be more explicit, there exists a Lagrange multiplier $\lambda \in \mathbb{R}^1$ such that

$$\nabla f(x_0) + \lambda \nabla g(x_0) = 0.$$

The Lagrange multiplier has a clear geometric meaning: if $\mathcal{M} = g^{-1}(1)$ is a differentiable manifold, then $\nabla f(x)$ is parallel to the outward normal vector at the point $(x, g(x))$, whenever x is the constraint minimizer. We restrict f on $\mathcal{M}$ and denote it by $\tilde{f} = f|_{\mathcal{M}}$, whose differential

$$d\tilde{f}(x) = \nabla f(x) - \frac{\nabla f(x) \cdot \nabla g(x)}{\|\nabla g(x)\|^2} \nabla g(x).$$

At an extreme point, it must satisfy

$$d\tilde{f}(x) = 0 \iff \nabla f(x) + \lambda \nabla g(x) = 0,$$

where

$$\lambda = -\frac{\nabla f(x) \cdot \nabla g(x)}{\|\nabla g(x)\|^2}$$

is the projection of $-\nabla f(x)$ onto the unit outward normal $\nabla g(x)$ (when $\|\nabla g(x)\| = 1$).

This indicates by means of the Lagrange multiplier, the solution of the constrained extreme value problem becomes a stationary point (critical point) of the adjusted function $f + \lambda g$.

Constrained extreme value problem of a functional can also be turned into an unconstrained extreme value problem of another functional via the Lagrange multiplier. We have the following:

Theorem 7.1 *Given* $L, G \in C^2(\bar{\Omega} \times \mathbb{R}^N \times \mathbb{R}^{nN})$, $\rho \in C^1(\partial\Omega, \mathbb{R}^N)$, *and* $\mathcal{M} = \{u \in C^1(\bar{\Omega}, \mathbb{R}^N) \,|\, u|_{\partial\Omega} = \rho\}$. *Define on* $\mathcal{M}$ *the functionals*

$$I(u) = \int_\Omega L(x, u(x), \nabla u(x)) dx$$

and

$$N(u) = \int_\Omega G(x, u(x), \nabla u(x)) dx.$$

Suppose c is a constant such that $N^{-1}(c) \cap \mathcal{M} \neq \emptyset$. Suppose $u_0 \in \mathcal{M}$ is a weak minimum of I under the constraint $N(u) = c$, i.e.

$$I(u_0) = \min_{u \in \mathcal{M} \cap N^{-1}(c)} I(u),$$

and suppose $\exists \, \varphi_0 \in C_0^1(\Omega, \mathbb{R}^N)$ such that $\delta N(u, \varphi_0) \neq 0$, then $\exists \, \lambda \in \mathbb{R}^1$ satisfying

$$\delta I(u_0, \varphi) + \lambda \delta N(u_0, \varphi) = 0, \quad \forall \, \varphi \in C_0^1(\Omega, \mathbb{R}^N).$$

Namely, if let

$$Q = L + \lambda G$$

be the adjusted Lagrangian, then u_0 satisfies the corresponding E-L equation

$$\sum_{\alpha=1}^n \partial_\alpha Q_{p_\alpha^i}(x, u_0(x), \nabla u_0(x)) = Q_{u^i}(x, u_0(x), \nabla u_0(x)), \quad i = 1, \ldots, N. \quad (7.1)$$

Proof Note the mapping $\varphi \mapsto \delta N(u_0, \varphi)$ is linear, so without loss of generality, we may assume $\delta N(u_0, \varphi_0) = 1$. We regard $N^{-1}(c) \cap \mathcal{M}$ as a hypersurface in a function space. For any $\varphi \in C_0^1(\Omega, \mathbb{R}^N)$ linearly independent of φ_0, consider the plane π at u_0 spanned by the vectors φ_0 and φ (see Figure 7.1)

$$\pi = \{u_0 + \epsilon\varphi + \tau\varphi_0 \mid (\epsilon, \tau) \in \mathbb{R}^2\},$$

and the two functions Φ and Ψ on π:

$$\Phi(\epsilon, \tau) = I(u_0 + \epsilon\varphi + \tau\varphi_0)$$

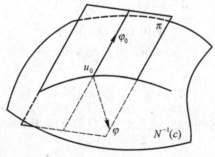

Fig. 7.1

and

$$\Psi(\epsilon, \tau) = N(u_0 + \epsilon\varphi + \tau\varphi_0).$$

Notice

$$\Psi(0,0) = N(u_0) = c, \quad \Psi_\tau(0,0) = \delta N(u_0, \varphi_0) = 1,$$

for $\epsilon_0, \tau_0 > 0$ sufficiently small, we now apply the implicit function theorem on $R = (-\epsilon_0, \epsilon_0) \times (-\tau_0, \tau_0)$: $\exists\, \xi \in C^1(-\epsilon_0, \epsilon_0)$ such that $(\epsilon, \xi(\epsilon)) \in R$ is the unique solution of

$$\Psi(\epsilon, \tau) = c$$

inside R. This means

$$\Psi(\epsilon, \xi(\epsilon)) = c, \ \xi(0) = 0, \ \xi'(0) = -\Psi_\epsilon(0,0) = -\delta N(u_0, \varphi).$$

So inside a small enough R,

$$N^{-1}(c) \cap \pi = u_0 + \epsilon\varphi + \xi(\epsilon)\varphi_0.$$

Let

$$g(\epsilon) = I(u_0 + \epsilon\varphi + \xi(\epsilon)\varphi_0).$$

Since $v = u_0 + \epsilon\varphi + \tau\varphi_0 \in \mathcal{M}$,

$$\|v - u_0\|_{C^1} \le |\epsilon| \|\varphi\|_{C^1} + |\tau| \|\varphi_0\|_{C^1}.$$

However, u_0 is a weak minimum of I on $\mathcal{M} \cap N^{-1}(c)$, so for ϵ_0, τ_0 sufficiently small,

$$\Phi(\epsilon, \xi(\epsilon)) \ge \Phi(0,0).$$

Thus, 0 is a minimum of g.

From this, we deduce that

$$
\begin{aligned}
0 &= g'(0)\\
&= \frac{d}{d\epsilon}\Phi(\epsilon, \xi(\tau))|_{\epsilon=0}\\
&= \Phi_\epsilon(0,0) + \Phi_\tau(0,0)\xi'(0)\\
&= \delta I(u_0, \varphi) + \lambda\delta N(u_0, \varphi),
\end{aligned}
\tag{7.2}
$$

where

$$\lambda = -\Phi_\tau(0,0)$$

is independent of φ. Letting $Q = L + \lambda G$, (7.2) is the **E-L** equation associated to Q:

$$\operatorname{div} Q_p(x, u_0(x), \nabla u_0(x)) = Q_u(x, u_0(x), \nabla u_0(x)). \qquad \square$$

Remark 7.1 We still call λ a Lagrange multiplier.

Remark 7.2 We can also consider the constrained extreme value problem with multiple constraints: given $L, G^1, \ldots, G^m \in C^2(\bar{\Omega} \times \mathbb{R}^N \times \mathbb{R}^{nN})$, $\mathcal{M} \subset C^1(\bar{\Omega}, \mathbb{R}^N)$, and constants $c_1, c_2, \ldots, c_m$. Let

$$I(u) = \int_\Omega L(x, u(x), \nabla u(x))dx,$$

$$N_j(u) = \int_\Omega G^j(x, u(x), \nabla u(x))dx, \quad j = 1, 2, \ldots, m,$$

if $u \in \mathcal{M}$ is a minimum of the functional I under the constraints

$$\int_\Omega G^j(x, u(x), \nabla u(x))dx = c_j, \quad j = 1, 2, \ldots, m$$

and if $\exists \, \varphi_k \in C_0^1(\Omega, \mathbb{R}^N)$ $(1 \le k \le m)$ such that $\det(a_{jk}) \neq 0$, where

$$a_{jk} = \delta N_j(u, \varphi_k)$$
$$= \int_\Omega [\langle G_u^j(x, u(x), \nabla u(x), \varphi_k(x)\rangle + \langle G_p^j(x, u(x), \nabla u(x)), \nabla \varphi_k(x)\rangle] dx,$$

$$j, k = 1, \ldots, m,$$

then there exist Lagrange multipliers $\lambda_1, \ldots, \lambda_m$ such that u satisfies the E-L equation of the adjusted Lagrangian

$$Q = L + \lambda_1 N_1 + \cdots + \lambda_m G_m.$$

Example 7.1 (The isoperimetric problem continued) By Theorem 7.1, we introduce the Lagrange multiplier λ and consider the E-L equation of the adjusted functional

$$I(x, y) = A(x, y) + \lambda L(x, y),$$

it follows that

$$\begin{cases} -x' = \lambda \left(\dfrac{y'}{\sqrt{x'^2 + y'^2}} \right)', \\ y' = \lambda \left(\dfrac{x'}{\sqrt{x'^2 + y'^2}} \right)'. \end{cases}$$

Upon solving, we arrive at

$$\begin{cases} x - c_1 = -\lambda \dfrac{y'}{\sqrt{x'^2 + y'^2}}, \\ y - c_2 = \lambda \dfrac{x'}{\sqrt{x'^2 + y'^2}}, \end{cases}$$

where c_1 and c_2 are constants. This is exactly the standard circle equation

$$(x - c_1)^2 + (y - c_2)^2 = \lambda^2,$$

whose radius $r = \lambda = \frac{1}{2\pi}$ and centered at (c_1, c_2).

Example 7.2 (The eigenvalue problem continued) In Lecture 1, we asked the following constrained extreme value problem: for a given domain $\Omega \subset \mathbb{R}^n$ and a bounded continuous function $q \in C(\bar{\Omega})$, define the functionals

$$I(u) = \int_\Omega |\nabla u|^2 dx, \ N(u) = \int_\Omega q(x)|u(x)|^2 dx, \quad \mathcal{M} = C_0^1(\Omega).$$

Find

$$\min\{I(u) \,|\, u \in \mathcal{M}, \ N(u) = 1\}.$$

According to Theorem 7.1, we introduce the Lagrange multiplier λ. The E-L equation of the adjusted Lagrangian $Q = p^2 - \lambda q(x)u^2$ is

$$-\Delta u = \lambda q u.$$

Here λ precisely coincides with the eigenvalues of the Laplace operator $-\Delta$ with respect to the weight function q.

If $u_0 \in \mathcal{M}$ is a minimum satisfying the constraint $N(u) = \int_\Omega q(x)|u(x)|^2 dx = 1$, then $\lambda = \int_\Omega |\nabla u_0(x)|^2 dx \neq 0$ such that

$$-\Delta u_0(x) = \lambda q(x) u_0(x).$$

This minimum u_0 is an eigenfunction of the Laplace operator $-\Delta$ with respect to the weight function q, and the Lagrange multiplier λ is its corresponding eigenvalue.

In fact, by introducing the Lagrange multiplier λ, all critical points of the functional

$$\int_\Omega [|\nabla u(x)|^2 - \lambda q(x) u^2(x)] dx$$

are eigenfunctions.

7.2 Pointwise constraints

The constraint appeared in the isoperimetric problem is of integral form, there is however another kind of constraint, which is given pointwise.

For instance, given a function $M \in C^1(\bar{\Omega} \times \mathbb{R}^N \times \mathbb{R}^{nN}, \mathbb{R}^1)$, we want to find the extrema in $\mathcal{M}$, under the pointwise constraint

$$M(x, u(x), \nabla u(x)) = 0, \quad \forall \, x \in \Omega,$$

of the functional

$$I(u) = \int_\Omega L(x, u(x), \nabla u(x)) dx.$$

It is natural to ask, is there a Lagrange multiplier-like method available? In the following, we will only address this question in the case of holonomic constraints, namely M only depends on u (but independent of p).

Theorem 7.2 *Let $\bar{\Omega} \subset \mathbb{R}^n$ be a closed and bounded set. Let $L \in C^2(\bar{\Omega} \times \mathbb{R}^N \times \mathbb{R}^{nN}, \mathbb{R}^1)$, $M \in C^2(\mathbb{R}^N, \mathbb{R}^1)$, $\rho \in C^1(\partial\Omega, \mathbb{R}^N)$, and*

$$\mathcal{M} = \{u \in PWC^1(\bar{\Omega}, \mathbb{R}^N) \,|\, u|_{\partial\Omega} = \rho\}.$$

Suppose $u_0 \in \mathcal{M}$ is a local minimum under the above constraint and it is C^2 outside finitely many $(n-1)$ dimensional piecewise C^1 hypersurfaces. If $\forall\, x \in \bar{\Omega}$, $\nabla M(u_0(x)) \neq 0$, then there exists a continuous function $\lambda \in C(\bar{\Omega})$ such that u_0 satisfies the E-L equation of the adjusted Lagrangian $Q = L + \lambda M$:

$$\boxed{L_{u^i} + \lambda M_{u^i} = \sum_{\alpha=1}^n \frac{\partial}{\partial x_\alpha} L_{p_\alpha^i}, \quad 1 \leq i \leq N. \qquad (7.3)}$$

Proof We will first construct such λ to be continuous locally on each piece, and then glue them together to a globally defined continuous function on $\bar{\Omega}$.

1. $\forall\, x_0 \in \Omega$, there exists a ball $B_r(x_0) \subset \Omega$ such that $\nabla_x(M(u_0(x))) \neq 0$, $\forall\, x \in B_r(x_0)$. Since $\nabla_u M(u_0(x)) \cdot \nabla u_0(x) = \nabla_x(M(u_0(x)))$, $\nabla_u M(u) \neq 0$, $\forall\, u \in u_0(B_r(x_0))$, $\nabla u_0(x) \neq 0$, $\forall\, x \in B_r(x_0)$. Without loss of generality, we may assume $M_{u^N}(u_0(x)) \neq 0$, $\forall\, x \in B_r(x_0)$. If we adopt the notation $\tilde{u} = (u^1, \ldots, u^{N-1})$, then we can solve and obtain $u^N = U(\tilde{u})$, where U is a C^2 function (see Figure 7.2).

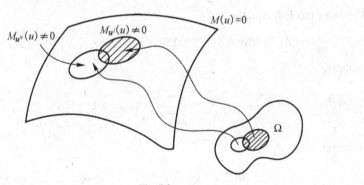

$M_{u^N}(u) \neq 0 \qquad M_{u^i}(u) \neq 0 \qquad M(u) = 0$

Ω

Fig. 7.2

We further adopt the notation

$$\tilde{p} = (p_\alpha^i)_{\substack{1 \leq i \leq N-1 \\ 1 \leq \alpha \leq n}},$$

$$p^N = \left(\sum_{i=1}^{N-1} U_{u^i} p_\alpha^i\right)_{1 \leq \alpha \leq n},$$

and

$$P^N = \sum_{i=1}^{N-1} U_{u^i} \nabla u^i.$$

2. Since u_0 is a minimum, when the domain of the integral of I is restricted to $B_r(x_0)$, $u_0|_{B_r(x_0)}$ is still a minimum. This can be justified as follows: suppose not, then $\exists\, v \in PWC^1(\bar{B}_r(x_0))$ such that $v|_{\partial B_r(x_0)} = u_0|_{\partial B_r(x_0)}$ and

$$\int_{\bar{B}_r(x_0)} L(x, v(x), \nabla v(x))dx < \int_{\bar{B}_r(x_0)} L(x, u_0(x), \nabla u_0(x))dx.$$

Now, if we replace $u_0(x)$ by $v(x)$ on $\bar{B}_r(x_0)$, then we will end up with a new piecewise differentiable function whose functional value is strictly less than $I(u_0)$, which contradicts the fact that u_0 is a minimum.

Let

$$\Lambda(x, \tilde{u}, \tilde{p}) = L(x, \tilde{u}, U(\tilde{u}), \tilde{p}, p^N),$$

since u_0 is the minimum of I under the constraint $M(u) = 0$, $\tilde{u}_0$ must be a minimum of

$$J(\tilde{u}) = \int_{B_r(x_0)} \Lambda(x, \tilde{u}(x), \nabla \tilde{u}(x))dx.$$

The latter has E-L equation

$$\Lambda_{u^i} = \operatorname{div} \Lambda_{p^i}(x, \tilde{u}(x), \nabla \tilde{u}(x)), \ i = 1, \ldots, N-1.$$

Namely,

$$L_{u^i} + L_{u^N} U_{u^i} + \sum_{\alpha=1}^{n} L_{p_\alpha^N} \frac{\partial p_\alpha^N}{\partial u^i} = \sum_{\alpha=1}^{n} \frac{\partial}{\partial x_\alpha}(L_{p_\alpha^i} + L_{p_\alpha^N} U_{u^i}), \ i = 1, \ldots, N-1. \tag{7.4}$$

However,

$$\frac{\partial p_\alpha^N}{\partial u^i} = \sum_{j=1}^{N-1} \frac{\partial^2 U}{\partial u^i \partial u^j} u_\alpha^j = \frac{\partial U_{u^i}}{\partial x_\alpha}. \tag{7.5}$$

By (7.5), the right-hand side of (7.4) is equal to

$$\sum_{\alpha=1}^{n} \left(\frac{\partial}{\partial x_\alpha} L_{p_\alpha^i} + U_{u^i} \frac{\partial}{\partial x_\alpha} L_{p_\alpha^N} + L_{p_\alpha^N} \frac{\partial}{\partial x_\alpha} U_{u^i} \right)$$

$$= \sum_{\alpha=1}^{n} \left(\frac{\partial}{\partial x_\alpha} L_{p_\alpha^i} + U_{u^i} \frac{\partial}{\partial x_\alpha} L_{p_\alpha^N} + L_{p_\alpha^N} \frac{\partial p_\alpha^N}{\partial u^i} \right).$$

Hence, (7.4) is reduced to

$$L_{u^i} + U_{u^i}\left(L_{u^N} - \sum_{\alpha=1}^{N-1} \frac{\partial}{\partial x_\alpha} L_{p_\alpha^N}\right) = \sum_{\alpha=1}^{n} \frac{\partial}{\partial x_\alpha} L_{p_\alpha^i}. \tag{7.6}$$

Noting

$$M(\tilde{u}_0, U(\tilde{u}_0)) = 0,$$

upon differentiating, we obtain

$$M_{u^i} + M_{u^N} U_{u^i} = 0,$$

i.e.

$$U_{u^i} = -\frac{M_{u^i}}{M_{u^N}}. \tag{7.7}$$

We now define on $B_r(x_0)$ the function

$$\lambda = \frac{1}{M_{u^N}}(\operatorname{div} L_{p^N} - L_{u^N}). \tag{7.8}$$

Substituting (7.7) and (7.8) into (7.6), it follows that

$$L_{u^i} - \operatorname{div} L_{p^i} + \lambda M_{u^i} = 0, \; i = 1, \ldots, N-1. \tag{7.9}$$

Combining (7.8) and (7.9) gives the local E-L equation (7.3) on $B_r(x_0)$.

3. Since $\bar{\Omega}$ is compact, we have a finite open sub-covering of the covering by $\{B_r(x_0) \,|\, x_0 \in \Omega, \, r = r(x_0)\}$, together with the already defined functions $\lambda|_{B_r(x_0)}$. Suppose on $B_{r_1}(x_1)$, $M_{u^N} \neq 0$ and on $B_{r_2}(x_2)$, $M_{u^j} \neq 0$, then we have $\lambda|_{B_{r_2}(x_2)} = \frac{1}{M_{u^j}}(\operatorname{div} L_{p^j} - L_{u^j})$ and $\lambda|_{B_{r_1}(x_1)}$ is the λ in (7.8). Thus, when any two such small balls have a nonempty intersection $(B_{r_1}(x_1) \cap B_{r_2}(x_2) \neq \emptyset)$, by (7.9), it is immediate that

$$\lambda|_{B_{r_1}(x_1)} = \lambda|_{B_{r_2}(x_2)}.$$

We now glue these $\lambda|_{B_r(x_0)}$ together to create a globally continuous function λ as desired. $\qquad\square$

Remark 7.3 Just like the isoperimetric problem, we can also consider multiple pointwise constraints. As long as the constraints $M_1(u) = M_2(u) = \cdots = M_s(u) = 0$ satisfy

$$\operatorname{rank}\left(\frac{\partial M_i(u(x))}{\partial u^j}\right) = s, \quad \forall\, x \in \Omega.$$

Now the adjusted Lagrangian is

$$Q = L + \sum_{i=1}^{s} \lambda_i M_i,$$

where $\lambda_i \in C(\Omega)$, $i = 1, \ldots, s$.

Remark 7.4 If the constraint functions also depend on other variables, then we call them non-holonomic constraints. As to variational problems with non-holonomic constraints, is there still a Lagrange multiplier method? The answer to this question is far more complicated; we refer the interested readers to the book by Giaquinta and Hidelbrandt [GH]. However, some special cases are also known to be optimal control problems, which we will address in the third part of this book.

Example 7.3 (Revisit of geodesics on spheres) In our earlier discussions, we have studied the geodesics problems from an unconstrained extreme value problem point of view. We now provide a different viewpoint. Regard the sphere as the constraint

$$x^2 + y^2 + z^2 = 1.$$

Find the extrema of the functional

$$I(u) = \int_a^b \sqrt{\dot{x}^2 + \dot{y}^2 + \dot{z}^2} dt$$

given this pointwise constraint.

Let $u = (x, y, z)$ and $p = (\xi, \eta, \zeta)$. Introducing the Lagrange multiplier $\lambda(t)$, we have the adjusted Lagrangian

$$Q = \sqrt{\xi^2 + \eta^2 + \zeta^2} + \lambda(x^2 + y^2 + z^2).$$

Its E-L equation is

$$\frac{d}{dt}\frac{\dot{u}}{|\dot{u}|} = 2\lambda u;$$

furthermore, it also satisfies $|u| = 1$. Since

$$\frac{d}{dt}\left(\frac{\dot{u}}{|\dot{u}|} \times u\right) = \frac{d}{dt}\left(\frac{\dot{u}}{|\dot{u}|}\right) \times u + \frac{\dot{u}}{|\dot{u}|} \times \frac{du}{dt} = 2\lambda u \times u + \frac{\dot{u}}{|\dot{u}|} \times \dot{u} = 0,$$

$v = \frac{\dot{u}}{|\dot{u}|} \times u$ must be a constant vector. Consequently, $u \perp v$, i.e. the geodesic u must lie in a plane orthogonal to the constant vector v, hence this curve must be a piece of a great circle.

Example 7.4 On the slant plane $x = y$, find the equation of the trajectory of a moving particle of unit mass by gravity. We denote the coordinate of the particle by (x, y). The Lagrangian is

$$L = \frac{1}{2}(p^2 + q^2) - gy,$$

where g is the gravitational constant and the constraint is

$$M(x, y) = y - x = 0.$$

The adjusted Lagrangian is

$$Q = L + \lambda M = \frac{1}{2}(p^2 + q^2) - gy + \lambda(y - x)$$

whose E-L equation is

$$\begin{cases} -\lambda = \dfrac{\partial Q}{\partial x} = \dfrac{d}{dt}Q_p = \ddot{x}, \\[2mm] -g + \lambda = \dfrac{\partial Q}{\partial y} = \dfrac{d}{dt}Q_q = \ddot{y}. \end{cases}$$

Since $y = x$, $\lambda = g/2$, from which we solve to get

$$x = y = -\frac{1}{2}\lambda t^2 + \dot{x}(0)t + x(0),$$

where $\lambda = g/2$.

Example 7.5 (Harmonic mappings to spheres) Let Ω be the unit ball in $\mathbb{R}^3$, S^2 be the unit sphere in $\mathbb{R}^3$, and $u = (u_1, u_2, u_3) : \Omega \to S^2$. If u is a solution of the constrained extreme value problem

$$\min\{I(u) \mid M(u) = 0\},$$

where

$$I(u) = \int_\Omega |\nabla u|^2 dx = \int_\Omega \sum_{i=1}^{3} \sum_{\alpha=1}^{3} |\partial_\alpha u^i|^2 dx,$$

$$M(u) = |u|^2 - 1 = u_1^2 + u_2^2 + u_3^2 - 1,$$

then we call u a harmonic mapping from the solid unit ball to the unit sphere. Find the differential equation for which such a harmonic mapping satisfies.

Solution The E-L equation of the adjusted Lagrangian is

$$-\Delta u_i = \lambda u_i,$$

where Δ is the Laplace operator and the Lagrange multiplier λ is a continuous function. From

$$u_1^2 + u_2^2 + u_3^2 = 1,$$

by differentiating, it yields

$$\langle u, \nabla u \rangle = 0,$$

where $\langle \cdot, \cdot \rangle$ denotes the standard inner product on $\mathbb{R}^3$. Differentiating once more, it yields

$$\langle u, \Delta u \rangle + |\nabla u|^2 = 0.$$

Multiplying by u on both sides of the above E-L equation, it is immediate

$$\lambda = -\langle \Delta u, u \rangle = |\nabla u|^2,$$

i.e.

$$-\Delta u = u|\nabla u|^2.$$

$\square$

7.3 Variational inequalities

In addition to constraints presented in terms of equalities, there are also variational problems with inequality constraints.

Example 7.6 (The Obstacle problem) Let $\Omega \subset \mathbb{R}^2$ be a bounded open subset.
As for the boundary, there is a given function $\varphi \in C^1(\partial\Omega)$.
As for the obstacle, there is a given function $\psi \in C^1(\bar{\Omega})$.
As for the external force, there is a given function $f \in C(\bar{\Omega})$.

Over Ω, we consider a thin membrane, whose boundary is fixed, $u|_{\partial\Omega} = \varphi$, while applying the external force, it cannot move beyond the obstacle, i.e. $u(x) \leq \psi(x), \forall x \in \bar{\Omega}$.

We propose this as the following variational problem: find the equilibrium position

$$u \in M = \{u \in PWC^1(\bar{\Omega}) \mid u|_{\partial\Omega} = \varphi\},$$

subject to the inequality constraint

$$u(x) \leq \psi(x), \quad \forall x \in \Omega, \quad u \in M,$$

such that the energy I of the thin membrane achieves its minimum:

$$I(u) = \int_\Omega \left[\frac{1}{2}|\nabla u(x)|^2 - f(x)u(x) \right] dx.$$

Generally speaking, the domain M is a given set of functions defined on $\bar{\Omega} \subset \mathbb{R}^n$, for a given convex subset C of M and a Lagrangian L, we ask to

$$\min \left\{ I(u) = \int_\Omega L(x, u(x), \nabla u(x))dx \mid u \in C \right\}.$$

Suppose $u \in C$ is a minimum, we want to derive a formula similar to the E-L equation. In fact, since C is convex, $\forall v \in C$, $tv + (1-t)u \in C$, $\forall t \in [0, 1]$, hence

$$I(tv + (1-t)u) \geq I(u), \quad \forall t \in [0, 1].$$

Consequently,

$$\delta I(u, v-u) = \lim_{t \to 0^+} [I(tv + (1-t)u) - I(u)] \geq 0,$$

i.e. $\forall v \in C$,

$$\boxed{\begin{aligned}
\int_\Omega &[L_u(x, u(x), \nabla u(x))(v(x) - u(x)) \\
&+ L_p(x, u(x), \nabla u(x)) \cdot (\nabla v(x) - \nabla u(x))]dx \geq 0.
\end{aligned}}$$

Note the only difference between the E-L equation and this expression is that the former is an equality, whereas the latter is an inequality. This is why we shall call it a variational inequality.

Returning to the obstacle problem, $C = \{u \in PWC^1(\bar{\Omega}) \mid u(x) \leq \psi(x), \forall x \in \bar{\Omega}, u|_{\partial\Omega} = \varphi\}$ is a convex set, so the resulting variational inequality is

$$\int_\Omega [\nabla u \nabla(v-u) - f(v-u)]dx \geq 0, \quad \forall v \in C.$$

Exercises

1. Find
$$\min\{I(u) \mid u \in C^1[0,1], u(0) = 0, u(1) = 2, N(u) = L\},$$

   where
$$I(u) = \int_0^1 \dot{u}^2(t)dt, \quad N(u) = \int_0^1 u(t)dt.$$

2. (The Dido problem) Find
$$\max\{I(u) \mid u \in C_0^1(0,b), \ N(u) = L\},$$

   where
$$I(u) = \int_0^b u(t)dt, \quad N(u) = \int_0^b (1 + \dot{u}(t)^2)^{\frac{1}{2}}dt.$$

3. Find the E-L equation of
$$I(u) = \int_a^b [(u'')^2(t) - p(t)(u')^2(t) + q(t)u^2(t)]dt,$$

   subject to the isoperimetric constraint $\int_a^b r(t)u^2(t)dt = 1$, where p, q, r are continuous functions on $[a, b]$, and u satisfies the boundary condition: $u(a) = \dot{u}(a) = u(b) = \dot{u}(b) = 0$.

4. Find the extreme values of the functional
$$I(u) = \int_0^1 (\dot{u}_1^2 + \dot{u}_2^2)dt, \quad M = \{(u_1, u_2) \in C_0^1((0,1), \mathbb{R}^2)\},$$

   under the constraint $u_2^2 + (u_1 - t) = 0$.

5. Let $\Omega \subset \mathbb{R}^n$, $u : \Omega \to \mathbb{R}^N$. Suppose $G \in C^1(\mathbb{R}^N)$ such that $\nabla G(u) \neq 0$, $\forall u \in G^{-1}(1)$. Let

$$I(u) = \int_\Omega |\nabla u(x)|^2 dx.$$

Prove that the E-L equation of the constrained variational problem

$$\min\{I(u) \,|\, G(u(x)) = 1 \,\forall\, x \in \Omega\}$$

is

$$-\Delta u^i = \lambda \partial_{u^i} G, \quad i = 1, \ldots, N,$$

$$\lambda = \frac{\sum_{k,j=1}^{N} \sum_{\alpha=1}^{n} \partial_{u^k u^j}^2 G(u) \partial_\alpha u^k \partial_\alpha u^j}{|\nabla_u G(u)|^2}.$$

This is the harmonic mapping equation of the hypersurface $G(u) = 1$.

6. Let $\Omega \subset \mathbb{R}^n$ be a bounded open subset and $u : \bar\Omega \to \mathbb{R}^1$ be a non-parametric equation of a surface. For a given boundary value $u|_{\partial\Omega} = \psi$ (ψ is a continuous function), its area $A(u) = \int_\Omega (1 + |\nabla u(x)|^2)^{\frac{1}{2}} dx$ with enclosed volume $V(u) = \int_\Omega u(x) dx$. Determine the surface equation whose area is a minimum but whose volume is the given constant V_0.

7. Let $X = (X_1(u,v), X_2(u,v), X_3(u,v))$, $\forall\, (u,v) \in B := \{(u,v) \,|\, u^2 + v^2 = 1\}$ be the parametric equation of a surface S. Define

$$D(X) = \frac{1}{2} \int_B \sum_{i=1}^{3} |\nabla X^i|^2 dudv$$

and

$$V(X) = \frac{1}{3} \int_B X \cdot (X_u \wedge X_v) dudv.$$

Determine the equation of the surface S such that $D(X)$ achieves its minimum under the constraint $V(X) = V_0$.

Lecture 8

The conservation law and Noether's theorem

In physics and mechanics, we frequently encounter various kinds of conservation laws, such as conservation of energy, conservation of momentum, and conservation of angular momentum, etc. E. Noether found out that the reason for such conservation laws is because the Lagrangian has invariant property under certain group actions.

8.1 One parameter diffeomorphisms and Noether's theorem

1. A special one parameter family of functions

Given a bounded open domain $\Omega \subset \mathbb{R}^n$, a function $u \in C^1(\bar{\Omega}, \mathbb{R}^N)$, and a Lagrangian $L \in C^2(\bar{\Omega} \times \mathbb{R}^N \times \mathbb{R}^{nN})$. We now introduce a 1-parameter family of diffeomorphisms $\eta_\varepsilon : \Omega \to \Omega_\varepsilon$, $\varepsilon \in (-\varepsilon_0, \varepsilon_0)$, where $\Omega_\varepsilon = \eta_\varepsilon(\Omega) \subset \mathbb{R}^n$ is the family of deformation of Ω under η_ε and $\eta_0 = \mathrm{id}$ (see Figure 8.1).

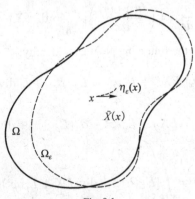

Fig. 8.1

Suppose

$$\frac{\partial}{\partial\varepsilon}\eta_\varepsilon|_{\varepsilon=0} = \bar{X}(x).$$

Then the deformation

$$y = \eta_\varepsilon(x) = x + \varepsilon\bar{X}(x) + o(\varepsilon).$$

Moreover, we define $v : \mathbb{R}^n \times (-\varepsilon_0, \varepsilon_0) \to \mathbb{R}^N$ satisfying $v(x, 0) = u(x)$. This induces a family of functions $v_\varepsilon = v(\cdot, \epsilon) : \Omega_\varepsilon \to \mathbb{R}^N, (y, \varepsilon) \in \Omega_\varepsilon \times (-\varepsilon_0, \varepsilon_0)$. Suppose

$$\frac{\partial}{\partial\varepsilon}v_\varepsilon(x)\bigg|_{\varepsilon=0} = \varphi(x).$$

In our previously discussed variational problems, in the same domain of the functional, all functions are defined over the same underlying region Ω. However, this requirement is unnecessary, we can indeed allow the functions in M to take on different domain Ω. To emphasize the functional dependence on the region, we shall denote

$$I(u, \Omega) = \int_\Omega L(x, u(x), \nabla u(x))dx.$$

Consequently, on the family of functions $\{v_\epsilon, \Omega_\epsilon\}$, I takes values

$$\Phi(\varepsilon) = I(v_\varepsilon, \Omega_\varepsilon) = \int_{\Omega_\varepsilon} L(y, v(y), \nabla v_\varepsilon(y))dy$$

$$= \int_\Omega L(\eta_\varepsilon(x), v_\varepsilon(\eta_\varepsilon(x)), \nabla_y v_\varepsilon(\eta_\varepsilon(x))) \det\left(\frac{\partial(\eta_\varepsilon(x))}{\partial x}\right)dx.$$

Since

$$\frac{d}{d\epsilon}\det\left(\frac{\partial(\eta_\varepsilon(x))}{\partial x}\right)\bigg|_{\epsilon=0} = \mathrm{div}(\bar{X}),$$

when $u \in C^2$, we have the following Noether's identity:

$$\Phi'(0) = \int_\Omega [\partial_{x_\alpha}L\bar{X}^\alpha + L_{u^i}\varphi^i + L_{p_\alpha^i}\varphi_{x^\alpha}^i + L\ \mathrm{div}\ \bar{X}]dx$$

$$= \int_\Omega [\mathrm{div}(L\bar{X}) + L_{u^i}\varphi^i - \partial_\alpha(L_{p_\alpha^i})\varphi^i + \mathrm{div}(L_{p^i}\varphi^i)]dx,$$

i.e.

$$\Phi'(0) = \int_\Omega [E_L(u)\varphi + \mathrm{div}(L\bar{X} + L_{p^i}\varphi^i)]dx. \tag{8.1}$$

2. General local 1-parameter transformation group

We consider deformation in the "phase space" (the space where both x and u are changing simultaneously). Let $\Omega \subset \mathbb{R}^n$. Let $\{\phi_\varepsilon\} : \Omega \times \mathbb{R}^N \to \mathbb{R}^n \times \mathbb{R}^N, |\epsilon| < \epsilon_0$, be a family of mappings satisfying

$$\begin{cases} Y(x, u, 0) = x, \\ W(x, u, 0) = u. \end{cases}$$

We call $(x, u) \mapsto (Y(x, u, \varepsilon), W(x, u, \varepsilon))$ a local 1-parameter transformation group. Its generating vector field is

$$\left. \frac{d\phi_\varepsilon}{d\varepsilon} \right|_{\varepsilon=0} = \sum_{\alpha=1}^{n} X^\alpha(x, u) \frac{\partial}{\partial x^\alpha} + \sum_{i=1}^{N} U^i(x, u) \frac{\partial}{\partial u^i}. \quad (8.2)$$

Thus,

$$\begin{cases} X(x, u) = \left. \frac{\partial}{\partial \varepsilon} Y(x, u, \varepsilon) \right|_{\varepsilon=0}, \\ U(x, u) = \left. \frac{\partial}{\partial \varepsilon} W(x, u, \varepsilon) \right|_{\varepsilon=0}. \end{cases}$$

For any $u \in C^1(\bar{\Omega}, \mathbb{R}^N)$, we want to convert it into the special 1-parameter family of functions as given above. Let

$$\begin{cases} \eta(x, \varepsilon) = Y(x, u(x), \varepsilon), \\ \omega(x, \varepsilon) = W(x, u(x), \varepsilon), \end{cases}$$

then

$$\begin{cases} \eta(x, 0) = x, \\ \omega(x, 0) = u(x). \end{cases}$$

We also let

$$\begin{aligned} \bar{X}(x) &= \left. \frac{\partial \eta(x, \varepsilon)}{\partial \varepsilon} \right|_{\varepsilon=0} = X(x, u(x)), \\ \bar{U}(x) &= \left. \frac{\partial \omega(x, \varepsilon)}{\partial \varepsilon} \right|_{\varepsilon=0} = U(x, u(x)). \end{aligned} \quad (8.3)$$

Introducing the variable

$$y = \eta_\varepsilon(x) = \eta(x, \varepsilon) = x + \varepsilon \bar{X}(x) + o(\varepsilon)$$

as well as the deformed region $\Omega_\varepsilon = \eta_\varepsilon(\Omega)$, then $\eta_\varepsilon : \bar{\Omega} \to \bar{\Omega}_\varepsilon$ is a diffeomorphism.

Denote the inverse mapping $\xi_\varepsilon = \eta_\varepsilon^{-1}$, then we have

$$x = \xi_\varepsilon(y) = y - \varepsilon \bar{X}(y) + o(\varepsilon) \qquad \forall y \in \Omega_\varepsilon$$

and another family of mappings

$$v_\varepsilon(y) = v(y, \varepsilon) = \omega(\xi_\varepsilon(y), \varepsilon) = \omega(x, \varepsilon) \qquad \forall\, y \in \Omega_\varepsilon.$$

Let

$$\varphi(x) = \left.\frac{\partial v(x, \varepsilon)}{\partial \varepsilon}\right|_{\varepsilon=0} \qquad \forall\, x \in \bar\Omega,$$

it follows that

$$\bar U = \frac{\partial v(\eta(x, \varepsilon), \varepsilon)}{\partial \varepsilon}\Big|_{\varepsilon=0} = \varphi(x) + \sum_{\alpha=1}^{n} \partial_\alpha u(x) \bar X^\alpha(x),$$

i.e.

$$\boxed{\varphi(x) = \bar U - \sum_{\alpha=1}^{n} u_{x_\alpha} \bar X^\alpha.} \qquad (8.4)$$

That said, for a given $u \in C^2(\Omega, \mathbb{R}^N)$, we can turn the locally 1-parameter transformation group ϕ_ε, whose generating vector field is (8.2), into a special 1-parameter family of functions. Note $\bar X$ and φ are determined by (8.3) and (8.4). Substituting into (8.1), we obtain the Noether's identity of a general local 1-parameter transformation group.

It is worth noting the representation of $I(u, \Omega)$ under the transformation $\{\phi_\varepsilon\}$ is

$$I(v_\varepsilon, \Omega_\varepsilon) = \int_{\Omega_\varepsilon} L(y, v_\varepsilon(y), \nabla v_\varepsilon(y))\,dy.$$

Definition 8.1 $\forall\, \Omega_0 \subset \bar\Omega_0 \subset \Omega$. Let $(\Omega_0)_\varepsilon = \eta(\Omega_0, \varepsilon)$. If

$$I(v_\varepsilon, (\Omega_0)_\varepsilon) = \text{const. (independent of } \varepsilon), \quad \forall\, u \in C^1(\bar\Omega_0, \mathbb{R}^N),$$

then I is invariant under $\{\phi_\varepsilon\}$.

In fact, if at each point, the following holds:

$$L(\eta_\varepsilon(x), v_\varepsilon(\eta_\varepsilon(x)), \nabla_y v_\varepsilon(\eta_\varepsilon(x)))\det(\eta_\varepsilon(x)) = L(x, u(x), \nabla u(x)),$$

then I is invariant under $\{\phi_\varepsilon\}$. In summary, we arrive at

Theorem 8.1 (Noether) *Suppose the local 1-parameter transformation group* $\{\phi_\varepsilon\}$ *is generated by the vector field* (8.2). *Let*

$$I(u) = \int_\Omega L(x, u(x), \nabla u(x))\,dx.$$

Then $\forall\, u \in C^2(\Omega, \mathbb{R}^N)$, *the Noether identity* (8.1) *holds, where*

$$\begin{cases} \bar X(x) = X(x, u(x)), \\ \bar U(x) = U(x, u(x)), \\ \varphi^i(x) = \bar U^i(x) - \displaystyle\sum_{\alpha=1}^{n} \frac{\partial u^i}{\partial x^\alpha} \bar X^\alpha(x). \end{cases}$$

If the functional I is invariant under $\{\phi_\varepsilon\}$, then

$$\boxed{E_L(u)\varphi + \operatorname{div}(L\bar{X} + L_{p^i}\varphi^i) = 0.}$$

Here we should substitute $(u(x), \nabla u(x))$ into (u, p) in L and L_p.

Furthermore, if $u \in C^2$ is a weak minimum of I, then the $(n-1)$-form

$$\nu = \sum_{\alpha=1}^{n}\left[L\bar{X}^\alpha + \sum_{i=1}^{N} L_{p^i_\alpha}\left(\bar{U}^i - \sum_{\beta=1}^{n} u^i_{x_\beta}\bar{X}^\beta\right)\right] dx^1 \wedge \cdots \wedge \hat{dx}^\alpha \wedge \cdots dx^n$$

is closed, i.e.

$$d\nu = 0.$$

Corollary 8.1 *When $n = 1$, let $\{\phi_\varepsilon\}$ be a local 1-parameter transformation group on $\mathbb{R}^N$, whose generating vector field is U (this means $X = 0$). If I is invariant under $\{\phi_\varepsilon\}$ and $u \in C^2$ is a weak minimum of I, then*

$$\sum_{i=1}^{N} \bar{U}^i(u(t))L_{p^i}(t, u(t), \dot{u}(t)) = \text{const.}$$

Corollary 8.2 *When $n = 1$, if L is autonomous (i.e. L is independent of t), then for the solution u of the E-L equation of I, we have*

$$\left(L - \sum_{i=1}^{N} L_{p^i}p^i\right)\Bigg|_{(u,p)=(u(t),\dot{u}(t))} = \text{const.}$$

Proof In fact, for the local 1-parameter transformation group $\{\phi_\varepsilon\}$, its generating vector fields are $X = 1$, $U = 0$, whence $\varphi = \frac{du}{dt}$. $\qquad\square$

8.2 The energy–momentum tensor and Noether's theorem

When $n = 1$, the Legendre transform of the Lagrangian L is the Hamiltonian H. The roles of L and H are usually symmetric in formulas. However, when $n > 1$, since p is no longer a vector but a tensor instead, we introduce the following Hamilton energy–momentum tensor (or the energy–momentum tensor for short) to reflect such symmetry:

$$T(x, u, p) = (T^\beta_\alpha(x, u, p)),$$

where

$$T^\beta_\alpha = p^i_\alpha L_{p^i_\beta} - \delta^\beta_\alpha L.$$

In other words, every component of T is the Legendre transform of L. In this sense, the energy–momentum tensor is the generalization of the Hamiltonian in high dimensions.

Example 8.1 Let

$$L = \frac{1}{2}\left(\sum_{i,j=0}^{3} g_{ij}p_i p_i - M^2 u^2\right) = \frac{1}{2}(p_0^2 - p_1^2 - p_2^2 - p_3^2 - M^2 u^2),$$

where $g_{00} = 1$, $g_{ii} = -1$, $i = 1, 2, 3$, and $g_{\beta\gamma} = 0$ for $\beta \neq \gamma$. Then

$$T_\beta^\alpha = \sum_{\gamma=0}^{3} g_{\alpha\gamma}p_\beta p_\gamma - \delta_{\alpha\beta}L.$$

In particular,

$$T_0^0 = \frac{1}{2}(p_0^2 + p_1^2 + p_2^2 + p_3^2 + M^2 u^2), \quad T_\beta^0 = -p_0 p_\beta, \quad \beta = 1, 2, 3.$$

Using the energy–momentum tensor, we can rewrite Noether's theorem as follows.

Theorem 8.2 *Let* $L \in C^2(\Omega \times \mathbb{R}^N \times \mathbb{R}^{nN}, \mathbb{R}^1)$. *Suppose*

$$I(u) = \int_\Omega L(x, u(x), \nabla u(x)) dx$$

is invariant under local 1-parameter transformation group $\{\phi_\varepsilon\}$. *If* $u \in C^2(\bar{\Omega}, \mathbb{R}^N)$ *is a weak minimum of* I, *then*

$$\boxed{\sum_{\alpha=1}^{n}\left(\sum_{i=1}^{n} L_{p_\alpha^i}\bar{U}^i - \sum_{\beta=1}^{n} T_\beta^\alpha \bar{X}^\beta\right)_{x_\alpha} = 0.}$$

Or simply,

$$\operatorname{div}\left(L_{p^i}\bar{U}^i - T \cdot \bar{X}\right) = 0.$$

Example 8.2 Consider the system of l particles $m_1, \ldots, m_l$, whose space coordinates are $X = (X_1, \ldots, X_l)$, where $X_i = (x_i, y_i, z_i)$ $(1 \leq i \leq l)$ is the space coordinates of the ith particle. The energy is

$$T = \frac{1}{2}\sum m_j \left|\dot{X}_j(t)\right|^2 = \frac{1}{2}\sum m_j(\dot{x}_j^2 + \dot{y}_j^2 + \dot{z}_j^2).$$

The energy potential is

$$V = -k\sum_{i<j} \frac{m_i m_j}{|X_i - X_j|} = -k\sum_{i<j} \frac{m_i m_j}{[(x_i - x_j)^2 + (y_i - y_j)^2 + (z_i - z_j)^2]^{\frac{1}{2}}},$$

where k is a constant. The Lagrangian

$$L = T - V,$$

whose associated functional is

$$I(X) = \int_{t_0}^{t_1} L(X(t), \dot{X}(t)) dt.$$

• **The space translation group $\{S_\varepsilon\}$.**

Let $\{S_\varepsilon\}$ be a family of space coordinate transformations depending on the parameter ε:

$$\tilde{x}_i = x_i + \varepsilon, \quad \tilde{y}_i = y_i, \quad \tilde{z}_i = z_i, \quad 1 \le i \le l.$$

Since L is independent of ε, I is invariant under $\{S_\varepsilon\}$, whose generating vector fields $X = 0$, $U = (e_1, \ldots, e_1)$, $P_i^1 = m_i \dot{x}_i(t)$, where $e_1 = (1, 0, 0)$. It follows from Noether's theorem that

$$\sum_{i=1}^{l} P_i^1 = \text{const.}$$

Likewise, applying the translations to y and z respectively, we obtain $\sum_{i=1}^{l} P_i^2 = \text{const.}$ and $\sum_{i=1} P_i^3 = \text{const.}$ Denote $P = (P^1, P^2, P^3)$, then

$$\sum_{i=1}^{l} P_i = \sum_{i=1}^{l} m_i \dot{X}_i(t) = \text{const.},$$

which is the conservation of momentum.

• **The time translation group.**

Let $\{T_\varepsilon\}$ be a family of space-time coordinate transformations depending on the parameter ε:

$$\tilde{t} = t + \varepsilon, \quad \tilde{x}_i = x_i, \quad \tilde{y}_i = y_i, \quad \tilde{z}_i = z_i, \quad 1 \le i \le l.$$

Then I is invariant under $\{T_\varepsilon\}$. Now $X = 1$ and $U = 0$, it follows from Noether's theorem that

$$H = pL_p - L = \sum m_j |\dot{X}_j|^2$$
$$- \left[\frac{1}{2} \sum m_j |\dot{X}_j^2| + k \sum_{i<j} \frac{m_i m_j}{[(x_i - x_j)^2 + (y_i - y_j)^2 + (z_i - z_j)^2]^{\frac{1}{2}}} \right]$$
$$= T + V = \text{const.},$$

which is the conservation of energy.

• **The 1-parameter rotational group $\{R_\varepsilon\}$.**

Let $\{R_\varepsilon\}$ be a family of space-time coordinate transformations depending on the parameter ε:

$$\begin{cases} \tilde{t} = t, \\ \tilde{x}_i = x_i \cos\varepsilon + y_i \sin\varepsilon, \\ \tilde{y}_i = -x_i \sin\varepsilon + y_i \cos\varepsilon, \\ \tilde{z}_i = z_i \quad 1 \le i \le l. \end{cases}$$

Then I is invariant under $\{R_\varepsilon\}$. Now

$$X \doteq 0, \quad U = (Z_1, Z_2, \ldots, Z_l),$$

where $Z_i = (y_i, -x_i, 0)$, $1 \le i \le l$, then

$$\sum_{i=1}^{l} m_i(y_i\dot{x}_i - x_i\dot{y}_i) = \sum_{i=1}^{l} L_{\dot{X}_i} \cdot Z_i = \text{const.}$$

Likewise, for rotations in the yz-plane and the xz-plane, we also have similar identities. This is the conservation of angular momentum:

$$\sum_{i=1}^{l} m_i X_i \wedge \dot{X}_i = \text{const.}$$

Example 8.3 Consider a gravitational field $u : \mathbb{R}^1 \times \mathbb{R}^3 \to \mathbb{R}^N$, where $u = u(x)$ denotes the distribution of the field with $x = (x_0, x_1, x_2, x_3)$, $x_0 = t$ being the time coordinate, and $\bar{x} = (x_1, x_2, x_3)$ being the space coordinate. Given a Lagrangian $L : (\mathbb{R}^1 \times \mathbb{R}^3) \times \mathbb{R}^N \times \mathbb{R}^{4N}$, the corresponding functional is

$$I(u) = \int_\Omega L(x, u(x), \nabla u(x))dx.$$

For example, for $N = 1$

$$L(x, u, p) = \frac{1}{2}(p_0^2 - p_1^2 - p_2^2 - p_3^2 - M^2u^2).$$

This is the Klein–Gordon field.

According to the special relativity, the Lagrangian of any gravitational field remains invariant under a positive Lorentz transformation. A positive Lorentz transformation is a time orientation preserving linear transformation on the Minkowski space-time which leaves the quadratic form $x_0^2 - x_1^2 - x_2^2 - x_3^2$ invariant. All positive Lorentz transformations form a group, called the positive Lorentz transformation group.

Obviously, the Lagrangian L is invariant under the space-time translation group, which infinitesimal generators are

$$\bar{X}^\beta = (\delta_\gamma^\beta)_{0 \le \gamma \le 3}.$$

In addition, $U = 0$, it follows from Noether's theorem that

$$\sum_{\alpha=0}^{3} \left(T_\gamma^\alpha\right)_{x_\alpha} = 0, \ \gamma = 0, 1, 2, 3,$$

where

$$T_\gamma^\alpha = \sum_\beta g_{\alpha\beta}\partial_\beta u \partial_\gamma u - \delta_{\alpha\gamma}L(x, u, \nabla u).$$

Or simply,

$$\operatorname{div} T_\gamma = 0, \quad \gamma = 0, 1, 2, 3,$$

where $T_\gamma = (T_\gamma^\alpha)_{0 \leq \alpha \leq 3}$.

Choose any $[t_1, t_2] \times B_R(\theta) \subset \mathbb{R}^4$ as our domain of integration, then

$$\int_{B_R(\theta)} T_\gamma^0(t_2, \bar{x})d\bar{x} - \int_{B_R(\theta)} T_\gamma^0(t_1, \bar{x})d\bar{x} + \int_{[t_1,t_2]\times\partial B_R(\theta)} T_\gamma^0(t, \bar{x}) \cdot \frac{\bar{x}}{|\bar{x}|} dt d\sigma = 0,$$

where $d\sigma$ is the area element of the 2-sphere and $|\bar{x}|$ is the norm of $\bar{x}$. If as $R \to \infty$, $T_\gamma^0(t, x)$, $|x| = R$, tend to zero uniformly, and $T_\gamma^0(t, \cdot)$ is integrable on $\mathbb{R}^3$, then

$$P_\gamma(t) = \int_{R^3} T_\gamma^0(t, \bar{x})d\bar{x} = \text{const.}$$

In particular, take $\gamma = 0$, then

$$P_0(t) = \int_{\mathbb{R}^3} T_0^0(t, \bar{x})d\bar{x}$$

$$= \int_{\mathbb{R}^3} [L_{p_0}u_{x_0} - L](t, \bar{x})d\bar{x}$$

$$= \int_{\mathbb{R}^3} \frac{1}{2}(|\partial_t u|^2 + |\nabla u|^2 + M^2 u^2)dx = \text{const.}, \quad \forall t \in \mathbb{R}^1.$$

This shows the conservation of energy. Next, take $\gamma = 1, 2, 3$, then

$$P_\gamma(t) = \int_{\mathbb{R}^3} T_\gamma^0(t, \bar{x})d\bar{x}$$

$$= \int_{\mathbb{R}^3} [L_{p_0}u_{x_\gamma}](t, \bar{x})d\bar{x}$$

$$= -\int_{\mathbb{R}^3} \partial_t u \partial_\gamma u = \text{const.}, \quad \forall t \in \mathbb{R}^1.$$

This shows the conservation of momentum.

The positive Lorentz transformation group includes the following six rotational generators:

$$\varepsilon_{\mu\nu} = -\varepsilon_{\nu\mu}, \quad 0 \leq \mu \leq \nu \leq 3.$$

Consider the transformation

$$y_\mu = x_\mu + \sum_0^3 g_{\nu\nu}\epsilon_{\mu\nu}x_\nu,$$

whose corresponding vector field is

$$X_\mu^{\alpha\beta} = \frac{dy_\mu}{d\varepsilon_{\alpha\beta}}\Big|_{\varepsilon_{\alpha\beta}=0}$$

$$= g_{\beta\beta}x_\beta\delta_{\mu\alpha} - g_{\alpha\alpha}x_\alpha\delta_{\mu\beta}, \quad 0 \le \beta < \alpha \le 3.$$

We call

$$M_{\alpha\beta\nu} = \sum_{\mu=0}^{3}(L_{p_\nu}u_{x_\mu} - \delta_{\mu\nu}L)X_\mu^{\alpha\beta} = \sum_{\mu=0}^{3}T_\mu^\nu X_\mu^{\alpha\beta}$$

angular momentum. According to Noether's theorem,

$$\sum_{\mu,\nu=0}^{3}(T_\mu^\nu X_\mu^{\alpha\beta})_{x_\nu} = \sum_{\nu=0}^{3}\partial_\nu(g_{\beta\beta}T_\alpha^\nu x_\beta - g_{\alpha\alpha}T_\beta^\nu x_\alpha) = 0.$$

Hence,

$$\int_{\mathbb{R}^3} M_{\alpha\beta,0}(t)d\bar{x} = \text{const.},$$

$\forall\, (\alpha,\beta), 0 \le \alpha < \beta \le 3$. This shows the conservation of angular momentum.

Remark 8.1 For electromagnetic field, complex vector fields, and Dirac field, etc, their conservation laws can be deduced in a similar fashion using Noether's theorem. It is clear that Noether's theorem is of fundamental importance.

8.3　Interior minima

In this section, we discuss another necessary condition for a functional to attain its minimum. In our earlier discussion, for a given functional, we started with the dependent variable u to derive a necessary condition for which the functional achieves its minimum: the E-L equation. However, we can also view this from a different angle: fix u and let x vary. As a result, we will end up with different functions, hence the functional varies accordingly. We attempt to describe a necessary condition for which the functional achieves its minimum from this point of view.

We adopt our earlier notation, let $\eta_\epsilon : \bar{\Omega} \to \bar{\Omega}$ be a self-diffeomorphism,

$$y = \eta_\epsilon(x) = x + \epsilon\bar{X}(x) + o(\epsilon),$$

where $\bar{X}|_{\partial\Omega} = 0$. η_ϵ has inverse mapping

$$x = \xi_\epsilon(y).$$

Given $u \in C^1(\bar{\Omega}, \mathbb{R}^N)$, let

$$v_\epsilon(y) = u(x),$$

i.e.

$$v_\epsilon = u \circ \xi_\epsilon = u \circ \eta_\epsilon^{-1}(y).$$

Hence,

$$I(v_\epsilon, \Omega) = \int_\Omega L(y, v_\epsilon(y), \nabla v_\epsilon(y)) dy$$

$$= \int_\Omega L\left(\eta_\epsilon(x), u(x), \frac{\partial \xi_\epsilon}{\partial y} \nabla u(x)\right) \det\left(\frac{\partial \eta_\epsilon}{\partial x}\right) dx.$$

Noting

$$\left(\frac{\partial \xi_\epsilon}{\partial y}\right) = \left(\frac{\partial \eta_\epsilon}{\partial x}\right)^{-1} = I - \epsilon \frac{\partial X}{\partial x} + o(\epsilon),$$

so for $u \in C^2$, we have

$$\frac{d}{d\varepsilon} I(v_\varepsilon, \Omega)|_{\varepsilon=0} = \int_\Omega \sum_{\alpha=1}^n \left[-\sum_{\beta=1}^n \sum_{i=1}^N L_{p_\beta^i} \frac{\partial u^i}{\partial x_\alpha} \frac{\partial \bar{X}^\alpha}{\partial x_\beta} + L_{x_\alpha} \bar{X}^\alpha + L \frac{\partial \bar{X}^\alpha}{\partial x_\alpha} \right] dx$$

$$= \int_\Omega \left\{ \sum_{\alpha=1}^n \left[L_{x_\alpha} - \partial_{x_\alpha}(L(x, u(x), \nabla u(x))) + \sum_{\beta=1}^n \partial_{x_\beta}\left(L_{p_\beta^i} u_{x_\alpha}^i\right) \right] \bar{X}^\alpha \right\} dx$$

$$= \int_\Omega E_L(u) \cdot \left(-\sum_{\alpha=1}^n \frac{\partial u}{\partial x_\alpha} \bar{X}^\alpha \right) dx. \tag{8.5}$$

We now introduce the following.

Defintion 8.2 $u \in C^1(\bar{\Omega}, \mathbb{R}^N)$ is said to be an interior minimum of I, if $\forall \bar{X} \in C_0^1(\Omega, \mathbb{R}^n)$, the functional I satisfies the equation under the transformation v_ε:

$$\frac{d}{d\epsilon} I(v_\varepsilon, \Omega)|_{\varepsilon=0} = 0.$$

Consequently, when $u \in C^2$ is an interior minimum of I, by (8.5), we have

$$\boxed{E_L(u) \frac{\partial u}{\partial x_\alpha} = 0.}$$

It follows immediately that

Corollary 8.3 *A C^2 weak minimum is an interior minimum.*

It is also worth noting that a necessary condition for having an interior minimum in C^2 can be expressed via the energy–momentum tensor. $\forall u \in C^2(\bar{\Omega}, \mathbb{R}^N)$, $\forall X \in C_0^1(\Omega, \mathbb{R}^N)$,

$$\int_\Omega \sum_{\alpha=1}^n \left[\sum_{i=1}^N L_{p_\beta^i} \frac{\partial u^i}{\partial x_\alpha} \frac{\partial \bar{X}^\alpha}{\partial x_\beta} - L_{x_\alpha} \bar{X}^\alpha - L \frac{\partial \bar{X}^\alpha}{\partial x_\alpha} \right] dx$$

$$= -\int_\Omega \sum_{\alpha=1}^n \left[\sum_{\beta=1}^n \frac{\partial T_\alpha^\beta(x, u(x), \nabla u(x))}{\partial x_\beta} + L_{x_\alpha}(x, u(x), \nabla u(x)) \right] \bar{X}^\alpha dx.$$

Thus,

$$\sum_{\beta=1}^n \frac{\partial T_\alpha^\beta(x, u(x), \nabla u(x))}{\partial x_\beta} + L_{x_\alpha}(x, u(x), \nabla u(x)) = 0.$$

As an application of the above necessary condition, we have the following.

Example 8.4 (Conformal mapping condition) Let Ω be a planar region. Suppose $u \in C^2(\Omega, \mathbb{R}^2)$ is a weak minimum of $D(u) = \int_\Omega |\nabla u|^2 dx dy$, then

$$\phi(z) := |\partial_x u|^2 - |\partial_y u|^2 + 2i \partial_x u \partial_y u$$

is an analytic function of $z = x + iy$ $((x,y) \in \Omega)$.

Proof $\forall X = (X^1, X^2) \in C_0^1(\Omega, \mathbb{R}^2)$, notice that $L = \sum_{\alpha=1}^2 |p_\alpha|^2$ is independent of z and u. Since u is an interior minimum, according to the first equality in (8.5),

$$\int_\Omega \left[L \operatorname{div} X - \sum_{\alpha=1}^2 \sum_{\beta=1}^2 L_{p_\alpha^i} \partial_\beta u \partial_\alpha X^\beta \right] dx dy$$

$$= \int_\Omega [|\nabla u|^2 (\partial_x X^1 + \partial_y X^2) - \partial_x u (\partial_x u \partial_x X^1 + \partial_y u \partial_x X^2)$$

$$- \partial_y u (\partial_x u \partial_y X^1 + \partial_y u \partial_y X^2)] dx dy$$

$$= \int_\Omega [(|\partial_x u|^2 - |\partial_y u|^2)(\partial_x X^1 - \partial_y X^2) + 2\partial_x u \cdot \partial_y u (\partial_x X^2 + \partial_y X^1)] dx dy$$

$$= 0.$$

Let $\xi = |\partial_x u|^2 - |\partial_y u|^2$, $\eta = 2\partial_x u \cdot \partial_y u$, then

$$\int_\Omega [-(\partial_x \xi + \partial_y \eta) X^1 + (\partial_y \xi - \partial_x \eta) X^2] dx dy = 0.$$

Since $X \in C_0^1(\Omega, \mathbb{R}^2)$ is arbitrary, we have:

$$\begin{cases} \partial_x \xi + \partial_y \eta = 0, \\ \partial_y \xi - \partial_x \eta = 0. \end{cases}$$

This is precisely the Cauchy–Riemann equation. Hence, ϕ is analytic. $\qquad \square$

8.4* Applications

Example 8.5 (Clairaut's Theorem) Let l be a geodesic on the smooth surface of revolution S. $\forall\, P \in l$, denote $r(P)$ the radius of the cross-section at point P and $\alpha(P)$ the angle between l and the meridian at P, then

$$r(P)\sin\alpha(P) = \text{const.}$$

Proof We parametrize S by

$$(x, y, z) = (r\cos\theta, r\sin\theta, f(r)). \tag{8.6}$$

A curve on S can be represented by $r = r(\theta)$. Since the arclength functional

$$L(r) = \int_{\theta_1}^{\theta_2} \sqrt{r^2 + (1 + f'(r)^2)\dot{r}^2}\, d\theta$$

is independent of θ, we have the conservation law:

$$\frac{r^2}{\sqrt{r^2 + (1 + f'(r)^2)\dot{r}^2}} = \text{const.}$$

Notice the differential of the arclength is

$$ds = \sqrt{r^2 + (1 + f'(r)^2)\dot{r}^2}\, d\theta.$$

The conservation law can therefore be rewritten as

$$r^2 \frac{d\theta}{ds} = \text{const.} \tag{8.7}$$

We re-parametrize S by the arclength s: $r = r(s)$, $\theta = \theta(s)$, the equation of l can then be obtained by substituting $r(s)$ and $\theta(s)$ into the r and s of (8.6), the tangent vector of l is

$$a = (\cos\theta, \sin\theta, f')\dot{r} + (-\sin\theta, \cos\theta, 0)r\dot{\theta}.$$

The cross-section's equation is

$$(x, y, z) = (r\cos\theta, r\sin\theta, f(r)),$$

where $r = \text{const.}$ and whose tangent vector is

$$b = (-r\sin\theta, r\cos\theta, 0).$$

The cosine value of the angle between these two vectors is

$$a \cdot b = r(s)^2\dot{\theta}(s). \tag{8.8}$$

However, the angle between l and the cross-section is $\beta(s) = \frac{\pi}{2} - \alpha(s)$, i.e.

$$a \cdot b = r(s)\cos(\beta(s)) = r(s)\sin(\alpha(s)).$$

Combining (8.7) and (8.8), we have

$$r(s)\sin(\alpha(s)) = \text{const.}$$

$\square$

Example 8.6 (Pohozaev's identity) Given a region $\Omega \subset \mathbb{R}^3$ with smooth boundary and $g \in C(\mathbb{R}^1)$. Consider the following nonlinear elliptic equation

$$\begin{cases} -\Delta u = g(u) & \text{in } \Omega \\ u = 0 & \text{on } \partial\Omega. \end{cases} \tag{8.9}$$

We now prove when $n \geq 3$, its (C^1 weak) solution satisfies the identity

$$\frac{n-2}{2}\int_\Omega |\nabla u|^2 - n\int_\Omega G(u) + \frac{1}{2}\int_{\partial\Omega} \left|\frac{\partial u}{\partial\nu}\right|^2 (x\cdot\nu)d\sigma = 0, \tag{8.10}$$

where G is an anti-derivative of g satisfying $G(0) = 0$, $d\sigma$ is the area element of $\partial\Omega$, and ν is the unit outward normal of $\partial\Omega$.

Proof The Lagrangian is

$$L(u,p) = \frac{1}{2}p^2 - G(u)$$

and $M = C_0^1(\Omega)$, whose corresponding E-L equation is exactly (8.9).

Without loss of generality, we may assume the origin $\theta \in \Omega$. Let $u \in C_0^1(\Omega)$ and consider $\Omega_\varepsilon = (1+\varepsilon)\Omega$ together with the 1-parameter family of diffeomorphisms $\eta_\varepsilon : \Omega \to \Omega_\varepsilon$ for $\eta_\varepsilon(x) = (1+\varepsilon)x$ (see Figure 8.2).

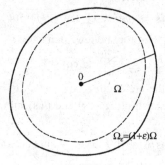

Fig. 8.2

Let $v_\varepsilon : \Omega_\varepsilon \to \mathbb{R}^1$,

$$v_\varepsilon(y) = u((1+\varepsilon)^{-1}y),$$

then $\bar{X} = x$ and

$$\varphi(x) = \frac{d}{d\varepsilon}u(1+\varepsilon)x)|_{\varepsilon=0} = -x\cdot\nabla u.$$

By Noether's theorem, we have

$$\frac{d}{d\varepsilon}I(v_\varepsilon, \Omega_\varepsilon)|_{\varepsilon=0}$$

$$= \int_\Omega \text{div}(L\bar{X} + L_p\varphi)dx$$

$$= \int_\Omega \left[\text{div}\left(\frac{1}{2}|\nabla u|^2 - G(u)\right)x - \nabla u(x\cdot\nabla u)\right]dx. \tag{8.11}$$

On one hand, from (8.9), it is equal to

$$\int_\Omega \left[\frac{n}{2}|\nabla u|^2 - nG(u) + x \cdot \nabla\left(\frac{1}{2}|\nabla u|^2 - G(u)\right)\right.$$
$$\left. - \Delta u(x \cdot \nabla u) - \nabla u \cdot \nabla(x \cdot \nabla u)\right] dx$$

$$= \int_\Omega \left[\frac{n}{2}|\nabla u|^2 - nG(u) + \frac{x}{2} \cdot \nabla(|\nabla u|^2) - \frac{x}{2} \cdot \nabla(|\nabla u|^2) - |\nabla u|^2\right] dx$$

$$= \int_\Omega \left[\frac{n-2}{2}|\nabla u|^2 - nG(u)\right] dx.$$

On the other hand, by Green's formula, since $u|_{\partial\Omega} = 0$, it follows that

$$\frac{d}{d\varepsilon}I(v_\varepsilon, \Omega_\varepsilon)|_{\varepsilon=0} = -\frac{1}{2}\int_{\partial\Omega}\left|\frac{\partial u}{\partial \nu}\right|^2 (x \cdot \nu)d\sigma.$$

Consequently,

$$\frac{n-2}{2}\int_\Omega |\nabla u|^2 - n\int_\Omega G(u) + \frac{1}{2}\int_{\partial\Omega}\left|\frac{\partial u}{\partial \nu}\right|^2 (x \cdot \nu)d\sigma = 0.$$

As a concrete application, let us take a star-shaped region $\Omega \subset \mathbb{R}^n$ ($n \geq 3$), that is, every ray emanating from the origin θ intersects $\partial\Omega$ once and only once. Let

$$g(u) = u^{\frac{n+2}{n-2}}.$$

The equation

$$\begin{cases} -\Delta u = u^{\frac{n+2}{n-2}} & x \in \Omega, \\ u|_{\partial\Omega} = 0 \end{cases} \tag{8.12}$$

has no nontrivial solution.

In fact, let $G(u) = \frac{n-2}{2n}u^{\frac{2n}{n-2}}$, if u is a solution of (8.12), then by (8.10),

$$\int_{\partial\Omega}\left|\frac{\partial u}{\partial \nu}\right|^2 (x \cdot \nu)d\sigma = 0.$$

Since Ω is star-shaped region, $x \cdot \nu > 0$. Hence, $u|_{\partial\Omega} = \frac{\partial u}{\partial \nu}|_{\partial\Omega} = 0$. By the uniqueness of the initial value problem of Laplace equation, $u \equiv 0$.

Exercises

1. Let $L(t, u, p) = t^2(p^2 - \frac{1}{3}u^6)$. Suppose $\varphi_\varepsilon : \mathbb{R}^1 \times \mathbb{R}^N \to \mathbb{R}^1 \times \mathbb{R}^N$ and define

$$Y(t, u, \epsilon) = (1 + \epsilon)t, \quad W(t, u, \epsilon) = \frac{u}{(1 + \epsilon)^{\frac{1}{2}}}.$$

Prove:

(1)

$$I(u) = \int_0^1 t^2 \left(\dot{u}^2 - \frac{1}{3} u^6 \right) dt$$

is invariant under $\{\varphi_\varepsilon\}$.

(2) If u is a solution of the E-L equation of I, then

$$\frac{t^3}{3} u^6 + t^3 \dot{u}^2 + t^2 u\dot{u} = \text{const.}$$

2. Let $L = (p + ku)^2$, where k is a constant. Let $\{\varphi_\varepsilon\}$ be defined such that

$$Y(t, u, \varepsilon) = t + \varepsilon, \quad W(t, u, \varepsilon) = u + \varepsilon\alpha e^{-kt}, \quad \alpha \in \mathbb{R}^1.$$

(1) Verify $I(u) = \int_0^1 (\dot{u} + ku)^2 dt$ is invariant under $\{\varphi_\varepsilon\}$.

(2) If u is a solution of the E-L equation of I, find the conservation law for which u satisfies.

(3) Solve for u by means of Hamiltonian system.

(4) For this u, verify the conclusion of (2).

Lecture 9

Direct methods

9.1 The Dirichlet's principle and minimization method

Before the 20th century, calculus of variations was largely based on the studies of E-L equations. Just like in mathematical analysis, finding the extrema of a function is usually turned into a problem of solving equation of critical points, finding the extrema of a functional can also be turned into solving the corresponding E-L equation.

E-L equations are differential equations. When $n = 1$, these are ordinary differential equations (or systems of ordinary differential equations); only under some special circumstances, one can find their analytic solutions. When $n > 1$, E-L equations are partial differential equations, the circumstances for which one will be able to find analytic solutions are extremely rare.

In the 19th century, driven by the studies of electromagnetism and complex variables, people were seeking solutions of the Laplace equation

$$\begin{cases} \triangle u = 0 \\ u|_{\partial\Omega} = \varphi, \end{cases} \text{ in } \Omega, \tag{9.1}$$

and the Poisson equation

$$\begin{cases} \triangle u = f \\ u|_{\partial\Omega} = \varphi. \end{cases} \text{ in } \Omega. \tag{9.2}$$

The Riemann (conformal) mapping theorem is a famous example: for a non-empty simply connected open domain $\Omega \subset \mathbb{C}$, there exists a biholomorphic mapping f (i.e. a bijective holomorphic mapping whose inverse is also holomorphic) from Ω onto the open unit disk. The fact f is biholomorphic implies that it is conformal. To establish the existence of the conformal mapping, Riemann turned this into a boundary value problem of the harmonic equation (9.1). He noticed that (9.1) is the E-L equation of the Dirichlet integral (regarded as a functional)

$$D(u) = \int_\Omega |\nabla u(x)|^2 dx \tag{9.3}$$

on the set $M = \{u \in C^1(\bar{\Omega}) \mid u|_{\partial\Omega} = \varphi\}$, the solution of (9.1) can thus be obtained by finding the minimum of D.

This idea opened up a new path in solving partial differential equations. If a differential equation is the E-L equation of some functional, then solving the partial differential equation can be turned into finding the extrema of the corresponding functional.

In order to prove the existence of solutions to the boundary value problem of the harmonic equation (9.1), it is led to prove the Dirichlet integral achieves its minimum on M. However, why does such minimum exist? Riemann's argument was based on the following Dirichlet's principle, a widely accepted statement back in the mid-19th century.

Dirichlet's principle Since D is bounded below, it "must" attain its infimum. That is, "there exists" u such that D achieves its minimum at u.

"Rationale": Choose a sequence $\{u_n\} \subset M$ such that $D(u_n) \to \inf_{u \in M} D(u)$. Since $\{u_n\}$ is bounded, there exists a convergent subsequence $u_{n_k} \to u_0$. This u_0 is then the desired solution $D(u_0) = \min_{u \in M} D(u)$.

This argument is clearly flawed in a modern reader's eyes. Shortly after publishing the Riemann mapping theorem, a heated debate began about the validity of the "Dirichlet's principle". In 1870, Weierstrass constructed the following counterexample. Consider the following extreme-value problem:

$$I(u) = \int_{-1}^1 x^2 u'^2 dx, \quad M = \{u \in C^1[-1,1] \mid u(-1) = -1, \ u(1) = 1\}.$$

(1) $\inf_{u \in M} I = 0$. In fact, $I \geq 0$. Letting

$$u_\epsilon = \frac{\arctan \frac{x}{\epsilon}}{\arctan \frac{1}{\epsilon}}, \quad \forall \epsilon > 0,$$

it follows that

$$I(u_\epsilon) < \int_{-1}^1 \frac{\epsilon^2 (x^2 + \epsilon^2)^{-1}}{(\arctan \frac{1}{\epsilon})^2} dx = \frac{2\epsilon}{\arctan \frac{1}{\epsilon}} \to 0, \quad \text{as } \epsilon \to 0.$$

(2) If u_0 is a minimum of I on M, then $u_0' \equiv 0$, so $u_0 = $ const. This contradicts the boundary condition!

Any reader with some rigorous background in mathematical analysis understands: it is in general not true that every minimizing sequence contains a subsequence converging to the minimal value.

Given a topological space X, suppose $f : X \to \mathbb{R}^1$ is bounded below, that is, $\exists\, M > 0$ such that $f(x) > -M$, hence there exists $m = \inf_{x \in X} f(x)$. Let $\{x_j\} \subset X$ be a minimizing sequence such that $f(x_j) \to m$. Under what condition does $\{x_j\}$ have a subsequence converging to the minimum?

Denote $f_t = \{x \in X | f(x) \leq t\},\ \forall t \in \mathbb{R}^1$. If

$$\exists\, t > m \text{ such that } f_t \text{ is “sequentially compact”}, \qquad (9.4)$$

then for N sufficiently large, $\{x_j \,|\, j \geq N\} \subset f_t$. Hence, there exists a subsequence $x_{j_k} \longrightarrow x_0 \in f_t$.

As to whether such x_0 is a minimum, we require additionally

$$f(x_0) \leq \underline{\lim} f(x_{j_k}).$$

The condition

$$x_n \to x_0 \Longrightarrow f(x_0) \leq \underline{\lim} f(x_n)$$

is called the sequentially lower semicontinuity of f. (Sometimes, without confusion, we simply refer to as lower semicontinuity.)

In summary, if $f : X \to \mathbb{R}^1$ is sequentially lower semicontinuous and if $\exists\, t > m$ and $\exists$ a sequentially compact set $K_t \supset f_t := \{x \in X | f(x) \leq t\}$, then f achieves its minimal value on X.

Next, we will apply the above abstract theorem to solve variational problems.

In finite dimensional Euclidean spaces, for a lower semicontinuous function f, if we impose the coercive condition:

$$f(x) \to +\infty, \qquad \text{as } \|x\| \to \infty, \qquad (9.5)$$

it then follows that $\exists\, t > m$ such that f_t is “sequentially compact”, which is well known in mathematical analysis.

This is however a completely different matter in infinite dimensional spaces. For example, in an infinite dimensional Hilbert space, consider the norm-square function $f(x) = \|x\|^2,\ \forall t > 0$, it is clearly coercive, but the set

$$f_t = \{x | \|x\| \leq \sqrt{t}\}$$

is not sequentially compact with respect to the norm-topology! As an example, $X = L^2[0, \pi]$, f_t is not sequentially compact, since the sequence $\{\sqrt{\frac{2t}{\pi}} \sin nx\}_1^\infty$ has no convergent subsequence whatsoever.

If we equip M with the C^1-topology, then on one hand, from the boundedness of $\{D(u_n)\}$, it is not possible to conclude $\|u_n\|_{C^1}$ is bounded. On the other hand, it is not difficult to construct an example where the sequence is C^1-bounded, but has no C^1-convergent subsequence. This is the reason why the previous "rationale" is invalid.

9.2 Weak convergence and weak-* convergence

In finite dimensional Euclidean space $\mathbb{R}^n$, we usually write $x_n = (\xi_1^n, \xi_2^n, \ldots, \xi_m^n) \to x = (\xi_1, \xi_2, \ldots, \xi_m)$ for x_n converging to x. The meaning of convergence ("$\to$") is understood in the sense of norm convergence:

$$\|x_n - x\| = \left[\sum_{i=1}^m (x_i^n - x_i)^2 \right]^{\frac{1}{2}} \to 0.$$

This can also be interpreted as coordinate-wise convergence: $x_i^n \to x_i$, $i = 1, \ldots, m$, since these two notions of convergence are equivalent.

However, in infinite dimensional spaces, these two notions of convergence are vastly different! For example, in l^2, given $x_n = (\xi_1^n, \xi_2^n, \ldots)$, $x = (\xi_1, \xi_2, \ldots)$, where $\sum_{i=1}^\infty |\xi_i^n|^2 < \infty$ and $\sum_{i=1}^\infty |\xi_i|^2 < \infty$. We say x_n converges to x in norm (denoted $x_n \to x$) if

$$\|x_n - x\| = \left[\sum_{i=1}^\infty (x_i^n - x_i)^2 \right]^{\frac{1}{2}} \to 0,$$

whereas x_n converges to x coordinate-wise means

$$x_i^n \to x_i, \ i = 1, 2, 3, \ldots.$$

If we take $\xi_i^n = \delta_{in}$, $n = 1, 2, \ldots$, then x_n converges to 0 coordinate-wise, but it does not converge in norm.

If in l^2, we use "coordinate-wise convergence" to define convergence, then every bounded sequence indeed contains a convergent subsequence. This is the same as in finite dimensional spaces, whose proof is based on the Cantor's "diagonal method". In fact, let $\{x_n\}$ be bounded, i.e. $\exists M > 0$ such that $\sum_{i=1}^\infty (\xi_i^n)^2 \leq M^2$. It follows that $|\xi_i^n| \leq M \ \forall i, \ \forall n$. Hence,

$$\exists \{n_k^1 \mid k \in \mathbb{N}\} \subset \mathbb{N} \text{ such that } \xi_1^{n_k^1} \to \xi_1^0,$$
$$\exists \{n_k^2 \mid k \in \mathbb{N}\} \subset \{n_k^1 \mid k \in \mathbb{N}\} \text{ such that } \xi_2^{n_k^2} \to \xi_2^0,$$
$$\cdots,$$
$$\exists \{n_k^l \mid k \in \mathbb{N}\} \subset \{n_k^{l-1} \mid k \in \mathbb{N}\} \text{ such that } \xi_l^{n_k^l} \to \xi_l^0,$$
$$\cdots$$

When $k \to \infty$, since $\forall N$, $\exists K = K(N)$ such that for all $k > K$,

$$\sum_{i=1}^N |\xi_i^0|^2 \leq \sum_{i=1}^N |\xi_i^{n_k^i}|^2 + 1 \leq M^2 + 1,$$

which implies

$$\sum_{i=1}^{\infty} |\xi_i^0|^2 \leq M^2 + 1.$$

We have thus established the diagonal subsequence x_{n_k} "coordinate-wise converges to" $x_0 = (\xi_1^0, \xi_2^0, \ldots, \xi_k^0, \ldots) \in l^2$.

Weak convergence and weak-$*$ convergence both stem from the idea of "coordinate-wise convergence".

Definition 9.1 Let X be a normed linear space, the sequence $\{x_n\} \subset X$ is said to converge weakly to x, denoted $x_n \rightharpoonup x$, if for any $x^* \in X^*$, we have $\langle x^*, x_n - x \rangle \to 0$, where X^* is the dual space of X.

Let X^* be the dual space of a normed linear space X, a sequence $\{x_n^*\} \subset X^*$ is said to converge to x^* in the weak-$*$-topology, denoted $x_n^* \rightharpoonup^* x^*$, if for any $x \in X$, we have $\langle x_n^* - x^*, x \rangle \to 0$.

Remark 9.1 In fact, on X^*, we have both the notion of weak convergence and the notion of weak-$*$ convergence. By weak convergence $x_n^* \rightharpoonup x^*$, we mean for any $x^{**} \in X^{**}$, $\langle x^{**}, x_n^* - x^* \rangle \to 0$; by weak-$*$ convergence, we mean for any $x \in X$, $\langle x_n^* - x^*, x \rangle \to 0$. Since we have the continuous embedding $X \hookrightarrow X^{**}$, weak convergence implies weak-$*$ convergence.

It is evident that norm convergence implies both weak and weak-$*$ convergence, but not vice versa.

Example 9.1 In $L^2(-\infty, \infty)$, choose any nonzero function $\varphi(t)$ with compact support, let $\varphi_n(t) = \varphi(t + n)$, then $\varphi_n \rightharpoonup 0$, but $\|\varphi_n\| = \|\varphi\| \neq 0$.

Example 9.2 Note that $L^2[0, 2\pi]$ is self-dual, its dual space is again $L^2[0, 2\pi]$. Consider the sequence $\{\sin(nt)\} \subset L^2[0, 2\pi]$. According to the Riemann–Lebesgue lemma, $\forall f \in L^1[0, 2\pi]$,

$$\int_0^{2\pi} f(t) \sin(nt) dt \to 0,$$

from which we may conclude:

(1) Take $X = L^p[0, 2\pi]$, $1 \leq p < \infty$ and regard the sequence $\{\sin(nt)\} \subset L^{p'}[0, 2\pi] = (L^p[0, 2\pi])^*$, then

$$\sin(nt) \rightharpoonup^* 0, \quad \text{as } n \to \infty.$$

(2) Take $X = L^p[0, 2\pi]$, $1 \leq p < \infty$ and regard the sequence $\{\sin(nt)\} \subset L^\infty[0, 2\pi] \subset X$, $X^* = L^{p'}[0, 2\pi] \subset L^1[0, 2\pi]$, $1 < p' \leq \infty$, then

$$\sin(nt) \rightharpoonup 0, \quad \text{as } n \to \infty.$$

For $1 < p < \infty$, $L^p[0, 2\pi]$ is reflexive, weak and weak-$*$ convergence coincide.

More generally, we have

Example 9.3 Let $D = [0,1]^N$ be the unit hypercube in $\mathbb{R}^N$. Let $\varphi \in L^p(D)$, $1 \leq p \leq \infty$ and make its periodic continuation. Let $\varphi_n(x) = \varphi(nx)$, $\forall\, n$ and

$$\bar{\varphi} = \int_D \varphi(x)dx,$$

then

$$\varphi_n \rightharpoonup \bar{\varphi} \quad \text{in} \quad L^p(D), \quad 1 \leq p < \infty,$$

and

$$\varphi_n \stackrel{*}{\rightharpoonup} \bar{\varphi} \quad \text{in} \quad L^\infty(D).$$

Proof First, without loss of generality, we may assume $\bar{\varphi} = 0$; for otherwise, we can replace φ by $\tilde{\varphi} = \varphi - \bar{\varphi}$.

Next, notice

$$\|\varphi_n\|_p^p = \int_D |\varphi(nx)|^p dx = \frac{1}{n^N} \int_{nD} |\varphi(y)|^p dy = \|\varphi\|_p^p \,\forall\, 1 \leq p \leq \infty.$$

We now define the set-valued function $\Phi(E) = \int_E \varphi(x)dx$, where E is any measurable set. Φ is σ-additive, and since φ is periodic, $\Phi(x + D) = 0$, $\forall\, x \in \mathbb{R}^N$.

For any rectangular hypercube $Q = \Pi_1^N(c_i, d_i)$, by translating D inside nQ without overlapping, we obtain

$$\left| \int_D \chi_Q \varphi_n dx \right| = \left| \int_Q \varphi_n dx \right| = \frac{1}{n^N}|\Phi(nQ)| \leq \frac{N}{n} \int_D |\varphi| dx.$$

Thus, for the simple function $\xi = \sum_i \alpha_i \chi_{Q_i}$, $Q_i \cap Q_j = \emptyset$, $i \neq j$, we have

$$\int_D \varphi_n \xi dx \to 0, \qquad n \to \infty.$$

For $1 < p \leq \infty$, the above simple functions form a dense subset in $L^{p'}(D)$ ($p' = \frac{p}{p-1}$). That is, $\forall\, f \in L^{p'}(D)$, $\exists\, \xi$ a simple function such that $\|f - \xi\|_{p'} < \frac{\epsilon}{2\|\varphi\|_p}$. For n sufficiently large,

$$\left| \int_D \varphi_n f dx \right| \leq \|\varphi_n\|_p \|f - \xi\|_{p'} + \left| \int_D \xi \varphi_n \right| \leq \epsilon.$$

The proof for $p = 1$ can be modified from the above argument, we omit the proof, leaving as an exercise. $\qquad\qquad\qquad\square$

9.3 Weak-* sequential compactness

In calculus of variations, we hope to utilize weak convergence or weak-* convergence to deduce weak sequential compactness or weak-* sequential compactness. In the previous section, we have discussed the fact in l^2, a bounded sequence has a "coordinate-wise" convergent subsequence. We now extend this statement as well as its proof to be a more abstract theorem.

Theorem 9.1 (Banach-Alaoglu) *Let X^* be the dual space of a separable normed linear space X. Suppose $\{x_n^* \mid n = 1, 2, \ldots\} \subset X^*$ is a norm-bounded sequence: $M = \sup \|x_n^*\| < \infty$, then it has a weak-* convergent subsequence.*

Proof Since X is separable, it has a countable dense subset $\{x_k \mid k = 1, 2, \ldots\}$.

For x_1, since $|\langle x_n^*, x_1 \rangle|$ is a bounded sequence, it has a subsequence $x_{n_j^1}^*$ such that $\langle x_{n_j^1}^*, x_1 \rangle$ converges.

For x_2, since $|\langle x_{n_j^1}^*, x_1 \rangle|$ is a bounded sequence, it has a subsequence $x_{n_j^2}^*$ such that $\langle x_{n_j^2}^*, x_1 \rangle$ converges.

Continuing in this fashion and applying the diagonal method, we can choose a subsequence $\{x_{n_j}^*\}$ such that $\langle x_{n_j}^*, x_k \rangle$ converges, $\forall k = 1, 2, \ldots$.

However, since $\{x_k \mid k = 1, 2, \ldots\}$ is dense and $\{x_n^* \mid n = 1, 2, \ldots\} \subset X^*$ is bounded in norm, for any $x \in X$, the sequence $\{\langle x_{n_j}^*, x \rangle\}$ converges.

Define
$$f(x) = \lim_{j \to \infty} \langle x_{n_j}^*, x \rangle.$$

It is clear that $f(x)$ is linear and continuous,
$$|f(x)| \le \sup_j \|x_{n_j}^*\| \|x\| \le M\|x\| \qquad \forall x \in X.$$

Thus, $\exists x^* \in X^*$ such that $f(x) = \langle x^*, x \rangle, \forall x \in X$, i.e.
$$x_{n_j}^* \rightharpoonup^* x^*.$$

$\square$

Consequently, we have the following fundamental result.

Theorem 9.2 *Let X be the dual space of a separable Banach space (e.g. a reflexive Banach space). Let $E \subset X$ be a non-empty weak-* sequentially closed subset. If $f : E \to \mathbb{R}^1$ is sequentially weak-* lower semi-continuous (abbreviated s.w*. l.s.c), and if f is coercive ($\forall x \in E$, when $\|x\| \to \infty$, $f(x) \to +\infty$), then f attains its minimum on E.*

Proof Choose a minimizing sequence $\{x_n\} \subset E$ of f,
$$\lim f(x_n) = \inf_{x \in E} f(x).$$

Since f is coercive, $\{x_n\}$ is bounded. By Theorem 9.1, $\{x_n\}$ contains a weak-* convergent subsequence

$$x_{n_k} \rightharpoonup^* x_0.$$

By assumption, E is weak-* sequentially closed, $x_0 \in E$.

Next, since f is s.w*. l.s.c.,

$$f(x_0) \leq \liminf f(x_{n_k}).$$

Thus,

$$f(x_0) = \inf_{x \in E} f(x).$$

$\square$

We now return to the Dirichlet integral. On the one hand, from the boundedness of the Dirichlet integral $D(u)$, it is impossible to deduce its boundedness in the C^1-norm; on the other hand, we do not know whether $C^1(\bar{\Omega})$ is the dual space of some normed linear space. Because of this, in order to verify the validity of the Dirichlet's Principle, the space $C^1(\bar{\Omega})$ is not a proper choice.

Closely related to the Dirichlet integral $D(u)$ is the following semi-norm:

$$\| u \| = \left(\int_{\Omega} |\nabla u|^2 dx \right)^{\frac{1}{2}},$$

and the norm

$$\|u\|_1 = \left[\int_{\Omega} (|\nabla u|^2 + |u|^2) dx \right]^{\frac{1}{2}}.$$

This norm corresponds to the inner product

$$(u, v) = \int_{\Omega} (\nabla u \cdot \nabla v + uv) dx.$$

Unfortunately, $C^1(\bar{\Omega})$ is not complete with respect to such a norm. We denote its completion (with respect to the above norm) by $H^1(\Omega)$, which is a Hilbert space.

Consequently, the Dirichlet inner product associated with the semi-norm $D(u)$ is

$$D(u, v) = \int_{\Omega} \nabla u \cdot \nabla v \, dx;$$

they are related via $D(u) = D(u, u)$.

When Ω is a bounded open set, on $C_0^1(\bar{\Omega})$, we can extend Poincaré's inequality encountered in Lecture 3 from one-variable functions to multivariable functions.

Lemma 9.1 (Poincaré's inequality) *Let $\Omega \subset \mathbb{R}^n$ be a bounded open set, $u \in C_0^1(\bar{\Omega})$, then $\exists\, C = C(\Omega)$ such that*

$$\int_\Omega |u|^2 dx \leq C \int_\Omega |\nabla u|^2 dx.$$

Proof Choose a hypercube $D \subset \mathbb{R}^n$ such that $\Omega \subset D$, $\forall\, \varphi \in C_0^\infty(\Omega)$, let

$$\tilde{\varphi}(x) = \begin{cases} \varphi(x), & x \in \Omega \\ 0, & x \notin \bar{\Omega}. \end{cases}$$

Then

$$\int_\Omega |\varphi|^2 = \int_D |\tilde{\varphi}|^2, \quad \int_\Omega |\nabla \varphi|^2 = \int_D |\nabla \tilde{\varphi}|^2.$$

Denote $x = (x_1, \tilde{x})$. By the single variable Poincaré's inequality, we have

$$\int_J |\tilde{\varphi}(x_1, \tilde{x})|^2 dx_1 \leq C \int_J |\partial_{x_1} \tilde{\varphi}(x_1, \tilde{x})|^2 dx_1,$$

where J is the projection of D in the direction of x_1. Integrating with respect to $\tilde{x}$, it yields

$$\int_D |\tilde{\varphi}|^2 dx \leq C \int_D |\nabla \tilde{\varphi}|^2 dx,$$

i.e.

$$\int_\Omega |\varphi|^2 dx \leq C \int_\Omega |\nabla \varphi|^2 dx.$$

By taking the limit, the inequality holds $\forall\, u \in C_0^1(\bar{\Omega})$. $\qquad\square$

This shows that $D(u)$ is a norm on $C_0^1(\bar{\Omega})$. We denote the closure of $C_0^1(\bar{\Omega})$ in $H^1(\Omega)$ by $H_0^1(\Omega)$. It is a closed subspace of the Hilbert space $H^1(\Omega)$. We can therefore regard $D(v)$, $D(v_0, v)$, $\forall\, v \in H_0^1(\Omega)$ as the continuous extension of the Dirichlet integral and the Dirichlet inner product respectively. For a given $v_0 \in C^1(\bar{\Omega})$, on $M = v_0 + H_0^1(\Omega)$, the Dirichlet integral is defined as

$$D(u) = D(v_0) + 2D(v_0, v) + D(v),$$

where $u = v_0 + v \in M$.

If we regard the so-defined $D(u)$ as a functional on $H_0^1(\Omega)$, then

1. $D(u)$ is coercive.

We wish to prove

$$D(v) \to \infty \Rightarrow D(u) \to \infty.$$

It follows from Schwarz's inequality and Young's inequality that

$$|D(v_0, v)| \leq D(v_0) + \frac{1}{4}D(v),$$

it implies

$$D(u) \geq \frac{1}{2}D(v) - D(v_0).$$

This confirms coerciveness.

2. $D(u)$ is sequentially weakly semi-lower continuous.

Let $u_j = v_0 + v_j$, then

$$u_j \rightharpoonup u \ (H^1(\Omega)) \Leftrightarrow v_j \rightharpoonup v \ (H_0^1(\Omega)).$$

Notice

$$|D(v_0, \varphi)| \leq D(v_0)^{\frac{1}{2}} D(\varphi)^{\frac{1}{2}} \ \forall \varphi \in H_0^1(\Omega),$$

$\varphi \mapsto D(v_0, \varphi)$ is a continuous linear functional on $H_0^1(\Omega)$. Thus,

$$D(v_0, v_j) \to D(v_0, v).$$

Likewise,

$$D(v, v_j) \to D(v, v).$$

By Schwarz's inequality, we obtain

$$D(v) = \lim D(v, v_j) \leq \liminf D(v)^{\frac{1}{2}} D(v_j)^{\frac{1}{2}},$$

i.e.

$$D(v) \leq \liminf D(v_j).$$

That is,

$$D(u) \leq \liminf D(u_j).$$

$H_0^1(\Omega)$ is a Hilbert space, hence is self-dual.

We can also directly verify $H_0^1(\Omega)$ is self-dual, $H_0^1(\Omega)$ is the dual space of the Banach space $H_0^1(\Omega)$. Furthermore, $H_0^1(\Omega)$ is separable (the detailed proof will be given in the next lecture). We now apply Theorem 9.2 to deduce the Dirichlet integral indeed attains its minimum on $H_0^1(\Omega)$, hence affirms Dirichlet's principle.

Remark 9.2 To verify Dirichlet's Principle, there is also a more direct method - orthogonal projection. Geometrically speaking, it is equivalent to minimizing the distance from a hyperplane to a given point outside the hyperplane (we refer to Lecture 12). From which, we will derive the Riesz Representation Theorem and the self-duality of a Hilbert space.

Remark 9.3 In the above example, we obtain the solution $u_0 \in H^1(\Omega)$. However, we do not yet know if it is differentiable, nor do we know if it belongs to C^2. In our on-going discussions, we must address in what sense will u_0 be a solution of the harmonic equation.

As for a more general functional I and its boundary conditions, in order to establish the existence of solutions via the minimizing sequence method, we must provide:

1. a suitable function space, which is a reflexive Banach space, or the dual space of a separable Banach space.

2. the functional I is sequentially weak-* lower semi-continuous with respect to the underlying topology,

3. I is coercive with respect to the topology.

4. in what sense the so-obtained minimal solution will indeed be a solution of the original equation (satisfying the weak form of the E-L equation).

5. whether the minimal solution would satisfy the differentiability rendered by the equation.

9.4* Reflexive spaces and the Eberlein–Šmulian theorem

In functional analysis, there is in-depth study of weak sequential compactness. It is possible to deduce the weak sequential compactness directly by the reflexiveness, avoiding the separable assumption. (In this occasion, there is no difference between being weakly sequentially compact and being weakly-* sequentially compact.)

We recall the definition of a reflexive space: the dual space X^* of a Banach space X is also a Banach space. We can also consider the dual space $(X^*)^* = X^{**}$ of X^*, and we call it the second dual of X.

Notice $\forall \, x \in X$, we can define a functional on X^* via

$$\langle F_x, x^* \rangle = \langle x^*, x \rangle \qquad \forall \, x^* \in X^*.$$

F_x is linear on X^* and satisfies

$$|\langle F_x, x^* \rangle| \le \|x^*\| \, \|x\|.$$

We call $T : x \mapsto F_x$ a natural mapping. In fact, $T : X \to X^{**}$ is a continuous embedding. By the Hahn–Banach theorem, $\exists \, x^* \in X^*$ such that $\|x^*\| = 1$, $\langle x^*, x \rangle = \|x\|$, it follows that

$$\|x\| = \langle x^*, x \rangle = \langle F_x, x^* \rangle \le \|F_x\|.$$

Thus, T is an isometry, that is, X is isometric to a closed linear subspace of its second dual X^{**}.

Definition 9.2 A Banach space is said to be reflexive if the above isometry T is surjective.

The following result on separability is due to Banach.

Theorem 9.3 (Banach) *Let X be a normed linear space. If the dual space X^* is separable, then so is X itself.*

Proof 1. Denote S_1^* the unit sphere in X^*, then S_1^* is separable. In fact, let $\{x_n^*\} \subset X^*$ be a countable dense subset. Let $y_n^* = \frac{x_n^*}{\|x_n^*\|}$, then $\{y_n^*\}$ is a countable dense subset of S_1^*. To see this, $\forall\, x^* \in S_1^*$, $\exists\, x_{n_j}^*$ such that $\|x_{n_j}^* - x^*\| \to 0$, hence $\|x_{n_j}^*\| \to 1$ and

$$\|y_{n_j}^* - x^*\| \leq \left|1 - \frac{1}{\|x_{n_j}^*\|}\right| \|x_{n_j}^*\| + \|x_{n_j}^* - x^*\| \to 0.$$

This shows $\{y_n^*\}$ is dense in S_1^*.

2. By definition, there exists $x_n \in X$ such that $\|x_n\| = 1$ and $\langle y_n^*, x_n \rangle \geq 1/2$. Let $X_0 = \overline{\text{span}}\{x_1, x_2, \ldots, x_n, \ldots\}$. We want to show $X_0 = X$. Suppose not, by the Hahn–Banach theorem, there exists $y^* \in S_1^*$ such that $\langle y^*, x \rangle = 0$, $\forall\, x \in X_0$. However,

$$\|y^* - y_n^*\| \geq |\langle y^* - y_n^*, x_n \rangle| \geq \frac{1}{2},$$

which is impossible. Thus, $X_0 = X$, i.e. X is separable. $\qquad\square$

As a consequence of Theorems 9.1 and 9.3, we have the following corollary.

Corollary 9.1 *If X is a separable, reflexive Banach space, then every bounded sequence has a weakly convergent subsequence.*

Moreover, the separable assumption can be removed, but we will need the assistance of the following theorem.

Theorem 9.4 (Pettis) *A closed linear subspace X_0 of a reflexive Banach space X is reflexive.*

Proof We want to show that $\forall\, x_0^{**} \in X_0^{**}$, $\exists\, x \in X_0$ such that

$$\langle x_0^{**}, x_0^* \rangle = \langle x_0^*, x \rangle, \qquad \forall\, x_0^* \in X_0^*. \tag{9.6}$$

1. Define the mapping $T : X^* \to X_0^*$ by $Tx^* = x^*|_{X_0}$,

$$\langle Tx^*, x_0 \rangle = \langle x^*|_{X_0}, x_0 \rangle \ \forall\, x_0 \in X_0.$$

T is linear and continuous. Its dual mapping $T^* \in L(X_0^{**}, X^{**})$ satisfying $\forall\, x_0^{**} \in X_0^{**}, T^* x_0^{**} \in X^{**}$.

2. Since X is reflexive, $\exists\, x \in X$ such that

$$\langle x^*, x \rangle = \langle T^* x_0^{**}, x^* \rangle = \langle x_0^{**}, x^*|_{X_0} \rangle \ \ \forall\, x^* \in X^*. \tag{9.7}$$

We want to show $x \in X_0$. We argue by contradiction. Suppose $x \notin X_0$, then applying the Hahn–Banach theorem to the closed linear subspace X_0, $\exists\, x_1^* \in X^*$ such that

$$\langle x_1^*, x \rangle \neq 0, \qquad x_1^*|_{X_0} = 0.$$

But
$$\langle x_1^*, x \rangle = \langle x_0^{**}, x_1^*|_{X_0} \rangle = 0,$$

which is a contradiction.

3. We now prove (9.6), i.e. x_0^{**} is the image of $x \in X_0$ under the natural mapping. According to the Hanh–Banach theorem, $\forall x_0^* \in X_0^*, \exists x^* \in X^*$ such that $x^*|_{X_0} = x_0^*$. Noting (9.7) implies

$$\langle x_0^{**}, x_0^* \rangle = \langle x^*, x \rangle = \langle x_0^*, x \rangle \quad \forall x_0^* \in X_0^*.$$

This shows X_0 is reflexive. $\qquad\square$

Corollary 9.2 *A Banach space X is reflexive if and only if X^* is reflexive.*

Proof "$\Rightarrow$" $(X^*)^{**} = (X^{**})^* = X^*$.

"$\Leftarrow$" Suppose X^* is reflexive. Using the forward implication, X^{**} is reflexive. However, since X is a closed linear subspace of X^{**}, by the Pettis Theorem, X is reflexive. $\qquad\square$

As a consequence, we have:

Theorem 9.5 (Eberlein–Šmulian) *Every bounded sequence $\{x_n\}$ of a reflexive Banach space X has a weakly convergent subsequence.*

Corollary 9.3 *Every bounded sequence $\{x_n\}$ of a Hilbert space has a weakly convergent subsequence.*

Besides Hilbert spaces, we consider the following reflexive Banach spaces.

Consider the Lebesgue function space $L^p(\Omega)$, where $\Omega \subset \mathbb{R}^n$ and $1 \le p < \infty$. We know

$$(L^p(\Omega))^* = L^{p'}(\Omega), \quad 1 \le p < \infty, \quad \frac{1}{p} + \frac{1}{p'} = 1.$$

This means, for every continuous linear functional F on $L^p(\Omega)$, there exists, in the sense of a.e., a unique function $v \in L^{p'}(\Omega)$ such that it can be represented by

$$F(u) = \int_\Omega v(x) u(x) dx$$

and

$$\|F\|_{(L^p)^*} = \left(\int_\Omega |v(x)|^{p'} dx \right)^{\frac{1}{p'}} = \|v\|_{p'}.$$

For all $u \in L^p(\Omega), 1 < p < \infty,$

$$v \mapsto \int_\Omega u(x) v(x) dx$$

can be viewed as a continuous linear functional G on $L^{p'}(\Omega)$. Likewise, there exists $w_u \in L^p(\Omega)$ such that

$$G(v) = \int_\Omega w_u(x)v(x)dx \quad \forall v \in L^{p'}(\Omega),$$

i.e.

$$\int_\Omega u(x)v(x)dx = G(v) = \int_\Omega w_u(x)v(x)dx \quad \forall v \in L^{p'}(\Omega).$$

Thus, $w_u(x) = u(x)$ a.e.

Following the above line of thought, when $1 < p < \infty$,

$$(L^p(\Omega))^{**} = (L^{p'}(\Omega))^* = L^p(\Omega),$$

whence space $L^p(\Omega)$ is reflexive.

Exercises

1. Prove that for a bounded sequence in l^2, coordinate-wise convergence $\Leftrightarrow$ weak convergence.
2. Given a family of functions

$$u_\epsilon = \frac{\arctan \frac{x}{\epsilon}}{\arctan \frac{1}{\epsilon}}, \quad \forall \epsilon > 0.$$

   In the space $L^2(-1,1)$, determine whether it converges in norm or converges weakly. In the space $H^1(-1,1)$, determine whether it converges in norm or converges weakly.
3. In $L^2([0,2\pi])$, find

$$w - \lim_{n\to\infty} \sin nt \quad \text{and} \quad w - \lim_{n\to\infty} \sin^2 nt.$$

4. In $H^1(0,1)$, given a sequence of functions

$$u_j(t) = \begin{cases} t - \dfrac{k}{j}, & t \in \left[\dfrac{k}{j}, \dfrac{2k+1}{2j}\right] \\[2mm] -t + \dfrac{k+1}{j}, & t \in \left[\dfrac{2k+1}{2j}, \dfrac{k+1}{j}\right] \end{cases} \qquad k = 0,1,2,\ldots,j-1,$$

$j = 1,2,\ldots$ Determine whether it converges in norm or converges weakly.

Lecture 10

Sobolev spaces

We have pointed out in our earlier discussions that neither C^1 nor C_0^1 is an appropriate space for verifying the Dirichlet's principle, instead, one should consider the spaces H^1 and H_0^1.

Such scenario frequently occurs when applying direct methods to solve variational problems. This is because the functionals are usually variational integrals involving derivatives, and the C-norm associated with the C-space consisting of the same order derivatives is determined by the maximal value of the pointwise norm. However, the C-norm cannot be controlled by such variational integrals. Moreover, in order to possess the weakly sequential compactness, the underlying space must be the dual space of a normed linear space, since such space is at least complete. The Sobolev spaces introduced in this lecture satisfy the above requirements.

10.1 Generalized derivatives

Let $u, v \in L^1_{\text{loc}}(\Omega)$, $\forall i = 1, 2, \ldots, n$, we call v the generalized derivative of u with respect to x_i, denoted $v = D_{x_i} u$, if $\forall \varphi \in C_0^\infty(\Omega)$,

$$\int_\Omega v\varphi dx = -\int_\Omega u\partial_{x_i}\varphi dx.$$

More generally, for a multi-index $\alpha = (\alpha_1, \ldots, \alpha_n)$, we denote

$$|\alpha| = \sum_{i=1}^n \alpha_i, \qquad \partial^\alpha = \partial_{x_1}^{\alpha_1} \cdots \partial_{x_n}^{\alpha_n}.$$

Definition 10.1 We call v the αth-order generalized derivative of u, if $\forall \varphi \in C_0^\infty(\Omega)$,

$$\int_\Omega v\varphi dx = (-1)^{|\alpha|} \int_\Omega u\partial^\alpha\varphi dx.$$

Denote $v = D^\alpha u$.

Example 10.1 Let $n = 1$, $J = (-1, 1)$, and $u(x) = |x|$. Since

$$\int_{-1}^{1} |x| \varphi'(x) dx = \int_{0}^{1} x\varphi'(x) dx + \int_{-1}^{0} (-x)\varphi'(x) dx$$

$$= -\int_{-1}^{1} \text{sgn } x \varphi(x) dx,$$

we have $D(|x|) = \text{sgn } x$.

Example 10.2 If $u \in C^k(\Omega)$, then $D^\alpha u = \partial^\alpha u$, $\forall \alpha$, $|\alpha| \le k$. This is because

$$\int_{\Omega} \partial^\alpha u \varphi dx = (-1)^{|\alpha|} \int_{\Omega} u \partial^\alpha \varphi dx.$$

10.2 The space $W^{m,p}(\Omega)$

Definition 10.2 (The space $W^{m,p}(\Omega)$) Suppose $p \in [1, \infty]$ and $m \in \mathbb{N}$, let

$$W^{m,p}(\Omega) := \{u \in L^p(\Omega) \,|\, D^\alpha u \in L^p(\Omega),\, \forall \alpha, |\alpha| \le m\},$$

on which, we define the norm by

$$\|u\|_{m,p} = \left(\sum_{|\alpha| \le m} \int_{\Omega} |D^\alpha u(x)|^p dx \right)^{\frac{1}{p}}, \qquad 1 \le p < \infty,$$

$$\|u\|_{m,\infty} = \text{esssup}_{x \in \Omega} \sum_{|\alpha| \le m} |D^\alpha u(x)|.$$

We call them Sobolev spaces.

Clearly, Sobolev spaces are normed linear spaces.

For a bounded region Ω, we have the following chains of containments:

$$W^{m,\infty}(\Omega) \subset W^{m,q}(\Omega) \subset W^{m,p}(\Omega) \subset W^{m,1}(\Omega), \quad 1 < p < q < \infty,$$

$$W^{m,p}(\Omega) \subset W^{l,p}(\Omega), \quad 0 \le l \le m,$$

also,

$$\text{if } \Omega_1 \subset \Omega_2, \ u \in W^{m,q}(\Omega_2) \text{ then } u|_{\Omega_1} \in W^{m,q}(\Omega_1).$$

Theorem 10.1 *The space $W^{m,p}(\Omega)$ is complete, hence is a Banach space.*

Proof Let $\{u_j\}$ be a Cauchy sequence, then $\forall \alpha, |\alpha| \le m, \{D^\alpha u_j\}$ is a Cauchy sequence in L^p. Hence, there exists $g_\alpha \in L^p(\Omega)$ such that $D^\alpha u_j \to g_\alpha$, $L^p(\Omega)$.

Now $\forall \varphi \in C_0^\infty(\Omega)$, we have

$$\langle D^\alpha u_j, \varphi \rangle = (-1)^{|\alpha|} \langle u_j, \partial^\alpha \varphi \rangle.$$

It follows that

$$\langle D^\alpha u_j, \varphi \rangle \to \langle g_\alpha, \varphi \rangle, \ \forall \alpha, \ |\alpha| \le m$$

and

$$\langle u_j, \partial^\alpha \varphi \rangle \to \langle g_0, \partial^\alpha \varphi \rangle, \ \forall \alpha, \ |\alpha| \le m.$$

That is,

$$\langle g_\alpha, \varphi \rangle = (-1)^{|\alpha|} \langle g_0, \partial^\alpha \varphi \rangle.$$

From which, we deduce that

$$g_\alpha = D^\alpha g_0$$

and

$$\| D^\alpha u_j - g_\alpha \|_p \to 0 \quad \forall \alpha, \ |\alpha| \le m.$$

Thus, $g_0 \in W^{m,p}(\Omega)$ and

$$\|u_j - g_0\|_{m,p} \to 0, \ j \to \infty. \qquad \square$$

It is worth mentioning that although the generalized derivatives are dually and globally defined, they are nevertheless closely related to the locally defined ordinary derivatives. In particular, for functions in the one dimensional Sobolev space, their generalized derivatives are the original derivatives almost everywhere! Assume $n = 1$ and $J = [a, b]$, we have:

Example 10.3 $W^{1,1}(J) = AC(J)$, the space of absolutely continuous functions defined on J and

$$Du(x) = u'(x), \qquad \text{a.e.} \qquad (10.1)$$

Proof $\forall u \in W^{1,1}(J)$, we show that

$$u(x) - u(a) = \int_a^x Du(t)dt, \ \forall x \in J.$$

$\forall n \in \mathbb{N}$, let

$$\varphi_n(t) = \begin{cases} n(t-a) & t \in [a, a + \frac{1}{n}], \\ 1 & t \in [a + \frac{1}{n}, x - \frac{1}{n}], \\ -n(x-t) & t \in [x - \frac{1}{n}, x] \\ 0 & t \in [x, b]. \end{cases}$$

Since $\exists \xi_{n,k} \in C_0^\infty(J), \|\xi_{n,k}\|_{C^1} \le 2n$ such that $|\varphi_n(t) - \xi_{n,k}(t)|$ converges to 0 uniformly on J and that $|\varphi_n'(t) - \xi_{n,k}'(t)| \to 0$ a.e. $t \in J$ as $k \to \infty$, from $\int_J u(t)\xi_{n,k}'(t)dt = -\int_J Du(t)\xi_{n,k}dt$, it follows that

$$\int_J u(t)\varphi_n'(t)dt = -\int_J Du(t)\varphi_n(t)dt.$$

Letting $n \to \infty$, it follows immediate

$$u(x) - u(a) = \int_a^x Du(t)dt, \quad \forall \, x \in J.$$

Thus, $u(x)$ is absolutely continuous on J; furthermore,

$$u'(x) = Du(x), \quad \text{a.e. } x \in J.$$

Conversely, $\forall \, u \in AC(J)$, $u'(x)$ exists a.e. and belongs to $L^1(J)$. It suffices to show

$$u'(x) = Du(x), \quad \text{a.e. } x \in J.$$

In fact, since $u(x) = \int_y^x u'(t)dt + u(y)$, $\forall \, x, y \in J$, we have

$$\int_J u(x)\varphi'(x)dx = -\int_J u'(x)\varphi(x)dx \quad \forall \varphi \in C_0^\infty(J). \qquad \square$$

This is (10.1).

Example 10.4 $W^{1,\infty}(J) = \text{Lip}(J)$, the space of Lipschitz functions on J.

Proof "$\supset$" Assume $u \in \text{Lip}(J)$, then u is absolutely continuous on J. Hence, u' exists almost everywhere and satisfies $u(y) - u(x) = \int_x^y u'(t)dt$ as well as

$$|u'(x)| \leq \sup_{y \in J} \frac{|u(y) - u(x)|}{|y - x|} \leq M.$$

From Example 10.3, we see that $Du \in L^\infty(J)$, i.e. $u \in W^{1,\infty}(J)$ and

$$\|u\|_{1,\infty} \leq \|u\|_{\text{Lip}}.$$

"$\subset$" Conversely, assume $u \in W^{1,\infty}(J)$, then by (10.1),

$$|u(y) - u(x)| \leq \int_x^y |u'(t)| \, dt \leq \|u'\|_\infty |y - x| \quad \forall \, x, y \in J.$$

Thus,

$$\|u\|_{\text{Lip}} \leq \|u\|_{W^{1,\infty}}. \qquad \square$$

Definition 10.3 We denote the closure of $C_0^\infty(\Omega)$ in $W^{m,p}(\Omega)$ by $W_0^{m,p}(\Omega)$.

Lemma 10.1 *If $u \in W^{m,p}(\Omega)$, $\psi \in C_0^\infty(\Omega)$, then $(\psi u) \in W_0^{m,p}(\Omega)$ and*

$$\text{supp}(D^\alpha(\psi u)) \subset \text{supp}(\psi), \quad |\alpha| \leq m.$$

Proof We only prove for $m = 1$, the rest can be proved by mathematical induction. In fact, $\forall\, \varphi \in C_0^\infty(\Omega)$,

$$\int_\Omega D_{x_i}(\psi u)\varphi dx = -\int_\Omega \psi u \partial_{x_i}\varphi dx$$

$$= -\int_\Omega u[\partial_{x_i}(\psi\varphi) - \partial_{x_i}\psi\varphi]dx$$

$$= \int_\Omega [\partial_{x_i}\psi u + \psi D_{x_i}u]\varphi dx.$$

Hence,

$$D_{x_i}(\psi u) = \psi(D_{x_i}u) + (\partial_{x_i}\psi)u. \qquad \square$$

10.3 Representations of functionals

In order to understand the weak topology on the Banach space $W^{m,p}(\Omega)$ (or $W_0^{m,p}(\Omega)$), we need to consider the representations of functionals on these spaces.

We know the dual space of $L^p(J)$ is

$$(L^p(\Omega))^* = L^{p'}(\Omega), \quad 1 \leq p < \infty, \quad \frac{1}{p} + \frac{1}{p'} = 1,$$

i.e. $\forall\, f \in (L^p(\Omega))^*,\, \exists\, v \in L^{p'}(\Omega)$ such that

$$\langle f, u\rangle = \int_\Omega u(x)v(x)dx \quad \forall u \in L^p(\Omega)$$

and

$$\|f\| = \sup_{\|u\|_p \leq 1} \langle f, u\rangle = \left(\int_\Omega |v|^{p'}\right)^{\frac{1}{p'}}.$$

In order to examine continuous linear functionals on $W^{m,p}(\Omega)$, we isometrically embed $W^{m,p}(\Omega)$ into the product space $\prod_{|\alpha| \leq m} L^p(\Omega)$ as a closed linear subspace:

$$i: \quad u \mapsto \{D^\alpha u,\, |\alpha| \leq m\},$$

$$W^{m,p}(\Omega) \to \prod_{|\alpha| \leq m} L^p(\Omega).$$

By the Hahn–Banach theorem, $f \in (W^{m,p}(\Omega))^*$ if and only if there exists $\{\psi_\alpha,\, |\alpha| \leq m\} \in \prod_{|\alpha| \leq m} L^{p'}(\Omega)$ such that

$$\langle f, u\rangle = \int_\Omega \left(\sum_{|\alpha| \leq m} D^\alpha u(x)\psi_\alpha(x)\right) dx.$$

Consequently, the weak convergence $u_j \rightharpoonup u$ in $W^{m,p}(\Omega)$ $(p \geq 1)$ can be expressed as

$$u_j \rightharpoonup u \text{ in } W^{m,p}(\Omega) \Leftrightarrow \int_\Omega \sum_{|\alpha| \leq m} D^\alpha(u_j - u)\psi_\alpha dx \to 0, \quad \forall \{\psi_\alpha\} \in \prod_{|\alpha| \leq m} L^{p'}(\Omega).$$

Theorem 10.2 $W^{m,p}(\Omega)$ $(1 < p < \infty)$ *is a reflexive Banach space.*

Proof We define the embedding $i : W^{m,p}(\Omega) \to \prod_{|\alpha| \leq m} L^p(\Omega)$ via

$$i : u \mapsto \{D^\alpha u, |\alpha| \leq m\}.$$

This is a closed map, therefore $i(W^{m,p}(\Omega)) \subset \prod_{|\alpha| \leq m} L^p(\Omega)$ is a closed linear subspace. Since $\prod_{|\alpha| \leq m} L^p(\Omega)$ is reflexive, by the Pettis theorem, $W^{m,p}(\Omega)$ is also reflexive. $\qquad \square$

10.4 Modifiers

In Lecture 6, in the proof of higher dimensional du Bois–Reymond lemma, we have introduced the "bump function"

$$\varphi(x) = \begin{cases} c_n^{-1} \exp\left(\dfrac{-1}{1 - |x|^2}\right) & |x| < 1, \\ 0 & |x| \geq 1, \end{cases}$$

where $|x| = (x_1^2 + \cdots + x_n^2)^{\frac{1}{2}}$ and c_n is such a constant that

$$\int_{R^n} \varphi(x) dx = 1.$$

$\forall \epsilon > 0$, let

$$\varphi_\epsilon(x) = \epsilon^{-n} \varphi\left(\frac{x}{\epsilon}\right),$$

then $\operatorname{supp} \varphi_\epsilon \subset \overline{B_\epsilon(\theta)}$.

We utilize this function to smooth out any given function. Let $u \in L^1(\Omega)$, $\operatorname{supp}(u) \subset \Omega_\delta := \{x \in \Omega \,|\, d(x, \partial\Omega) \geq \delta > 0\}$. We call the mapping $u \mapsto u_\epsilon$ $(0 < \epsilon < \delta)$ a modifier, where

$$u_\epsilon(x) = \int_\Omega \varphi_\epsilon(x - y) u(y) dy.$$

Modifiers have the following properties:

(1) $\operatorname{supp}(u_\epsilon) \subset (\operatorname{supp} u)_\epsilon := \{x \in \Omega \,|\, d(x, \operatorname{supp} u) \leq \epsilon\}$.
(2) $u_\epsilon \in C_0^\infty(\Omega)$ and

$$\partial^\alpha u_\epsilon(x) = \int_\Omega u(x) \partial^\alpha \varphi_\epsilon(x - y) dy.$$

(3) If $u \in C_0^m(\Omega)$, then $\partial^\alpha(u_\epsilon) = (\partial^\alpha u)_\epsilon$, $\forall \alpha \, |\alpha| \leq m$ for $\epsilon > 0$ sufficiently small.

(4) If $u \in C_0(\Omega)$, then $\|u - u_\epsilon\|_C \to 0$ as $\epsilon \to 0$.

(5) If $u \in L^p(\Omega)$, $1 \leq p < \infty$, then $\|u - u_\epsilon\|_{L^p} \to 0$ as $\epsilon \to 0$.

(6) $C_0^\infty(\Omega)$ is dense in $L^p(\Omega)$.

(7) If $u, \partial^\alpha u \in L^p(\Omega)$, $p \in [1, \infty]$, supp $u \subset \text{int}(\Omega)$, then for ϵ sufficiently small, we have $(D^\alpha u)_\epsilon = \partial^\alpha(u)_\epsilon$.

In fact,

$$\text{LHS} = \int_\Omega D^\alpha u(y) \varphi_\epsilon(x - y) dy$$

$$= (-1)^{|\alpha|} \int_\Omega u(y) \partial_y^\alpha \varphi_\epsilon(x - y) dy$$

$$= \int_\Omega u(y) \partial_x^\alpha \varphi_\epsilon(x - y) dy$$

$$= \partial_x^\alpha \int_\Omega u(y) \varphi_\epsilon(x - y) dy = \text{RHS}.$$

(8)

$$W_0^{m,p}(\mathbb{R}^n) = W^{m,p}(\mathbb{R}^n).$$

10.5 Some important properties of Sobolev spaces and embedding theorems

Sobolev spaces are fundamental function spaces, which play an important role in Harmonic Analysis, Partial Differential Equations, Functional Analysis, and Calculus of Variations. The important properties of Sobolev spaces are discussed in many textbooks. In this section, we only list some of the main results and refer the interested readers for the detailed proofs in the existing literature. However, in order to assist the readers' understanding of the significance of these results as well as the essence of the proofs, we would like to provide proofs for some particular or simplified cases.

Extension Theorem

Sobolev spaces are function spaces. When the domain of these functions is an arbitrary region $\Omega \subset \mathbb{R}^n$, the space is denoted $W^{m,p}(\Omega)$; when the domain of these functions is the whole $\mathbb{R}^n$, it is denoted $W^{m,p}(\mathbb{R}^n)$. A natural question comes to mind: can we extend each function $u \in W^{m,p}(\Omega)$ to be a function $\tilde{u} \in W^{m,p}(\mathbb{R}^n)$ such that $\tilde{u}|_\Omega = u$?

The answer is affirmative as long as $\Omega \subset \mathbb{R}^n$ has sufficiently smooth boundary.

Theorem 10.3 (Extension Theorem) *If Ω is a bounded region where $\partial\Omega$ is uniformly C^m, then $\forall\, 0 \le l \le m$, $\forall\, 1 \le p < \infty$, $\exists\, T \in L(W^{l,p}(\Omega), W^{l,p}(\mathbb{R}^n))$ such that*

$$Tu(x) = u(x), \qquad \text{a.e. } x \in \Omega.$$

A detailed proof can be found in [Ad] Theorem 4.26. This theorem is due to Lichtenstein, Hestenes, Seeley, and Calderon.

In contrast, for $W_0^{m,p}$ spaces, regardless of the choice of Ω, such extension is always possible, that is, $\exists\, T \in L(W_0^{l,p}(\Omega), W_0^{l,p}(\mathbb{R}^n))$ such that

$$Tu(x) = \begin{cases} u(x), & \text{a.e. } x \in \Omega, \\ 0, & x \notin \Omega. \end{cases}$$

This is because $C_0^\infty(\Omega)$ is a dense subspace of $W_0^{m,p}(\Omega)$ and $C_0^\infty(\Omega)$ is a subspace of $C_0^\infty(\mathbb{R}^n)$, whence $W_0^{l,p}(\Omega)$ is a closed subspace of $W_0^{l,p}(\mathbb{R}^n)$.

Approximation Theorem

We have previously defined $H^1(\Omega)$ as the metric completion of the space $C^1(\bar\Omega)$ with respect to the norm $\|u\| = (\int_\Omega(|\nabla u|^2 + |u|^2)dx)^{1/2}$. Additionally, $W^{1,2}(\Omega)$ is a complete metric space containing $C^1(\bar\Omega)$, whence $H^1(\Omega) \subset W^{1,2}(\Omega)$. It is natural to ask whether they are the same. We give a positive answer to this question via the following approximation theorem.

Theorem 10.4 (Serrin-Meyers Approximation Theorem) *If $p \in [1,\infty)$, then the set $S := C^\infty(\Omega) \cap W^{m,p}(\Omega)$ is dense in $W^{m,p}(\Omega)$,*

Proof Choose a sequence of open subsets $\{\Omega_j\}$ satisfying

$$\emptyset = \Omega_{-1} = \Omega_0 \subset \Omega_1 \subset \bar\Omega_1 \subset \Omega_2 \subset \bar\Omega_2 \subset \cdots \Omega_i \subset \bar\Omega_i \subset \Omega_{i+1} \subset \cdots,$$

$$\bigcup_{i=1}^\infty \Omega_i = \Omega.$$

For example, let $\Omega_i = \{x \in \Omega \,|\, \|x\| \le i, \operatorname{dist}(x, \partial\Omega) > \frac{1}{i}\}$, $i = 1,2,3,\ldots$. For an open covering of Ω

$$O = \{U_k = \Omega_{k+1} \backslash \Omega_{k-1} \,|\, k = 1,2,\ldots\},$$

the corresponding partition of unity $\{\psi_i\}$ subordinate to O satisfies:
  (1) $\psi_i \in C_0^\infty(\Omega)$,
  (2) $\operatorname{supp}\{\psi_i\} \subset \Omega_{i+1} \backslash \Omega_{i-1}$,
  (3) $\sum \psi_i(x) \equiv 1 \ \forall\, x \in \Omega$.
  Now $\forall\, u \in W^{m,p}(\Omega)$, $\forall\, \epsilon > 0$, choose $\delta_i < \operatorname{dist}(\Omega_i, \partial\Omega_{i+1})$ small enough such that

$$\|(\psi_i u)_{\delta_i} - \psi_i u\|_{m,q} < \frac{\epsilon}{2^i}. \tag{10.2}$$

This is possible due to properties (7), (5) in §10.4 and Lemma 10.1. $\forall \alpha, |\alpha| \leq m$, we have

$$\|\partial^\alpha(\psi_i u)_{\delta_i} - D^\alpha(\psi_i u)\|_q = \|(D^\alpha(\psi_i u))_{\delta_i} - D^\alpha(\psi_i u)\|_q \to 0.$$

Summing it up in (10.2), it follows that

$$\|\sum_{i=1}^\infty (\psi_i u)_{\delta_i} - u\|_{m,q} = \|\sum_{i=1}^\infty (\psi_i u)_{\delta_i} - \sum_{i=1}^\infty (\psi_i u)\|_{m,q} < \epsilon.$$

Since $\forall x \in \Omega$, $\sum_{i=1}^\infty (\psi_i u)_{\delta_i}(x)$ has finite sum, $v = \sum_{i=1}^\infty (\psi_i u)_{\delta_i}$ is infinitely many times differentiable.

Moreover, $\forall k \in \mathbb{N}$, on each Ω_k,

$$u(x) = \sum_{i=1}^{k+2} (\psi_i u)(x), \quad v(x) = \sum_{i=1}^{k+2} (\psi_i u)_{\delta_i}(x),$$

we have the estimate

$$\|u - v\|_{W^{m,p}(\Omega_k)} \leq \sum_{i=1}^{k+2} \|(\psi_i u)_{\delta_i} - \psi_i u\|_{m,p} < \epsilon.$$

Letting $k \to \infty$, it follows that $\|u - v\|_{m,p} < \epsilon$. It is evident $v \in S$ as desired. $\square$

Corollary 10.1 *If Ω is a bounded region where $\partial\Omega$ is uniformly C^m, then $\forall 1 \leq p < \infty$, $W^{m,p}(\Omega)$ is a separable Banach space.*

Proof Choose an open hypercube $D \subset \mathbb{R}^n$ such that $\Omega \subset \bar{\Omega} \subset D$. $\forall u \in W^{m,q}(\Omega), \exists \tilde{u} \in W^{m,q}(\mathbb{R}^n)$ such that

$$\|\tilde{u}\|_{W^{m,q}(\mathbb{R}^n)} \leq C\|u\|_{W^{m,q}(\Omega)}, \quad \tilde{u}|_\Omega = u.$$

For arbitrary $\epsilon > 0$, applying Theorem 10.4, we can find $v \in C^\infty(\mathbb{R}^n) \cap W^{m,q}(\mathbb{R}^n)$ such that

$$\|\tilde{u} - v\|_{W^{m,q}(\Omega)} \leq \|\tilde{u} - v\|_{W^{m,q}(D)} \leq \frac{\epsilon}{3}.$$

Choose $\psi \in C_0^\infty(D)$ satisfying $\psi(x) = 1, \forall x \in \Omega$, then $\psi v \in C_0^\infty(D)$. By the Weierstrass Approximation Theorem, there exists a polynomial P on $\bar{D}$ such that

$$\|\psi v - P\|_{W^{m,q}(D)} \leq \|\psi v - P\|_{C^m(\bar{D})} \leq \frac{\epsilon}{3}.$$

Since any polynomial can be approximated in the C^m sense by polynomials with rational coefficients, this asserts that $W^{m,p}(\Omega)$ has a countable dense subset. $\square$

Corollary 10.2 $H^1(\Omega) = W^{1,2}(\Omega)$.

Poincaré's Inequality

Poincaré's inequality in Lecture 9 can be extended to $W_0^{1,p}(\Omega)$, $1 \leq p < \infty$ as follows.

Poincaré's Inequality If $\Omega \subset \mathbb{R}^n$ is bounded, $u \in W^{1,p}(\Omega)$, $1 \leq p < \infty$, then $\exists \, C = C(p, \Omega)$ such that

$$\int_\Omega |u|^p dx \leq C \int_\Omega |\nabla u|^p dx.$$

Corollary 10.3 *If $\Omega \subset \mathbb{R}^n$ is bounded, then*

$$\|u\| = \left(\int_\Omega |\nabla u|^p \right)^{\frac{1}{p}}$$

defines an equivalent norm on $W_0^{1,p}(\Omega)$ ($1 \leq p < \infty$).

Since $W_0^{1,p}(\Omega)$ ($1 < p < \infty$) is a closed linear subspace of $W^{1,p}(\Omega)$, it is itself a reflexive Banach space.

Embedding Theorems

Theorem 10.5 (Sobolev) *Both embeddings*

$$W^{m,q}(\mathbb{R}^n) \hookrightarrow L^r(\mathbb{R}^n), \quad \frac{1}{r} = \frac{1}{q} - \frac{m}{n} \ (\text{if } mq < n),$$

and $\forall \, j \in \mathbb{N}$,

$$W^{m+j,q}(\mathbb{R}^n) \hookrightarrow C^{j,\lambda}(\mathbb{R}^n), \quad 0 < \lambda \leq m - \frac{n}{q} \ (\text{if } mq > n)$$

are continuous.

A proof can be found in [Ad] 5.4–5.10.

Combining the above theorem and the extension theorem, we arrive at

Theorem 10.6 (Sobolev Embedding Theorem) *Assume Ω is a bounded region with uniformly C^m boundary, $1 \leq q < \infty$, and $m \geq 0$ is an integer, then the embeddings*

$$W^{m,q}(\Omega) \hookrightarrow L^r(\Omega), \frac{1}{r} = \frac{1}{q} - \frac{m}{n} \ (\text{if } mq < n)$$

and $\forall \, j \in \mathbb{N}$,

$$W^{m+j,q}(\Omega) \hookrightarrow C^{j,\lambda}(\bar{\Omega}), \quad 0 < \lambda \leq m - \frac{n}{q} \ (\text{if } mq > n)$$

are both continuous.

The most frequently used version of this result is for $m = 1$, and we denote $q^* = \frac{nq}{n-q}$, then

$$W^{1,q}(\Omega) \hookrightarrow L^r(\Omega), \ r \leq q^* \text{ if } n > q$$

and

$$W^{1,q}(\Omega) \hookrightarrow C(\bar{\Omega}) \text{ if } q > n$$

are both continuous.

Remark 10.1 When $n = 1$, $\Omega = (a, b)$, the conclusion of the embedding theorem follows easily from Hölder's inequality and Example 10.3. Since in this case, the generalized derivatives coincide with the usual derivative functions almost everywhere, we have

$$|u(x) - u(y)| = \left| \int_y^x u'(t)dt \right| \le |x - y|^{\frac{1}{q'}} \left(\int_a^b |u'(t)|^q dt \right)^{\frac{1}{q}}, \qquad \forall\, x, y \in (a, b).$$

$\square$

Compact Embeddings

Theorem 10.7 (Rellich–Kondrachov compact embedding theorem) *Assume* Ω *is a bounded region with uniformly* C^m *boundary,* $1 \le q \le \infty$, *and* $m \ge 0$ *is an integer, then the embeddings*

$$W^{m,q}(\Omega) \hookrightarrow L^r(\Omega), \ 1 \le r < \frac{nq}{n - mq} \ (if\ m < n/q)$$

and

$$W^{m,q}(\Omega) \hookrightarrow C(\bar{\Omega}), \ (if\ m > n/q)$$

are both compact.

The proof can be found in [Ad] Theorem 6.2.

We now give a direct proof for a frequently used special case of the above result.

Theorem 10.8 (Rellich) *If* $\Omega \subset \mathbb{R}^n$ *is a bounded region,* $1 \le p < \infty$, *then a closed bounded ball in* $W_0^{1,p}(\Omega)$ *is sequentially compact in* $L^p(\Omega)$.

Proof It suffices to show the closed unit ball is sequentially compact. Denote B the closed unit ball in $W_0^{1,p}(\Omega)$ centered at $\mathbf{0}$. We proceed by showing $\forall\, \epsilon > 0$, under the L^p-norm, there exists a finite ϵ net of B.

The idea is to find a uniformly bounded and equicontinuous set of functions in an arbitrary L^p-neighborhood of B.

1. By definition, $C_0^\infty(\Omega)$ is dense in $W_0^{1,p}(\Omega)$. Denote $S = C_0^\infty(\Omega) \cap B$. For any $\delta > 0$, denote $S_\delta = \{v_\delta \mid v \in S\}$, where

$$v_\delta(x) = \int_\Omega v(y)\varphi_\delta(x - y)dy.$$

$\forall\, v \in S$,

$$\|v_\delta\|_{1,p} \le \|v\|_{1,p} \le 1.$$

Since

$$|v_\delta(x) - v(x)| = \left| \int_{|y| \le 1} \varphi(y)[v(x) - v(x - \delta y)]dy \right|$$

$$\le \int_{|y| \le 1} \varphi(y) \int_0^{\delta|y|} \left| \frac{\partial}{\partial r} v\left(x - r\frac{y}{|y|} \right) \right| dr dy,$$

by Hölder's inequality,

$$\|v_\delta - v\|_p^p \le \int_\Omega \left(\int_{|y|\le 1} \varphi(y)^{\frac{1}{p'}} \varphi(y)^{\frac{1}{p}} \int_0^{\delta|y|} \left| \frac{\partial}{\partial r} v\left(x - r\frac{y}{|y|} \right) \right| dr dy \right)^p dx$$

$$\le \int_{|y|\le 1} \varphi(y)\delta^p |y|^p \int_\Omega |\nabla v(x)|^p dx dy,$$

i.e.

$$\|v_\delta - v\|_p \le \delta \|\nabla v\|_p.$$

Thus, $\exists \delta_0 = \delta(\epsilon)$ such that

$$\|v - v_\delta\|_p \le \frac{\epsilon}{8}, \ \ \forall v \in S, \ \forall \delta \le \delta_0.$$

Fixing $\delta = \delta_0$, we have

$$|v_\delta(x)| = \left| \int_\Omega v(y)\varphi_\delta(x - y) dy \right| \le \frac{C}{\delta^n} \|v\|_p$$

and

$$|\nabla v_\delta(x)| = \left| \int_\Omega v(y)\nabla\varphi_\delta(x - y) dy \right| \le \frac{C}{\delta^{n+1}} \|v\|_p.$$

This shows $S_{\delta_0} \subset C(\bar\Omega)$ is uniformly bounded and equicontinuous. According to the Arzelà–Ascoli theorem, under the C-norm, S_{δ_0} has a finite $\frac{\epsilon}{4\text{mes}(\Omega)}$-net $\{w_1, w_2, \ldots, w_l\}$. That is, $\forall v \in S_{\delta_0}, \exists w_i \in S_{\delta_0}$ such that

$$\|v_\delta - w_i\|_C < \frac{\epsilon}{4\text{mes}(\Omega)}.$$

Thus,

$$\|v_\delta - w_i\|_p < \frac{\epsilon}{4}.$$

2. However, w_i may not lie in B, but for each w_i, there is a corresponding $v_{\delta_0}^i$, where $v^i \in S = C_0^\infty(\Omega) \cap B$. If $v_{\delta_0}^i \notin B$, its support must lie beyond $\bar\Omega$; we can then take $\delta_i \in (0, \delta_0]$ such that the support of $w_i' = v_{\delta_i}^i$ is confined in Ω. Consequently, we have $w_i' \in B$ and

$$\|w_i - w_i'\|_p \le \|w_i - v^i\|_p + \|w_i' - v^i\|_p \le \frac{\epsilon}{4}.$$

$\forall u \in B, \exists v \in S$ such that $\|u - v\|_{1,p} \le \frac{\epsilon}{4}$, hence

$$\|u - w_i'\|_p \le \|u - v\|_p + \|v - v_{\delta_0}\|_p + \|v_{\delta_0} - w_i\|_p + \|w_i - w_i'\|_p < \epsilon.$$

$\{w_1', \ldots, w_l'\}$ is the finite ϵ-net we are seeking. $\qquad\qquad\square$

10.6　The Euler–Lagrange equation

In this section, we extend the calculus of variation from previously discussed C^1-space to Sobolev spaces. Given a Lagrangian $L \in C(\Omega \times \mathbb{R}^N \times \mathbb{R}^{nN})$, $L = L(x, u, p)$ is differentiable, where $\Omega \subset \mathbb{R}^n$. In order to define the functional

$$I(u) = \int_\Omega L(x, u(x), \nabla u(x))dx$$

on some suitable Sobolev space $W^{1,q}(\Omega)$ and to make sure the E-L equation corresponding to the necessary condition of its extremal values actually makes sense, we must impose some additional assumptions on L as follows:

(1) L, L_u, L_p are continuous;
(2) $|L(x, u, p)| \leq C(1 + |u|^q + |p|^q)$;
(3) $|L_u(x, u, p)| + |L_p(x, u, p)| \leq C(1 + |u|^q + |p|^q)$.

In fact, by assumption (2), $\forall u \in W^{1,p}(\Omega, \mathbb{R}^N)$,

$$I(u) = \int_\Omega |L(x, u(x), \nabla u(x))|dx \leq C \int_\Omega (1 + |u(x)|^q + |\nabla u(x)|^q)dx < \infty.$$

Moreover, by assumption (3),

$$\int_\Omega |L_u(x, u(x), \nabla u(x))| + |L_p(x, u(x), \nabla u(x))|dx < \infty,$$

i.e. $\lambda(x) := L_u(x, u(x), \nabla u(x))$ and $\mu(x) := L_p(x, u(x), \nabla u(x)) \in L^1$.

Lemma 10.2 *Under the assumptions* (1), (2), *and* (3), $\forall u_0 \in W^{1,q}(\Omega)$, $\forall \varphi \in C_0^1(\Omega)$,

$$\delta I(u_0, \varphi) = \frac{d}{dt} I(u_0 + s\varphi)|_{s=0}$$

$$= \int_\Omega [L_u(x, u(x), \nabla u(x))\varphi(x) + L_p(x, u(x), \nabla u(x))\nabla\varphi(x)]dx. \quad (10.3)$$

Proof We need only be concerned with

$$s^{-1}[I(u + s\varphi) - I(u)] = \int_\Omega [\lambda_s(x)\varphi(x) + \mu_s(x)\nabla\varphi(x)]dx,$$

where

$$\lambda_s(x) = \int_0^1 L_u(x, u(x) + \theta s\varphi(x), \nabla(u(x) + \theta s\varphi(x)))d\theta,$$

$$\mu_s(x) = \int_0^1 L_p(x, u(x) + \theta s\varphi(x), \nabla(u(x) + \theta s\varphi(x)))d\theta.$$

By assumption (3), $\lambda_s, \mu_s \in L^1$, and as $s \to 0$, almost everywhere we have

$$\lambda_s(x) \to L_u(x, u(x), \nabla u(x)), \quad \mu_s(x) \to L_p(x, u(x), \nabla u(x)).$$

In addition, $\forall \varphi \in C_0^1(\Omega)$, the integrand in (10.3) is dominated by the integrable function

$$C(1 + |u(x)|^q + |\nabla u(x)|^q)(|\varphi(x)| + |\nabla \varphi(x)|).$$

Thus, the conclusion follows immediately from the Lebesgue Dominated Convergence Theorem. □

Remark 10.2 The continuity requirement in all three variables (x, u, p) in Lemma 10.2 assumption (1) can be replaced by the weaker Carathéodory condition:

$$\forall (u, p) \in \mathbb{R}^N \times \mathbb{R}^{nN}, \quad x \mapsto L(x, u, p) \text{ is Lebesgue measurable},$$

$$\text{for a.e. } x \in \Omega, \quad (u, p) \mapsto L(x, u, p) \text{ is continuous.} \qquad (10.4)$$

That is, assuming L, L_u, and L_p satisfy the Carathéodory condition.

Remark 10.3 Note that in the Sobolev Embedding Theorem, when $q < n$, $W^{1,q}(\Omega) \hookrightarrow L^r(\Omega)$, $r \leq q^* = \frac{nq}{n-q}$. The exponential growth requirement in Lemma 10.2 assumption (3) can be relaxed by:

$$(3') \ |L_u(x, u, p)| + |L_p(x, u, p)| \leq \begin{cases} \leq C(1 + |u|^r + |p|^q), & r \leq q^*, \ q < n \\ \leq C(1 + |u|^r + |p|^q), & r \geq 1, \quad q = n \\ \leq C(1 + |p|^q), & q > n. \end{cases}$$

As a consequence, we have:

Theorem 10.9 *Suppose the Carathéodory condition* (10.4), (2), *and* (3') *hold. If for any given $\rho \in W^{1,q}(\Omega, \mathbb{R}^N)$, $u_0 \in M = \rho + W_0^{1,q}(\Omega, \mathbb{R}^N)$ is a minimum of I in M, then u_0 satisfies the following E-L equation:*

$$\int_\Omega [L_u(x, u(x), \nabla u(x))\varphi(x) + L_p(x, u(x), \nabla u(x))\nabla \varphi(x)]dx = 0,$$

$$\forall \varphi \in C_0^\infty(\Omega, \mathbb{R}^N). \qquad (10.5)$$

In this sense, we call solutions of (10.5) the generalized solutions of E-L equation.

Remark 10.4 The generalization of the concept of directional derivatives in differential calculus on Banach spaces are the Gâteaux derivatives.

Let X be a Banach space, let $U \subset X$ be an open subset, and $f \in C(U, \mathbb{R}^1)$ be a function defined on U.

Definition 10.4 (The Gâteaux derivative) Let $x_0 \in U$. We say f is Gâteaux differentiable at x_0, if $\forall\, h \in X$, $\exists\, df(x_0, h) \in \mathbb{R}^1$ such that

$$|f(x_0 + th) - f(x_0) - tdf(x_0, h)| = o(t), \qquad (t \to 0).$$

We call $df(x_0, h)$ the Gâteaux derivative of f at x_0.

In calculus of variations, for a functional I in the space $W^{1,q}(\Omega, \mathbb{R}^N)$, the difference between the variation $\delta I(u_0, \varphi)$ and the Gâteaux derivative $dI(u_0, \varphi)$ is that, in the variation, $\varphi \in C_0^\infty(\Omega)$ or $C_0^1(\Omega)$; while in the Gâteaux derivative, $\varphi \in W_0^{1,q}(\Omega)$.

In order to insure the integral involved in $dI(u_0, \varphi)$ is well-defined, we must modify the exponential growth assumption (3) as follows:

$$(3'')\ |L_u(x, u, p)| + |L_p(x, u, p)| \leq \begin{cases} C(1 + |u|^{r-1} + |p|^{q-1}), & r < q^*, \quad q < n, \\ C(1 + |u|^r + |p|^{q-1}), & r \geq 1, \quad q = n, \\ C(1 + |p|^{q-1}), & q > n. \end{cases}$$

We have the following theorem regarding the Gâteaux derivative of I:

Theorem 10.10 *Suppose* (1), (2) *and* (3'') *hold. Given* $\rho \in W^{1,q}(\Omega, \mathbb{R}^N)$. *Let* $I(u) = \int_\Omega L(x, u(x), \nabla u(x))dx$ *be a functional defined on* $M = \rho + W_0^{1,q}(\Omega, \mathbb{R}^N)$. *If* I *attains its minimum at* $u_0 \in M$, *then* u_0 *is Gâteaux differentiable; furthermore,*

$$dI(u_0, \varphi) = \int_\Omega [L_u(x, u(x), \nabla u(x))\varphi(x) + L_p(x, u(x), \nabla u(x))\nabla\varphi(x)]dx,$$

$$\forall\, \varphi \in W_0^{1,q}(\Omega, \mathbb{R}^N).$$

Comparing to $(3')$, $(3'')$ is clearly stronger. So in calculus of variations, we often use variational derivatives, rather than the Gâteaux derivatives.

Remark 10.5 When $n = 1$ and $J = (a, b)$, we can derive the following integral form of the E-L equation via (10.5):

$$\int_a^t L_u(t, u^*(t), \dot{u}^*(t))dt = L_p(t, u^*(t), \dot{u}^*(t)) = c, \quad \text{a.e.} \tag{10.6}$$

In Lecture 2, Remark 2.2, we established the du Bois–Remond lemma on $L^\infty(J)$, we can now extend this to the Lebesgue space $L^1(J)$.

Lemma 10.3 *Suppose* $f \in L^1(J)$ *satisfies*

$$\int_J f(t)\varphi'(t)dt = 0, \quad \forall\, \varphi \in C_0^\infty(J),$$

then $f(t) = c$, *a.e.* $t \in J$.

Proof Fix any two Lebesgue points $a < t_0 < t_1 < b$ of f. Let $c = f(t_0)$. Choose $\varepsilon > 0$ such that $(t_0 - \varepsilon, t_1 + \varepsilon) \subset J$. We construct a piecewise linear function ψ as follows

$$\psi(t) = \begin{cases} 0, & t \notin [t_0 - \varepsilon, t_1 + \varepsilon] \\ 1, & t \in [t_0, t_1] \end{cases}$$

and connecting the rest with straight lines. Clearly, there exists $\{\varphi_n\} \subset C_0^\infty(J)$ such that $\varphi_n \to \psi$ in $W^{1,1}(J)$. Hence,

$$\int_J f(t)\varphi_n'(t)dt \to \int_J f(t)\psi'(t)dt;$$

consequently,

$$\varepsilon^{-1}\int_{t_0-\varepsilon}^{t_0} f(t)dt - \varepsilon^{-1}\int_{t_1}^{t_1+\varepsilon} f(t)dt = 0.$$

Letting $\varepsilon \to 0$, by taking the limit, $f(t_1) = f(t_0) = c$. Since t_1 can be an arbitrary Lebesgue point, it follows that for all Lebesgue point $t \in J$, $f(t) = c$. This completes the proof. $\square$

Thus, under the assumptions (1), (2) and (3'), the minimum $u^* \in u_0 + W^{1,q}(J, \mathbb{R}^N)$ of I satisfies the integral form of the E-L equation (10.6).

Exercises

1. Verify the properties (1)–(6) and (8) listed in §10.4.
2. Let $u \in W^{1,q}(\mathbb{R}^1)$. Show that

$$u \in \begin{cases} C^\lambda(\mathbb{R}^1), & \lambda = \dfrac{q-1}{q}, \quad 1 \le q < \infty, \\ \mathrm{Lip}(\mathbb{R}^1), & q = \infty. \end{cases}$$

Lecture 11

Weak lower semi-continuity

In Lecture 9, we have demonstrated that the weak sequential lower semi-continuity of a functional plays an important role in direct methods. In this lecture, we focus on the criteria for determining whether a functional is weakly sequentially lower semi-continuous.

11.1 Convex sets and convex functions

We shall investigate the lower semi-continuity of a function from the view-point of point-set topology. From definition, on a topological space X, a function $f : X \to \mathbb{R}^1$ is said to be lower semi-continuous if and only if its epigraph $f_t = \{x \in X \mid f(x) \leq t, \ \forall t \in \mathbb{R}^1\}$ is closed. On a normed linear space X, a function f is weakly sequentially lower semi-continuous if and only if the set f_t ($\forall t \in \mathbb{R}^1$) is weakly sequentially closed.

Generally speaking, being closed and weak closed for sets are quite different concepts: a weakly closed set must be closed; however, the converse may not be true.

A convex set in a Banach space has the following important property: if $C \subset X$ is a convex subset of a Banach space X, then

$$\text{closed} \iff \text{weakly sequentially closed.}$$

This is due to Mazur's Theorem: In a Banach space, if $x_n \rightharpoonup x$, then $\frac{1}{n} \sum_{i=1}^{n} x_i \to x$.

We can apply this result to determine whether a functional is weakly sequentially lower semi-continuous. Let us recall the definition of a convex function: given a function f defined on a convex subset C of a linear space E, if

$$f(\lambda x + (1 - \lambda)y) \leq \lambda f(x) + (1 - \lambda)f(y), \ \forall x, y \in C, \forall \lambda \in [0, 1],$$

149

then f is said to be convex. It is not difficult to see that

f is a convex function $\implies f_t := \{x \in X | f(x) \le t\}$ is a convex set, $\forall t \in \mathbb{R}^1$.

Combining these two statements, we obtain the following theorem.

Theorem 11.1 *Let X be a Banach space. If $f : X \to \mathbb{R}^1$ is a convex function, then*

f is sequentially lower semi-continuous $\iff$ f is weakly sequentially lower semi-continuous.

Proof We note that

f is lower semi-continuous $\iff \forall t \in \mathbb{R}^1$, f_t is closed.

f is weakly sequentially lower semi-continuous $\iff \forall t \in \mathbb{R}^1$, f_t is weakly sequentially closed.

Since $\forall t \in \mathbb{R}^1$, f_t is a convex set, these two notions are equivalent, our assertion follows. $\square$

Based on Theorem 11.1, we now determine the weak sequential lower semicontinuity of certain functionals.

From now on, without confusion, we will not distinguish the gradient ∇ and the generalized gradient D (in the sense of generalized derivatives), and denote both by ∇.

Example 11.1 Let $\Omega \subset \mathbb{R}^n$ be a bounded region, $f \in L^2(\Omega)$. On the Sobolev space $H_0^1(\Omega)$, the functional

$$I(u) = \int_\Omega \left(\frac{1}{2} |\nabla u(x)|^2 + f(x)u(x) \right) dx$$

is weakly sequentially lower semi-continuous.

Proof Since $u \longmapsto \int_\Omega f(x)u(x)dx$ is linear, and by Poincaré's inequality,

$$\int_\Omega f(x)u(x)dx \le \|f\|_{L^2} \|u\|_{L^2} \le C(f)\|u\|_{H_0^1}.$$

It is both continuous and convex, hence it is weakly sequentially lower semi-continuous.

Furthermore, since $u \to \int_\Omega |\nabla u(x)|^2 dx$ is continuous and convex, it is also weakly sequentially lower semi-continuous. $\square$

Example 11.2 Let $\Omega \subset \mathbb{R}^n$ be a bounded region. On the Sobolev space $W_0^{1,q}(\Omega)$ ($1 \le q < \infty$, $1 \le r \le q^* = \frac{nq}{n-q}$), the functional

$$I(u) = \int_\Omega (|\nabla u(x)|^q + |u(x)|^r)dx$$

is weakly sequentially lower semi-continuous. (The proof is similar and left as an exercise.)

If we replace the functional by

$$I(u) = \int_\Omega (|\nabla u(x)|^q + c(x)|u(x)|^r)dx,$$

where $c \in L^\infty(\Omega)$, $c \geq 0$, then the conclusion still holds and the proof remains the same. $\qquad\square$

However, if the non-negativity condition on c is removed, then in general, I is not necessarily convex, so it may not be weakly sequentially lower semi-continuous. But if we impose the condition $r < q^* = \frac{nq}{n-q}$, then I will be weakly sequentially lower semi-continuous.

In fact, if $u_j \rightharpoonup u$ in $W_0^{1,q}(\Omega)$, then

$$\liminf \int_\Omega |\nabla u_j(x)|^q \geq \int_\Omega |\nabla u(x)|^q.$$

We also have

$$\liminf \int_\Omega c(x)|u_n(x)|^r dx \geq \int_\Omega c(x)|u(x)|^r dx.$$

Suppose not, there exist $\epsilon > 0$ and a subsequence $\{u_{j'}\}$ such that

$$\int_\Omega c(x)|u_{j'}(x)|^r dx < \int_\Omega c(x)|u(x)|^r dx - \epsilon.$$

By the Rellich–Kondrachov embedding theorem, there exists a subsequence $\{u_{n_j}\} \subset \{u_{j'}\}$ such that $u_{n_j} \to u$ in $L^r(\Omega)$. Thus,

$$\int_\Omega c(x)|u_{n_j}(x)|^r dx \to \int_\Omega c(x)|u(x)|^r dx,$$

a contradiction. $\qquad\square$

11.2 Convexity and weak lower semi-continuity

In order to establish the variational integral $I(u) = \int_\Omega L(x, u(x), \nabla u(x))dx$ is convex in u, it would require the Lagrangian $L(x, u, p)$ to be convex in (u, p). But from Example 11.2, we see that this requirement is too strong. Using compact embedding, such convexity requirement for u can be replaced by some exponential growth conditions. We will carry out this idea in solving general variational problems.

Theorem 11.2 (Tonelli–Morrey) *Suppose $L : \bar{\Omega} \times \mathbb{R}^N \times \mathbb{R}^{nN} \to \mathbb{R}^1$ satisfies*

(1) $L \in C^1(\bar{\Omega} \times \mathbb{R}^n \times \mathbb{R}^{nN})$,

(2) $L \geq 0$,

(3) $\forall\, (x, u) \in \Omega \times \mathbb{R}^N$, $p \mapsto L(x, u, p)$ *is convex,*

then $I(u) = \int_{\Omega} L(x, u(x), \nabla u(x))dx$ *is weakly sequentially lower semi-continuous in* $W^{1,q}(\Omega, \mathbb{R}^N)$ $(1 \leq q < \infty)$.

We now need another important characteristic of convex functions.

Lemma 11.1 *Let X be a Banach space, $h \in X$, and $f : X \to \mathbb{R}^1$ be a convex function. If f has directional derivative $df(x_0, h)$ at x_0 in the direction of h, then*

$$f(x_0) + df(x_0, h) \leq f(x_0 + h).$$

This is a direct application of the well-known result of single variable convex functions.

Proof of Theorem 11.2 Let $u_j \rightharpoonup u$ $(W^{1,q})$, we want to show

$$\liminf \int_{\Omega} L(x, u_j(x), \nabla u_j(x))dx \geq \int_{\Omega} L(x, u(x), \nabla u(x))dx.$$

1. By the Rellich–Kondrachov theorem, we can find a subsequence, still denoted by $\{u_j\}$ such that

$$u_j \to u, \qquad L^q(\Omega, \mathbb{R}^N).$$

Then by the Riesz theorem, there exists a further subsequence, which is again denoted by $\{u_j\}$ such that

$$u_j(x) \to u(x) \quad \text{a.e. } x \in \Omega.$$

$\forall \varepsilon > 0$, there exists a compact subset $K \subset \Omega$ such that $\text{mes}(\Omega \backslash K) < \varepsilon$ and

(1) $u_j \to u$ uniformly on K (Egorov's theorem),

(2) u and ∇u are continuous on K (Luzin's theorem),

(3) if $\int_{\Omega} L(x, u(x), \nabla u(x))dx < +\infty$, then

$$\int_K L(x, u(x), \nabla u(x))dx \geq \int_{\Omega} L(x, u(x), \nabla u(x))dx - \varepsilon,$$

(the absolute continuity of Lebesgue integrals).

If $\int_{\Omega} L(x, u(x), \nabla u(x))dx = +\infty$, then we take

$$\int_K L(x, u(x), \nabla u(x))dx > \frac{1}{\varepsilon}.$$

2. Since L is convex in p, applying Lemma 11.1, we obtain

$$I(u_j) \geq \int_K L(x, u_j(x), \nabla u_j(x))dx$$

$$\geq \int_K L_p(x, u_j(x), \nabla u(x))(\nabla u_j(x) - \nabla u(x))dx + \int_K L(x, u_j(x), \nabla u(x))dx$$

$$= \int_K L(x, u_j(x), \nabla u(x))dx + \int_K L_p(x, u(x), \nabla u(x))(\nabla u_j(x) - \nabla u(x))dx$$

$$+ \int_K L_p(x, u_j(x), \nabla u(x)) - L_p(x, u(x), \nabla u(x))(\nabla u_j(x) - \nabla u(x))dx$$

$$= \mathrm{I} + \mathrm{II} + \mathrm{III}.$$

3. By (1), on K, $u_j \to u$ uniformly, and L is continuous, hence the first term

$$\mathrm{I} = \int_K L(x, u_j(x), \nabla u(x))dx \to \int_K L(x, u(x), \nabla u(x))dx.$$

By (2), $L_p(x, u(x), \nabla u(x)) \in L^\infty(K)$. Taking χ_K to be the characteristic function of K, it yields $\chi_K L_p(x, u(x), \nabla u(x)) \in L^\infty(\Omega) \subset L^{q'}(\Omega)$. Using the fact $u_j \rightharpoonup u$ (in $W^{1,q}(\Omega)$), we can deduce

$$\nabla u_j \rightharpoonup \nabla u \ (\text{in } L^q(\Omega, \mathbb{R}^{nN})).$$

Thus, the second term

$$\mathrm{II} = \lim_{j\to\infty} \int_K L_p(x, u(x), \nabla u(x))(\nabla u_j(x) - \nabla u(x))dx = 0.$$

Lastly, since a weakly convergent sequence is bounded,

$$\|\nabla u_j - \nabla u\|_1 \leq C_1(\|\nabla u_j - \nabla u\|_q) \leq C_1(\|\nabla u_j\|_q + \|\nabla u\|_q) \leq C_2;$$

furthermore, since $L_p(x, u_j(x), \nabla u(x)) \to L_p(x, u(x), \nabla u(x))$ uniformly on K, it follows that the third term

$$\mathrm{III} = \lim_{j\to\infty} \int_K (L_p(x, u_j(x), \nabla u(x)) - L(x, u(x), \nabla u(x)))(\nabla u_j - \nabla u)dx = 0.$$

In summary,

$$\varliminf_{j\to\infty} I(u_j) \geq \int_K L(x, u(x), \nabla u(x))dx \geq I(u) - \varepsilon.$$

Since $\varepsilon > 0$ is arbitrary,

$$\varliminf_{j\to\infty} I(u_j) \geq I(u). \qquad \square$$

Remark 11.1 It is worth noting that the theorem does not require the functional to be bounded from above, this is because we are only concerned with lower semi-continuity. Without the restriction on the growth of the functional, it is conceivable that $\exists u \in W^{1,q}(\Omega)$ such that $I(u) = +\infty$. However, this does not interfere with our discussion on lower semi-continuity.

Utilizing the concept of Carathéodory functions, assumption (1) in Theorem 11.2 can be replaced by

(1′): for a.e. (x, u), $p \mapsto L(x, u, p)$ is differentiable, L and L_p are both Carathéodory functions.

11.3 An existence theorem

Theorem 11.3 (The existence of extreme values) *Let $\Omega \subset \mathbb{R}^n$ be a bounded measurable set. If $u_0 \in W^{1,q}(\Omega, \mathbb{R}^N)$, $1 < q < \infty$ and if*

(1) *L and L_p are Carathéodory functions,*

(2′) *$\exists\, a \in L^1(\Omega)$, $b > 0$, such that $L(x, u, p) \geq -a(x) + b|p|^q$, $\forall\, (x, u, p) \in \Omega \times \mathbb{R}^N \times \mathbb{R}^{nN}$,*

(3) *$\forall\, (x, u) \in \Omega \times \mathbb{R}^N, p \mapsto L(x, u, p)$ is convex,*

then the functional

$$I(u) = \int_\Omega L(x, u(x), \nabla u(x))dx$$

attains its minimum on $u_0 + W_0^{1,q}(\Omega, \mathbb{R}^N)$.

Proof Consider the functional $v \mapsto I(u_0 + v)$ on the reflexive Banach space $W_0^{1,q}(\Omega, \mathbb{R}^N)$.

1. Although assumption (2′) and assumption (2) are not quite the same, we can consider the functional $I + \int_\Omega a(x)dx$ instead of I. Then applying the Tonelli–Morrey theorem to $W_0^{1,q}(\Omega, \mathbb{R}^N)$, it follows that I is weakly sequentially lower semi-continuous.

2. I is coercive, i.e.

$$\text{when } \|u\|_{1,q} \to \infty, \quad I(u) \to +\infty.$$

In fact, $\forall v \in W_0^{1,q}(\Omega, \mathbb{R}^N)$, by Poincaré's inequality, there exist $\alpha, \beta > 0$ such that

$$I(u_0 + v) \geq -\int_\Omega a(x)dx + b\int_\Omega |\nabla(u_0 + v)|^q dx \geq \alpha\|v\|_{1,q}^q - \beta. \qquad \square$$

As an example, we claim the Poisson equation has a generalized solution. Given a bounded $\Omega \subset \mathbb{R}^n$ and a function $f \in L^2(\Omega)$, there exists $u \in H_0^1(\Omega)$

such that

$$-\Delta u = f.$$

As a matter of fact, by Theorem 11.3, the functional

$$I(u) = \int_\Omega \left(\frac{1}{2} |\nabla u|^2 - fu \right) dx$$

attains its minimum on $H_0^1(\Omega)$.

11.4* Quasi-convexity

Although we have relaxed the convexity requirement to be $\forall\, (x, u) \in \Omega \times \mathbb{R}^N$, $p \mapsto L(x, u, p)$ is convex, this condition is yet still too strong in many applications.

For instance, the Jacobian determinant $\det\left(\frac{\partial u^i}{\partial x_j} \right)$ clearly possesses both geometric and mechanical meaning. When $\Omega \subset \mathbb{R}^n$, $u = (u^1, u^2, \ldots, u^n) : \Omega \to \mathbb{R}^n$, the Lagrangian containing the Jacobian determinant of u often occurs in both geometry and mechanics. For example, let $f \in C(\mathbb{R}^1, \mathbb{R}^1)$ be a convex function,

$$L : \mathbb{R}^{n^2} \to \mathbb{R}^1,$$

$$A \to L(A) = f(\det(A)),$$

where $A = \det\left(\frac{\partial u^i}{\partial x_j} \right)$. However, L is not convex in p.

It is natural to ask whether the assumption on L being convex in p is indeed necessary. We shall first take a look at the simple case where L only depends on p and $n = 1$. The more general cases will be inspired by this simple case.

Suppose $I(u) = \int_J L(u'(t))dt$ is sequentially weak-* lower semi-continuous, i.e.

$$u_j \rightharpoonup^* u \ (\text{in } W^{1,\infty}(J, \mathbb{R}^N)) \Longrightarrow \underline{\lim} I(u_j) \geq I(u).$$

We seek conditions for which L must satisfy.

Without loss of generality, we may assume $J = (0, 1)$; in particular, $\forall\, \eta \in \mathbb{R}^N$, $\forall\, \lambda \in (0, 1)$, take

$$\varphi(t) = \eta \begin{cases} t(1 - \lambda), & t \in [0, \lambda), \\ (1 - t)\lambda, & t \in [\lambda, 1), \end{cases}$$

then $\varphi \in W_0^{1,\infty}(J)$. We now periodically extend this function to the whole line, and let

$$\varphi_m(t) = \frac{1}{m} \varphi(mt).$$

For arbitrary $u_0, p_0 \in \mathbb{R}^N$, we construct the function

$$u(t) = u_0 + p_0 t$$

and the sequence

$$u_m(t) = u(t) + \varphi_m(t), \ m = 1, 2, \ldots.$$

We have:

$$|u_m(t) - u_m(s)| = |\varphi_m(t) - \varphi_m(s)| \leq \sup_{t \leq \theta \leq s} |\varphi'(m\theta)| \, |t - s|,$$

namely,

$$|u_m|_{1,\infty} \leq \|\varphi'\|_{L^\infty(J)}$$

and

$$u_m \to u \ (\text{in } L^\infty(J)).$$

By further applying the facts $\dot{\varphi}_m(t) = \varphi'(mt)$, $\varphi \in W_0^{1,\infty}(J, \mathbb{R}^N)$, $\bar{\varphi}' = 0$, and by Example 9.3 in Lecture 9, it follows that

$$\varphi_m' \rightharpoonup^* 0 \ (\text{in } L^\infty(J)).$$

Now $\forall \, \xi \in (W^{1,1}(J, \mathbb{R}^N))^*$, we have $\xi = (\xi_0, \xi_1) \in L^1(J, \mathbb{R}^{2N})$. Moreover, since $Du_m \rightharpoonup^* Du$ in $L^\infty(J)$, it follows that

$$\langle \xi, u_m - u \rangle = \int_J [\xi_0(u_m - u) + \xi_1 D(u_m - u)] dt \to 0.$$

Namely, $u_m \rightharpoonup^* u$ in $W^{1,\infty}(J, \mathbb{R}^N)$.

Notice that

$$I(u) = \int_J L(p_0) dt = L(p_0),$$

and by the periodicity of φ, we have

$$I(u_m) = \int_J L(p_0 + \varphi_m'(t)) dt$$

$$= \frac{1}{m} \int_{mJ} L(p_0 + \varphi'(t)) dt$$

$$= \int_J L(p_0 + \varphi'(t)) dt.$$

Suppose I is weakly sequentially lower semi-continuous, $\liminf I(u_m) \geq I(u)$, it yields that

$$\int_J L(p_0 + \varphi'(t)) dt \geq L(p_0).$$

Inspired by the above argument, we now consider the case $n > 1$.

Theorem 11.4 *Let $\Omega \subset \mathbb{R}^n$ be a region and $L \in C(\Omega, \mathbb{R}^{nN})$. If*

$$I(u) = \int_\Omega L(\nabla u(x))dx$$

is sequentially weak-* lower semi-continuous on $W^{1,\infty}(\Omega, \mathbb{R}^N)$, i.e.

$$u_j \rightharpoonup^* u \ (W^{1,\infty}(\Omega, \mathbb{R}^N)) \implies \underline{\lim} I(u_j) \geq I(u),$$

then for any hypercube $D \subset \bar{D} \subset \Omega$, $\forall A_0 \in \mathbb{R}^{nN}$ (an $n \times N$ matrix),

$$\int_D L(A_0 + \nabla\varphi(x))dx \geq L(A_0)\text{mes}(D), \ \forall \varphi \in W_0^{1,\infty}(\Omega, \mathbb{R}^N).$$

Proof 1. We first extend φ_m to high dimensions. Without loss of generality, we may assume $D = [0,1]^n$ is the unit hypercube. For any $k \in N$, we partition D into 2^k equal size subcubes, $D = \cup_{l=1}^{2^{kn}} D_l^k$, where each side of D_l^k has length 2^{-k} with center $c_l^k = 2^{-k}(y_1^l + \frac{1}{2}, y_2^l + \frac{1}{2}, \ldots, y_n^l + \frac{1}{2})$, where $(y_1^l, y_2^l, \ldots, y_n^l) l = 1, 2, \ldots, 2^{kn}$ range over the lattice points $(0, 1, 2, \ldots, 2^k - 1)^n$. For any $v \in W_0^{1,\infty}(D, \mathbb{R}^N)$, we extend it periodically to the entire $\mathbb{R}^n$. Let

$$w_k(x) = \frac{1}{2^k} v(2^k(x - c_l^k)), \ x \in D_l^k, \ \forall l = 1, 2, \ldots, 2^{kn},$$

then

$$\nabla w_k(x) = \nabla v(2^k(x - c_l^k)), \ x \in D_l^k, \ \forall l = 1, 2, \ldots, 2^{kn},$$

and

$$\begin{cases} w_k \to 0 & \text{in } L^\infty(D, \mathbb{R}^N), \\ \nabla w_k \rightharpoonup^* 0 & \text{in } L^\infty(D, \mathbb{R}^N). \end{cases}$$

2. Define

$$u_k(x) = A_0 x + w_k(x), \ k = 1, 2, \ldots,$$

where w_k is defined as above and it vanishes outside D. It is clear

$$u_k \rightharpoonup^* u = A_0 x.$$

Hence,

$$\text{mes}(\Omega)L(A_0) = I(u) \leq \liminf I(u_k)$$

and

$$I(u_k) = \int_D L(A_0 + \nabla w_k(x)dx + \int_{\Omega \backslash D} L(A_0)dx$$

$$= \sum_{l=1}^{2^{kn}} \int_{D_l^k} L(A_0 + \nabla v(2^k(x - c_l^k)))dx + L(A_0)\text{mes}(\Omega \backslash D)$$

$$= \int_D L(A_0 + \nabla v(x))dx + L(A_0)\text{mes}(\Omega \backslash D).$$

Our assertion now follows. $\qquad\square$

This is why Morrey introduced the following concept of quasi-convexity of a function.

Definition 11.1 A function L is said to be quasi-convex, if $\forall A \in \mathbb{R}^{nN}$ (an $n \times N$ matrix), $\forall$ hypercube $D \subset \mathbb{R}^n$, $\forall v \in W_0^{1,\infty}(D, \mathbb{R}^N)$, we have

$$\text{mes}(D)L(A) \leq \int_D L(A + \nabla v(x))dx.$$

The importance of quasi-convexity is to insure the weak sequential lower semi-continuity of a functional. The proof of the following theorem is rather lengthy, we refer the interested readers to [Da], pp. 156–167.

Theorem 11.5 (Morrey–Acerbi–Fusco) *When* $1 \leq p < \infty$, *if*

$$I(u) = \int_\Omega L(\nabla u)dx$$

is weakly sequentially lower semi-continuous on $W^{1,p}(\Omega, \mathbb{R}^N)$ *(or when* $p = \infty$, *I is weakly-* sequentially lower semi-continuous), then L is quasi-convex. Conversely, if we add the growth conditions:*

$$\begin{cases} |L(A) \leq \alpha(1 + |A|)| & p = 1 \\ -\alpha(1 + |A|^q) \leq L(A) \leq \alpha(1 + |A|^p) & 1 \leq q < p < \infty \\ |L(A)| \leq \eta(|A|) & p = \infty, \end{cases}$$

where $\alpha > 0$ *is a constant, and* η *is a non-decreasing continuous function, and if L is quasi-convex, then when* $1 \leq p < \infty$, *I is weakly sequentially lower semi-continuous on* $W^{1,p}(\Omega, \mathbb{R}^N)$ *(or when* $p = \infty$, *I is weak-* sequentially lower semi-continuous).*

What kind of functions are quasi-convex? Suppose $f : \mathbb{R}^1 \to \mathbb{R}^1$ is convex and $u \in L^1(\Omega)$, then from Jessen's inequality:

$$f\left(\frac{1}{\text{mes}(\Omega)} \int_\Omega u(x)\right)dx \leq \frac{1}{\text{mes}(\Omega)} \int_\Omega f(u(x))dx,$$

we see that if L is convex, then L must be quasi-convex. In fact, if $p \mapsto L(p)$ is convex, then

$$L(p) = L(\text{mes}(D)^{-1} \int_D (p + \nabla\varphi(x))dx)$$

$$\leq \text{mes}(D)^{-1} \int_D L(p + \nabla\varphi(x))dx, \quad \forall \varphi \in W_0^{1,\infty}(\Omega, \mathbb{R}^N).$$

It is worth noting when $n = 1$ or $N = 1$, quasi-convexity and convexity coincide.

We first verify the above statement by showing quasi-convexity implies convexity for $n = 1$.

Let $\xi, \eta \in \mathbb{R}^N, \forall \lambda \in [0,1]$. Let

$$\xi_1 = \xi + (1-\lambda)\eta,$$
$$\xi_2 = \xi - \lambda\eta.$$

Then

$$\xi = \lambda\xi_1 + (1-\lambda)\xi_2,$$
$$\eta = \xi_1 - \xi_2.$$

Define

$$\varphi(t) = \eta \begin{cases} t(1-\lambda), & t \in [0, \lambda), \\ (1-t)\lambda, & t \in [\lambda, 1), \end{cases}$$

and substituting it into the quasi-convexity assumption,

$$
\begin{aligned}
L(\xi) &\leq \int_0^1 L(\xi + \varphi'(t))dt \\
&= \int_0^\lambda L(\xi_1)dt + \int_\lambda^1 L(\xi_2)dt \\
&= \lambda L(\xi_1) + (1-\lambda)L(\xi_2),
\end{aligned}
$$

it shows that $p \mapsto L(p)$ is convex.

Next, we verify quasi-convexity implies convexity for $N = 1, n > 1$. That is, $\forall \xi_1, \xi_2 \in \mathbb{R}^n$, we want to show

$$L(\lambda\xi_1 + (1-\lambda)\xi_2) \leq \lambda L(\xi_1) + (1-\lambda)L(\xi_2).$$

Take a hypercube $D \subset \Omega$, without loss of generality, we may assume $D = [0,1]^n$. We continue to use the above defined function φ, $\forall x = (x_1, \ldots, x_n) \in D$, let

$$u_k(x) = \eta k^{-1}\varphi(kx_1),$$

then

$$\nabla u_k(x) = \eta \begin{cases} (1-\lambda), & \{kx_1\} \in [0, \lambda), \\ -\lambda, & \{kx_1\} \in [\lambda, 1), \end{cases}$$

where $\{y\}$ represents the fractional part of $y \in \mathbb{R}^1$ and $\eta = \xi_1 - \xi_2$. Furthermore, let

$$v_k(x) = \eta \min\{k^{-1}\varphi(kx_1), \text{dist}(x, \partial D)\},$$

where $\text{dist}(x, \partial D) = \inf\{\sup_{1 \leq i \leq n}\|x_i - y_i\| \mid y = (y_1, \ldots, y_n) \in \partial D\}$. Thus, $v_k|_{\partial D} = 0$, and there exists a constant $K > 0$ such that $|u_k(x) - v_k(x)| \leq K\|x - y\|$, whence, $v_k \in W_0^{1,p}(D), 1 < p < \infty$. In addition,

$$\text{mes}\{x \in D \mid \nabla u_k(x) \neq \nabla v_k(x)\} \to 0, \qquad k \to \infty.$$

If we take $D_1 = \{x \in D \mid \nabla u_k(x) = (1-\lambda)\eta\}$, $D_2 = \{x \in D \mid \nabla u_k(x) = -\lambda\eta\}$, then $D = D_1 \cup D_2$.

Since L is quasi-convex,

$$\text{mes}(D)L(\lambda\xi_1 + (1-\lambda)\xi_2) \le \int_D L(\lambda\xi_1 + (1-\lambda)\xi_2 + \nabla v_k(x))dx.$$

Let $\xi = \lambda\xi_1 + (1-\lambda)\xi_2$, then by the absolute continuity of integrals, we have

$$\lim \int_D L(\xi + \nabla v_k(x))dx = \lim \int_D L(\xi + \nabla u_k(x))dx$$

$$= \int_{D_1} L(\xi_1)dx + \int_{D_2} L(\xi_2)dx$$

$$= (\lambda L(\xi_1) + (1-\lambda)L(\xi_2))\text{mes}(D).$$

This proves our assertion. □

However, there are quasi-convex functions which are not convex. For example, the determinant function

$$A \mapsto \det(A).$$

We only verify this for the case $N = n = 2$. Given a matrix $A = (a_{ij})$. Let $u = (u_1, u_2)$, $P = (p_j^i)$, and $L(P) = \det(P)$. Since

$$\det(\nabla\varphi) = \partial_1(\varphi_1\partial_2\varphi_2) - \partial_2(\varphi_1\partial_1\varphi_2),$$

we have:

$$\int_D \det(\nabla\varphi)dx = 0, \ \forall \varphi \in W_0^{1,\infty}(\Omega).$$

Consequently,

$$\text{mes}(D)^{-1} \int_D \det(A + \nabla\varphi(x))dx$$

$$= \text{mes}(D)^{-1} \int_D [\det(A) + \det(\nabla\varphi(x)) + a_{11}\partial_2\varphi_2 + a_{22}\partial_1\varphi_1$$

$$- a_{12}\partial_1\varphi_2 - a_{21}\partial_2\varphi_1]dx$$

$$= \det(A).$$

Recall the Legendre–Hadamard condition from Lecture 6 in determining whether u_0 is a weak minimum,

$$L_{p_\alpha^j p_\beta^k}(x, u_0(x), \nabla u_0(x))\pi_\alpha^j \pi_\beta^k \ge 0, \ \forall \pi \in \mathbb{R}^{n \times N}, \ \text{rank}(\pi) = 1.$$

This is related to convexity. In fact, if $\forall (x, u) \in \Omega \times \mathbb{R}^N$, $p \mapsto L(x, u, p)$ is convex, and if $L \in C^2$, then for any u, the Legendre–Hadamard condition holds.

However, the Legendre–Hadamard condition does not require L to be convex for all $n \times N$ matrices, it only requires L to be convex for all rank 1 matrices. We shall call the Lagrangian L rank 1 convex (for brievity, we omit $(x, u) \in \Omega \times \mathbb{R}^N$) if it satisfies the following conditions:

$$L(\lambda B + (1 - \lambda)C) \leq \lambda L(B) + (1 - \lambda)L(C),$$
$$\forall \lambda \in [0, 1], \quad \forall B, C \in M^{n \times N}, \quad \text{rank}(B - C) = 1$$

In summary,

convex $\Longrightarrow$ quasi-convex $\Longrightarrow$ rank 1 convex $\overset{\text{for } n=1 \text{ or } N=1}{\Longrightarrow}$ convex.

Returning to the existence problem, with the assistance of the Morrey–Acerbi–Fusco theorem, we have a more general existence theorem.

Theorem 11.6 *Suppose* $L : M^{n \times N} \overset{C}{\longrightarrow} \mathbb{R}^1$ *is quasi-convex and there exist constants* $C_2 > C_1 > 0$ *such that*

$$C_1 |A|^p \leq L(A) \leq C_2(1 + |A|^p), \quad 1 < p < \infty.$$

If further, we assume $v \in W^{1,p}(\Omega, \mathbb{R}^N)$*, then in the set* $E = W_v^{1,p} = \{u \in W^{1,p}(\Omega, \mathbb{R}^N) \mid u|_{\partial\Omega} = v|_{\partial\Omega}\}$*,*

$$I(u) = \int_\Omega L(\nabla u(x)) \, dx$$

achieves its minimum.

Exercises

1. Suppose $\forall x \in \Omega$, $(u, p) \longmapsto L(x, u, p)$ is convex. Prove that the functional

$$I(u) = \int_\Omega L(x, u(x), \nabla u(x)) dx$$

   is convex in u.
2. Find the weak limit of $\varphi_n(x) = \frac{1}{n} \sin nx$ in $W^{1,q}(0, 2\pi)$, $1 \leq q < \infty$.
3. Let $\Omega \subset \mathbb{R}^n$ be a bounded region. Suppose in $W^{1,q}(\Omega)$ $(1 \leq q < \infty)$, the sequence $\{u_j\}$ converges weakly and $\{u_j\}$ converges almost everywhere to a function u_0. Does $\{u_j\}$ contain a convergent subsequence in $L^q(\Omega)$? If so, what does it converge to? Is $u_0 \in W^{1,q}(\Omega)$?
4. Let $J = (0, 1)$, $\lambda \in (0, 1)$, $\alpha, \beta \in \mathbb{R}^1$,

$$\varphi(x) = \begin{cases} \alpha, & x \in (0, \lambda), \\ \beta, & x \in (\lambda, 1), \end{cases}$$

   and let $\varphi_n(x) = \varphi(nx)$. Show that $\varphi_n \rightharpoonup \lambda\alpha + (1 - \lambda)\beta$.

5. Determine whether each of the following functionals is weakly sequentially lower semi-continuous on each specified spaces:

(1) Assume Φ is a single variable convex function, c is a continuous function on the bounded set $\bar{\Omega}$,

$$I(u) = \int_{\Omega} (\Phi(|\nabla u(x)|) + c(x)|u(x)|^4)dx \text{ in } W^{1,4}(\Omega).$$

(2) $I(u) = \int_{\Omega}[1 + |\nabla u(x)|^2]^{\frac{1}{2}}dx$ in $W^{1,1}(\Omega)$.

(3) $I(u) = \int_{-1}^{1} t^2 \dot{u}(t)^2 dt$ in $H^1(-1,1)$.

Lecture 12

Boundary value problems and eigenvalue problems of linear differential equations

12.1 Linear boundary value problems and orthogonal projections

In previous lectures, the methods used in analyzing the existence of extreme values of functionals not only rely on the weak sequential compactness of the domain but also the weak sequential lower semi-continuity of the functionals.

Interestingly, as mentioned in Lecture 9, for some special variational problems associated with a particular class of linear differential equations, the weak sequential compactness as well as the weak sequential lower semi-continuity can be replaced by the completeness of a Hilbert space, and we need only the orthogonal projections in a Hilbert space. In this lecture, we introduce the orthogonal projection method and its applications in variational problems.

Recall the Poisson equation in Lecture 11: let $\Omega \subset \mathbb{R}^n$ be a bounded region and $f \in L^2(\Omega)$, we want to find $u : \Omega \to \mathbb{R}^1$ satisfying the equation

$$\begin{cases} -\Delta u = f, & \text{in } \Omega, \\ u = 0, & \text{on } \partial\Omega. \end{cases}$$

In the last lecture, we obtain the minimum by means of proving that a minimizing sequence of the functional

$$I(u) = \int_\Omega \left(\frac{1}{2}|\nabla u|^2 - fu \right) dx$$

contains a convergent subsequence in $H_0^1(\Omega)$.

However, since $H_0^1(\Omega)$ is a Hilbert space equipped with the inner product

$$((u, v)) = \int_\Omega \nabla u \cdot \nabla v \, dx, \quad \forall\, u, v \in H_0^1(\Omega),$$

we note the first term in the functional I precisely corresponds to the norm induced by such inner product, whereas the second term can also be expressed via this inner product. To see this, we regard

$$\varphi \mapsto \int_\Omega f \cdot \varphi \, dx$$

as a linear functional on $H_0^1(\Omega)$. Using Schwarz's inequality, we have

$$\left| \int_\Omega f \cdot \varphi \, dx \right| \leq \|f\|_2 \cdot \|\varphi\|_2,$$

and Poincaré's inequality yields

$$\left| \int_\Omega f \cdot \varphi \, dx \right| \leq C \|f\|_2 \|\varphi\|_{H_0^1},$$

which implies the continuity of the functional. Hence, there exists $F \in (H_0^1(\Omega))^*$ such that

$$F(\varphi) = \int_\Omega f \cdot \varphi \, dx.$$

By the Riesz representation theorem from functional analysis, this continuous linear functional can be represented via the inner product as follows: $\exists \, u_0 \in H_0^1(\Omega)$ such that

$$((u_0, \varphi)) = F(\varphi) = \int_\Omega f\varphi \, dx.$$

Consequently, we can rewrite the E-L equation of I in terms of the inner product:

$$((u, \varphi)) = \int_\Omega f \cdot \varphi \, dx = ((u_0, \varphi)), \quad \forall \varphi \in C_0^\infty(\Omega).$$

Since $C_0^\infty(\Omega)$ is dense in $H_0^1(\Omega)$, $u = u_0$ is the solution.

In appearance this approach contains neither the minimizing sequence nor the weak sequential compactness and the weak sequential lower continuity. The solution is obtained directly. However, the proof of the Riesz representation theorem is itself a variational problem, namely, finding the distance from a point outside a given hyperplane to the hyperplane in a Hilbert space, along with the projection of the point onto the hyperplane.

Let us recall this proof. Denote

$$M = \{\eta \in H_0^1(\Omega) | F(\eta) = 0\}.$$

M is a closed linear subspace of $H_0^1(\Omega)$. For any $\varphi \in H_0^1(\Omega)$, we can find the orthogonal projection η of φ onto M, i.e. $\eta \in M$ and $\varphi - \eta = \xi \perp M$. If we take

$$v_0 = \begin{cases} F(\xi) \dfrac{\xi}{[\![\xi]\!]^2} & \xi \neq 0, \\ 0 & \xi = 0, \end{cases}$$

then $((v_0, \varphi)) = F(\varphi)$ as desired.

Note, if $\xi = 0$, then the above equality clearly holds. If $\xi \neq 0$, then $\forall \varphi \in H_0^1(\Omega)$,

$$F(\varphi) = F(\xi),$$

but

$$((v_0, \varphi)) = F(\xi)\left(\left(\frac{\xi}{[\![\xi]\!]^2}, \varphi\right)\right) = F(\xi),$$

which is exactly what we want.

If we take an equivalent inner product on the Hilbert space, specifically

$$[\![u]\!] = \left(\int_\Omega |\nabla u|^2 dx\right)^{\frac{1}{2}},$$

then the following argument precisely depicts the minimization process used in solving variational problems.

We now return to examine the existence of orthogonal projections (see Figure 12.1).

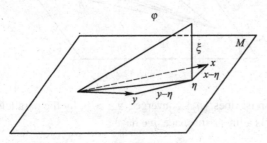

Fig. 12.1

This can be converted to a variational problem:

$$\min_{x \in M} [\![\varphi - x]\!].$$

In fact, if there exists $\eta \in M$ such that $[\![\varphi - \eta]\!] = \min_{x \in M} [\![\varphi - x]\!]$, then $\xi = \varphi - \eta$ satisfies

$$((\xi, x - \eta)) = 0, \quad \forall x \in M.$$

This is because $\forall x \in M$ and $\forall t \in [0, 1]$, by letting

$$g(t) = [\![\varphi - (tx + (1 - t)\eta)]\!]^2$$
$$= [\![\varphi - \eta]\!]^2 - 2t((\varphi - \eta, x - \eta)) + t^2 [\![x - \eta]\!]^2,$$

g becomes a quadratic function, which achieves its minimum at $t = 0$, whence,

$$g'(0) = -2((\varphi - \eta, x - \eta)) = 0, \tag{12.1}$$

namely,

$$((\xi, x - \eta)) = 0.$$

But why does this minimum exists? Let $m = \inf_{x \in M} [\![\varphi - x]\!]$, choose a minimizing sequence $\{\eta_j\} \subset M$ such that

$$[\![\eta_j - \varphi]\!] < m + \frac{1}{j}, \quad \forall j$$

(see Figure 12.2).

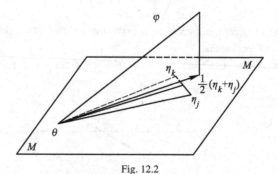

Fig. 12.2

The question is: does $\{\eta_j\}$ converge? $\forall \varepsilon > 0$, by the parallelogram law and the fact that M is a linear subspace, we have

$$[\![\eta_j - \eta_k]\!]^2 = 2([\![\eta_j - \varphi]\!]^2 + [\![\eta_k - \varphi]\!]^2) - 4 \left[\!\left[\frac{\eta_j + \eta_k}{2} - \varphi \right]\!\right]^2$$

$$\leq 4(m + \varepsilon)^2 - 4m^2, \quad \text{for } j, k \text{ sufficiently large.} \tag{12.2}$$

Hence, $\{\eta_j\}$ is a Cauchy sequence. By the completeness of $H_0^1(\Omega)$, $\{\eta_j\}$ is convergent, i.e. $\eta_j \to \eta$. Since M is closed, $\eta \in M$ and it achieves

$$[\![\eta - \varphi]\!] = \min_{x \in M} [\![\varphi - x]\!].$$

This argument provides a different angle to verify the Dirichlet's principle. By the peculiarity of the problem, it is unnecessary to employ weak convergence, while the metric completeness of the space plays a key role.

The same method can also be used in handling variational inequalities (cf. Lecture 7). For instance, the obstacle problem of a thin membrane: on the bounded region $\Omega \subset \mathbb{R}^n$, given a measurable function $\psi(x)$ as an obstacle and the external force function $f \in L^2(\Omega)$. We seek the equilibrium position of the thin membrane $u : \Omega \to \mathbb{R}^1$.

We restate this as a variational problem. Let $C = \{u \in H_0^1(\Omega) \mid u(x) \leq \psi(x) \text{ a.e.}\}$ be a closed convex subset of the Hilbert space $H_0^1(\Omega)$. Denote $(f, v) = \int_\Omega f \cdot v dx$, we wish to find $u \in C$ such that

$$((u, v - u)) - (f, v - u) \geq 0, \qquad \forall v \in C. \tag{12.3}$$

According to the above discussion, there exists $u_0 \in H_0^1(\Omega)$ such that

$$((u_0, v)) = (f, v), \qquad \forall v \in H_0^1(\Omega).$$

Thus, it becomes

$$((u - u_0, v - u)) \geq 0, \qquad \forall v \in C. \tag{12.4}$$

Based on the earlier argument, this can be accomplished by finding

$$\min_{v \in C} [\![u_0 - v]\!]. \tag{12.5}$$

In fact, suppose u is a solution to the problem, since $\forall v \in C$, $tv + (1 - t)u \in C$, (12.1) becomes

$$g'(0) \geq 0,$$

i.e. (12.4). As for the existence of a solution for (12.5), this can still be deduced by the minimizing sequence $\{\eta_j\} \subset C$. Since C is convex, in (12.2), $\frac{1}{2}(\eta_j + \eta_k) \in C$. By the parallelogram law, $\{\eta_j\}$ is a Cauchy sequence. Since C is closed, it must converge to a limit, at which the minimal distance is attained. Furthermore, we already know the solution of (12.5) is indeed the solution of (12.3).

12.2 The eigenvalue problems

Let $\Omega \subset \mathbb{R}^n$ be a region, we want to know for which values of $\lambda \in \mathbb{R}^1$, the boundary value problem

$$\begin{cases} -\Delta u = \lambda u, & \text{in } \Omega, \\ u = 0, & \text{on } \partial\Omega, \end{cases} \tag{12.6}$$

has a nonzero solution u. Similar to eigenvalue problems of matrices, we shall call this the eigenvalue problem of the differential operator $-\Delta$.

Those λ which correspond to nonzero solutions are called the "spectrum". For the nonzero solution $u \in L^2(\Omega)$, it is called an *eigenfunction*, and the corresponding λ is called an *eigenvalue*. This kind of eigenvalue problems is frequently encountered in geometry, mechanics, physics, and many other branches in mathematics.

Eigenvalue problems can be handled using constrained variational methods. As stated in Lecture 6, setting

$$I(u) = \int_\Omega |\nabla u|^2 \, dx$$

and

$$N(u) = \int_\Omega |u|^2 \, dx,$$

we wish to find

$$\min\{I(u) \,|\, u \in H_0^1(\Omega) \cap N^{-1}(1)\}. \tag{12.7}$$

If $\varphi_1 \in C^2(\bar{\Omega})$ indeed achieves such minimum, then by the Lagrange multipliers, for the adjusted E-L equation of the Lagrangian, there exists $\lambda_1 \in \mathbb{R}^1$ such that

$$-\Delta\varphi_1 = \lambda_1\varphi_1 \quad \text{in} \quad \Omega. \tag{12.8}$$

Since $N(\varphi_1) = \int_\Omega |\varphi_1|^2 \, dx = 1$, φ_1 is nonzero. Multiplying both sides of (12.8) by φ_1 and then integrating, it yields

$$I(\varphi_1) = \int_\Omega |\nabla\varphi_1|^2 \, dx = \lambda_1.$$

This means

$$\lambda_1 = \min\{I(u) \,|\, u \in H_0^1(\Omega) \cap N^{-1}(1)\}. \tag{12.9}$$

In the following, we shall verify the existence of a solution for the minimization problem (12.7). Suppose $\Omega \subset \mathbb{R}^n$ is bounded, we already know that I is coercive and weakly sequentially lower semi-continuous, it remains to show the set

$$M_1 = \{u \in H_0^1(\Omega) \,|\, N(u) = 1\}$$

is weakly sequentially closed. That is, if $\{u_j\} \subset M_1$, $u_j \rightharpoonup u$ in $H_0^1(\Omega)$, then $u \in M_1$.

By the Rellich–Kondrachov compact embedding theorem, for the bounded region Ω, $H_0^1(\Omega) \hookrightarrow L^2(\Omega)$ is a compact embedding. By assumption, $u_j \rightharpoonup u$ in $H_0^1(\Omega)$, it contains a subsequence $\{u_j'\}$ such that

$$u_j' \to u \quad \text{in} \quad L^2(\Omega).$$

Moreover, since

$$\int_\Omega u_j^2 \, dx = 1,$$

it follows that

$$\int_\Omega u^2 dx = 1,$$

i.e. $u \in M_1$, hence M_1 is weakly closed.

Once the first eigenvalue has been found, motivated by the eigenvalue problems of matrices, it is natural to ask: are there any other eigenvalues? If so, how to find them?

We continue to adopt the idea of constrained optimization. Consider the set

$$M_2 = \left\{ u \in M_1 \;\middle|\; \int_\Omega u \cdot \varphi_1 dx = 0 \right\},$$

we seek to

$$\min\{I(u) \mid u \in M_2\}.$$

If $\varphi_2 \in M_2$ is a minimum, then $\varphi_2 \neq 0$, and by the Lagrange multipliers, there exist $\lambda_2, \mu_2 \in \mathbb{R}^1$ such that

$$-\Delta \varphi_2 = \lambda_2 \varphi_2 + \mu_2 \varphi_1. \tag{12.10}$$

We now prove that $\mu_2 = 0$. First, multiplying both sides of (12.8) by φ_2 and then integrating, it yields

$$\int_\Omega \nabla\varphi_2 \nabla\varphi_1 dx = \int_\Omega \nabla\varphi_1 \nabla\varphi_2 = \lambda_1 \int_\Omega \varphi_1 \varphi_2 dx = 0.$$

Next, multiplying both sides of (12.10) by φ_1 and then integrating, it yields

$$\int_\Omega \nabla\varphi_2 \cdot \nabla\varphi_1 dx = \mu_2 \int_\Omega |\varphi_2|^2 dx.$$

Thus, $\mu_2 = 0$.

Consequently,

$$-\Delta \varphi_2 = \lambda_2 \varphi_2 \quad \text{in} \quad \Omega.$$

This confirms that λ_2 is indeed an eigenvalue with its corresponding eigenfunction φ_2, and $\varphi_2 \neq \varphi_1$. In addition,

$$\lambda_2 = I(\varphi_2) = \int_\Omega |\nabla\varphi_2|^2 \, dx = \min\{I(u) \mid u \in M_2\} \geq \lambda_1.$$

In order to show I attains its minimum on M_2, we must show M_2 is weakly closed. Suppose $u_j \rightharpoonup u$ (in $(H_0^1(\Omega))$, since $\{u_j\} \subset M_1$ and M_1 is weakly closed, $u \in M_1$. From $\int_\Omega u_j \varphi_1 = 0$, it follows that

$$\int_\Omega u \varphi_1 = 0.$$

Thus, $u \in M_2$, i.e. M_2 is weakly closed.

Continuing in this fashion, we let

$$M_n = \left\{ u \in M_{n-1} \;\middle|\; \int_\Omega u \varphi_{n-1} dx = 0 \right\}$$

and use a similar argument, we can show that each M_n is weakly closed. Hence, the constrained optimization problem

$$\min\{I(u) \mid u \in M_n\}$$

has a solution $\varphi_n \neq 0$ satisfying

$$-\Delta \varphi_n = \lambda_n \varphi_n + \sum_{j=1}^{n-1} \mu_j \varphi_j.$$

Likewise, using mathematical induction, we can show $\mu_1 = \cdots = \mu_{n-1} = 0$ and

$$\Delta \varphi_n = \lambda_n \varphi_n,$$

where

$$\lambda_n = I(\varphi_n) = \min\{I(u) \mid u \in M_n\} \geq \lambda_{n-1}.$$

As a consequence, we have the following theorem.

Theorem 12.1 *Suppose $\Omega \subset \mathbb{R}^n$ is a bounded region, then Eq. (12.6) has an increasing sequence of eigenvalues $0 < \lambda_1 \leq \lambda_2 \leq \cdots \leq \lambda_n \to +\infty$, with corresponding eigenfunctions $\{\varphi_1, \varphi_2, \ldots\} \subset H_0^1(\Omega)$ satisfying*

$$\begin{cases} \Delta \varphi_i = \lambda_i \varphi_i, \\ \displaystyle\int_\Omega |\varphi_i|^2 dx = 1, \end{cases} \quad i = 1, 2, \ldots$$

and

$$((\varphi_i, \varphi_j)) = \int_\Omega \nabla \varphi_i \cdot \nabla \varphi_j dx = \lambda_i \int_\Omega \varphi_i \varphi_j dx = 0, \quad \forall\, i \neq j.$$

Proof It suffices to show $\lambda_n \to +\infty$.

We argue by contradiction. Suppose $\exists\, C > 0$ such that $\lambda_n \leq C$, then

$$\int_\Omega |\nabla \varphi_j|^2 \, dx = \lambda_j \int_\Omega |\varphi_j|^2 \, dx = \lambda_j \leq C.$$

This implies $\{\varphi_j\}_1^\infty$ is a bounded sequence in $H_0^1(\Omega)$, hence there is a weakly convergent subsequence

$$\varphi_j' \rightharpoonup \varphi^* \; (\text{in } H_0^1(\Omega)).$$

On one hand, by the Rellich–Kondrachov compact embedding theorem, we have

$$\varphi_j' \to \varphi^* \; (\text{in } L^2(\Omega)). \tag{12.11}$$

On the other hand, since $\int_\Omega \varphi_i \varphi_j dx = 0$, $i \neq j$, we see that

$$\int_\Omega |\varphi_i - \varphi_j|^2 dx = \int_\Omega (\varphi_i^2 + \varphi_j^2) dx = 2, \quad i \neq j. \tag{12.12}$$

Substituting (12.11) into (12.12), it follows that

$$\int_\Omega |\varphi^* - \varphi_j|^2 = 2, \quad \forall j.$$

In particular, if we take $j = j'$, then $\int_\Omega |\varphi^* - \varphi_j'|^2 dx = 2$, which contradicts (12.11). $\qquad\square$

Remark 12.1 According to the regularity theory of elliptical equations, the eigenfunctions $\varphi_1, \varphi_2, \ldots$ are indeed infinitely differentiable on Ω. Furthermore, if the boundary of Ω is smooth, then they are infinitely differentiable on $\bar{\Omega}$.

12.3 The eigenfunction expansions

We have established the collection of eigenfunctions $\{\varphi_i\}_1^\infty$ is an orthogonal family not only in $H_0^1(\Omega)$ but also in $L^2(\Omega)$. Thus, they form an orthogonal basis for $L^2(\Omega)$. We now show that this family is a complete orthogonal basis for the Hilbert space $L^2(\Omega)$.

$\forall u \in L^2(\Omega)$, let

$$c_n = \int_\Omega u(x)\varphi_n(x) dx, \quad \forall n,$$

and we call them generalized Fourier coefficients of u. Consider the partial sum

$$s_m(x) = \sum_{n=1}^m c_n \varphi_n(x);$$

by the completeness of $\{\varphi_i\}_1^\infty$, we mean as $m \to \infty$,

$$s_m(x) \to u(x) \quad (\text{in } L^2(\Omega)).$$

On one hand, by orthogonality,

$$\int_\Omega |\nabla(u - s_m)|^2\, dx = \int_\Omega |\nabla u|^2\, dx - 2\int_\Omega \nabla u \cdot \nabla s_m dx + \int_\Omega |\nabla s_m|^2\, dx$$

$$= \|\nabla u\|_2^2 - 2\sum_{n=1}^m \lambda_n |c_n|^2 + \sum_{n=1}^m \lambda_n |c_n|^2$$

$$= \|\nabla u\|_2^2 - \sum_{n=1}^m \lambda_n |c_n|^2, \tag{12.13}$$

it follows that

$$\sum_{n=1}^m \lambda_n |c_n|^2 \le \int_\Omega |\nabla u|^2\, dx. \tag{12.14}$$

On the other hand, since $u - s_m \in M_{m+1}$, we have

$$\int_\Omega |\nabla(u - s_m)|^2\, dx \ge \lambda_{m+1} \int_\Omega |u - s_m|^2\, dx.$$

Using (12.13), we deduce that

$$\|u - s_m\|_2^2 \le \frac{1}{\lambda_{m+1}} \|\nabla u\|_2^2 \to 0, \quad m \to \infty.$$

Hence, we have the Fourier expansion

$$u(x) = \sum_{n=1}^\infty c_n \varphi_n(x) \quad \text{(in the sense of } L^2 \text{ norm)}$$

and the Parseval's identity,

$$\int_\Omega |u|^2 dx = \lim_{m\to\infty} \|s_m\|_2^2 = \lim_{m\to\infty} \sum_{i,j=1}^m c_i c_j \int_\Omega \varphi_i \varphi_j dx = \sum_{m=1}^\infty |c_m|^2.$$

In addition, we can also obtain

(1) $s_m \to u$ in $H_0^1(\Omega)$,

(2) $\int_\Omega |\nabla u|^2 dx = \sum_{n=1}^\infty \lambda_n |c_n|^2$.

To prove (1), combining (12.13) and (12.14) and letting $m > n \to \infty$, it yields that

$$\|\nabla(s_m - s_n)\|_2^2 = \sum_{j=n+1}^m \lambda_j |c_j|^2 \to 0.$$

Hence, $s_m \to u$ in $H_0^1(\Omega)$. In the meantime,

$$\|\nabla u\|_2^2 = \lim_{m\to\infty} \|\nabla s_m\|_2^2 = \sum_{j=1}^\infty \lambda_j |c_j|^2.$$

Remark 12.2 In the special case of $n = 1$, we consider the following Sturm–Liouville problem. On the interval $J = [a, b]$, given functions $p \in C^1(J)$ and $q \in C(J)$. Suppose there exists a constant $\alpha > 0$ such that

$$p(x) \geq \alpha.$$

By further assuming $q(x) \geq 0, \ \forall \, x \in J$, we define the Sturm–Liouville operator to be

$$Lu = -(pu')' + qu.$$

Consider the following eigenvalue problem:

$$\begin{cases} Lu = \lambda u & \text{in } J, \\ u(a) = u(b) = 0. \end{cases}$$

As before, on $H_0^1(J)$, consider the functional

$$I(u) = \int_J \frac{1}{2}(p(t)|u'(t)|^2 + q(t)|u(t)|^2)dt.$$

However, we define an equivalent norm and inner product on $H_0^1(J)$ via I as follows:

$$\|u\| = \left(\int_J p(t)|u'(t)|^2 + q(t)|u(t)|^2 dt \right)^{\frac{1}{2}},$$

$$((u, v)) = \int_J p(t)u'(t)v'(t) + q(t)u(t)v(t)dt.$$

Inductively, by introducing the constraints

$$M_1 = \left\{ u \in H_0^1(J) \,\middle|\, \int |u|^2 = 1 \right\}, \quad M_2 = \left\{ u \in M_1 \,\middle|\, \int_J u\varphi_1 dx = 0 \right\}, \ldots,$$

we obtain the eigenvalues and eigenfunctions

$$0 < \lambda_1 \leq \lambda_2 \leq \cdots \leq \lambda_n \to +\infty,$$

and

$$\varphi_1, \varphi_2, \ldots, \varphi_n, \ldots.$$

It is not difficult to verify $\varphi_n \in C^2(J), \ \forall \, n$.
They satisfy

$$\int_J \varphi_n \varphi_m dx = \delta_{nm}, \quad \forall \, n, m$$

and

$$L\varphi_n = \lambda_n \varphi_n, \quad \forall \, n.$$

Furthermore, $\forall\, u \in H_0^1(J)$, its generalized Fourier expansion is

$$u(x) = \sum_{n=1}^{\infty} c_n \varphi_n(x),$$

which converges in the sense of $H_0^1(J)$, where

$$c_n = \int_J u(x)\varphi_n(x)dx.$$

Example 12.1 Let $p = 1$, $q = 0$, and $J = [0, \pi]$, then

$$Lu = -\ddot{u}.$$

It is easy to see

$$\lambda_n = n^2, \quad \varphi_n(x) = \sin nx, \quad n = 1, 2, \ldots.$$

Remark 12.3 Besides the Dirichlet problem, we can also consider eigenvalue problems with other types of boundary conditions. As an example, we consider the Neumann problem:

$$\begin{cases} -(pu')' + qu = \lambda u & \text{in } (a, b), \\ u'(a) = u'(b) = 0. \end{cases}$$

As mentioned in Lectures 1 and 6, for the Neumann problem, we shall use the space $H^1(\Omega)$ instead of $H_0^1(\Omega)$. The reason is as follows: using integration by parts:

$$\int_a^b (pu')'\varphi dx = (pu')\varphi|_a^b - \int_a^b (pu')\varphi' dx,$$

the integral form of the E-L equation turns out to be

$$\int_a^b (pu'\varphi' + (q - \lambda)u\varphi)dx = 0, \quad \forall\, \varphi \in H^1(J).$$

Simultaneously, we derive the following ordinary differential equation

$$-(pu')' + qu = \lambda u,$$

with boundary condition

$$u'(a) = u'(b) = 0.$$

However, on the space $H^1(J)$, the functional

$$I(u) = \int_a^b (p|u'|^2 + q|u|^2)dx$$

is not coercive. We use the subspace

$$X = \left\{ u \in H^1(J) \;\middle|\; \int_a^b u\,dx = 0 \right\},$$

i.e. replacing $H_0^1(J)$ by the subspace which is orthogonal to all constant functions.

On the subspace X, Poincaré's inequality still remains valid. The proof is identical, since $\int_a^b u\,dx = 0$, there exists $\xi \in (a,b)$ such that $u(\xi) = 0$. Thus,

$$|u(x)| = \left| \int_\xi^x u'(t)\,dt \right| \le (b-a)^{\frac{1}{2}} \left(\int_a^b |u'|^2 dt \right)^{\frac{1}{2}}.$$

We can now minimize I on the subspace X. In other words, we insert, on $H^1(J)$, an integral constraint

$$\int_J u(t)\,dt = 0.$$

It is worth noting the Lagrange multiplier associated with this constraint will naturally disappear in the equation.

Note that when $q = 0$, a nonzero constant is itself an eigenfunction of the eigenvalue 0. Under the Neumann boundary condition, the Sturm–Liouville problem has eigenvalues $\{0, \lambda_1, \lambda_2, \ldots\}$ and eigenfunctions $\{1/\sqrt{(b-a)}, \varphi_1, \varphi_2, \ldots\}$.

Next, we consider the eigenvalue problem for the Laplace operator with Neumann boundary condition:

$$\begin{cases} \Delta u = \lambda u & \text{in } \Omega, \\ \left. \dfrac{\partial u}{\partial n} \right|_{\partial \Omega} = 0, \end{cases}$$

where n denotes the unit normal direction on $\partial \Omega$. First, we introduce the subspace

$$X = \left\{ u \in H^1(\Omega) \;\middle|\; \int_\Omega u(x)\,dx = 0 \right\}.$$

We then extend Poincaré's inequality to the subspace X; (after inserting an integral constraint), we minimize $I(u) = \int_\Omega |\nabla u|^2 dx$ on X.

Remark 12.4 The same method applies to the eigenvalue problem of the Laplace–Beltrami operator on a closed (no boundary) compact Riemannian manifold (M, g):

$$\Delta_g u = \frac{1}{\sqrt{g}} \sum_{ij} \partial_i (g^{ij} \sqrt{g} \partial_j u),$$

again we denote $g = \det(g_{ij})$, without any boundary condition.

In particular, it is worth noting the following eigenvalue problem with periodic boundary condition:

$$\begin{cases} -(pu')' + qu = \lambda u & \text{in } (a,b), \\ u(a) = u(b), \ u'(a) = u'(b), \end{cases}$$

can also be regarded as eigenvalue problem on the closed compact manifold S^1. In this case, we replace $H_0^1(J)$ by $H_{\text{per}}^1(J)$, the space of periodic functions; we replace Poincaré's inequality by Wirtinger's inequality (see Lemma 13.4 in Lecture 13), which can be used in verifying the coerciveness of a functional.

Remark 12.5 The above steps used in finding the eigenvalues and eigenfunctions coincide with the geometric approach of diagonalizing a quadratic form determined by a quadratic hyper-surface in $\mathbb{R}^n$. In which, $\lambda_1^{-1}, \lambda_2^{-1}, \ldots$ are the principal axes.

12.4 The minimax description of eigenvalues

Our previous description of the eigenvalues $\{\lambda_n\}$ is inductive; in other words, assuming the $n-1$ eigenfunctions $\varphi_1, \ldots, \varphi_{n-1}$ have been found, then we can determine λ_n. The following min-max theorem gives a direct approach to find λ_n.

Theorem 12.2 (Courant's Min-Max Theorem)

$$\lambda_n = \max_{E_{n-1}} \ \min_{u \in E_{n-1}^\perp \setminus \{\theta\}} \frac{\displaystyle\int_\Omega |\nabla u|^2 \, dx}{\displaystyle\int_\Omega |u|^2 \, dx},$$

where E_{n-1} is any $(n-1)$-dimensional linear subspace of $H_0^1(\Omega)$.

Proof Assume $v_1, \ldots, v_{n-1} \in H_0^1(\Omega)$ is a collection of linearly independent functions. Let

$$E_{n-1} = \text{span}\{v_1, v_2, \ldots, v_{n-1}\}$$

and

$$\mu(E_{n-1}) = \min_{u \in E_{n-1}^\perp \setminus \{\theta\}} \frac{\displaystyle\int_\Omega |\nabla u|^2 \, dx}{\displaystyle\int_\Omega |u|^2 \, dx}.$$

On one hand, we will prove

$$\mu(E_{n-1}) \leq \lambda_n.$$

Let $\{\varphi_1, \ldots, \varphi_n\}$ be the first n eigenfunctions, then $(E_{n-1}^{\perp} \backslash \{\theta\}) \cap \text{span}\{\varphi_1, \ldots, \varphi_n\} \neq \emptyset$, i.e. there exists a nonzero u such that

$$\begin{cases} u = \displaystyle\sum_{i=1}^{n} c_i \varphi_i, \\ \displaystyle\int_{\Omega} u v_j dx = 0, \end{cases} \quad j = 1, \ldots, n-1,$$

or

$$\sum_{i=1}^{n} c_i \int_{\Omega} \varphi_i v_j dx = 0 \quad j = 1, \ldots, n-1.$$

This system of $(n-1)$ linear equations in the n unknowns $c_1, \ldots, c_n$ with coefficients $\int_{\Omega} \varphi_i v_j dx$ must possess a nontrivial solution. From $\lambda_1 \leq \lambda_2 \leq \cdots \leq \lambda_n$, it follows that

$$\mu(E_{n-1}) \leq \frac{\displaystyle\int_{\Omega} |\nabla u|^2 dx}{\displaystyle\int_{\Omega} |u|^2 dx} = \frac{\displaystyle\sum_{i=1}^{n} \lambda_i |c_i|^2}{\displaystyle\sum_{i=1}^{n} |c_i|^2} \leq \lambda_n.$$

On the other hand, if we choose the particular subspace $\tilde{E}_{n-1} = \text{span}\{\varphi_1, \ldots, \varphi_{n-1}\}$, then

$$\max_{E_{n-1}} \mu(E_{n-1}) \geq \mu(\tilde{E}_{n-1}) = \lambda_n.$$

This completes the proof. $\qquad\qquad\qquad\qquad\qquad\qquad\qquad\qquad\qquad\square$

Exercises

1. Let $\Omega \subset \mathbb{R}^n$ be bounded. If $y \in C(\bar{\Omega})$ is a positive continuous function, prove that

$$\begin{cases} -\Delta u(x) = \lambda r(x) u(x), & x \in \Omega, \\ u(x) = 0, & x \in \partial\Omega, \end{cases}$$

has infinitely many eigenvalues and eigenfunctions. Furthermore, the eigenfunctions are mutually orthogonal with respect to the inner product

$$(u, v) = \int_{\Omega} u(x) v(x) r(x) dx.$$

2. We adopt the above notations. Assume $a_{ij} \in C^1(\bar{\Omega})$, $c \in C(\bar{\Omega})$, if there exists $\alpha > 0$ such that

$$\sum_{i,j=1}^{n} a_{i,j}(x) \xi_i \xi_j \geq \alpha |\xi|^2, \quad \forall \xi = (\xi_1, \ldots, \xi_n) \in \mathbb{R}^n,$$

prove that

$$\begin{cases} -\sum_{i,j=1}^{n} \partial_i(a_{i,j}(x)\partial_j u(x)) + c(x)u(x) = \lambda u(x), & x \in \Omega, \\ u(x) = 0, & x \in \partial\Omega, \end{cases}$$

has infinitely many eigenvalues and eigenfunctions.

3. Let H be a Hilbert space, $\alpha : H \times H \to \mathbb{R}^1$ be a bounded bilinear functional, and $V \subset H$ be a linear subspace. Suppose there exists $c > 0$ such that

$$a(v,v) \geq c\|v\|^2, \quad \forall v \in V,$$

prove that $(V, a(\cdot, \cdot))$ is a Hilbert space.

4. In exercise 2 above, assume $f \in L^2(\Omega)$, show that

$$\begin{cases} -\sum_{i,j=1}^{n} \partial_i(a_{i,j}(x)\partial_j u(x)) + c(x)u(x) = f(x), & x \in \Omega, \\ u(x) = 0, & x \in \partial\Omega, \end{cases}$$

has a unique solution $u \in H_0^1(\Omega)$.

Lecture 13

Existence and regularity

In the previous two lectures, we turned the existence of solutions of an E-L equation into a problem of finding extreme values of a functional. Under certain conditions, the minimum u can be obtained by a minimizing sequence in some Sobolev space $W^{1,q}(\Omega)$, which satisfies the following equation only in the generalized sense:

$$\int_\Omega L_u(x, u(x), \nabla u(x))\varphi(x) + L_p(x, u(x), \nabla u(x))\nabla\varphi(x)dx = 0, \ \forall\varphi \in C_0^\infty(\Omega).$$

That is, the minimum is a generalized solution of the corresponding E-L equation.

However, for a functional containing first order derivatives, its E-L equation is a second order differential equation. A generalized solution is a solution in the ordinary sense only if it is twice differentiable. From a differential equation perspective, we need also address whether such a generalized solution would have enough differentiability to fulfill the E-L differential equation. In other words, is it possible to deduce $u \in C^2$ from the generalized solution u? Or perhaps more differentiability, or even analyticity? This is the so-called regularity problem.

We call the problem of finding a generalized solution an "existence" problem, and call the problem of determining the differentiability of a generalized solution a "regularity" problem.

That being said, in direct methods, the "existence" and "regularity" are twin problems.

Among Hilbert's 23 problems, problem 19 inquires the analyticity of solutions of some regular variational problems (in fact, it is about elliptical equations), whereas problem 20 is related to the solvability of a regular variational problem with boundary conditions (namely, the existence).

179

13.1   Regularity ($n = 1$)

In the following, we shall henceforth assume the Lagrangian $L \in C^2$, we ask: when does the minimum u^* of functional I belong to $C^2(J)$?

Lemma 13.1 *Suppose $u^* \in C^1(J)$ is a minimum of $I(u) = \int_J L(t, u(t), \dot{u}(t))dt$. If $\det(L_{p_i p_j}(t, u^*(t), \dot{u}^*(t))) \neq 0, \forall t \in J$, then $u^* \in C^2(J)$.*

Proof Denote

$$q(t) = \int_a^t L_u(t, u^*(t), \dot{u}^*(t))dt - c,$$

where c is a constant. Using the integral form of the E-L equation, we have

$$L_p(t, u^*(t), \dot{u}^*(t)) = q(t).$$

Define $\varphi : \bar{J} \times \mathbb{R}^N \to \mathbb{R}^N$ via

$$\varphi(t, p) = L_p(t, u^*(t), p) - q(t).$$

We know that $\varphi \in C^1(\bar{J} \times \mathbb{R}^N, \mathbb{R}^N)$ satisfies

$$\det(\varphi_p(t, u^*(t))) = \det(L_{pp}(t, u^*(t), \dot{u}^*(t))) \neq 0$$

and

$$\varphi(t, \dot{u}^*(t)) = 0.$$

By the Implicit Function Theorem, the equation

$$\varphi(t, p) = 0$$

has a unique local C^1-solution, i.e. $\forall t_0 \in \bar{J}$, there exists a neighborhood $U = U(t_0)$ and a unique $\lambda \in C^1(U, \mathbb{R}^N)$ such that in the neighborhood of $(t, \dot{u}^*(t))$, $\varphi(t, \lambda(t)) = 0$.

Thus, $\dot{u}^*(t) = \lambda(t) \in C^1$, which implies $u^* \in C^2$. $\square$

The condition $\det(L_{p_i p_j}(t, u^*(t), \dot{u}^*(t))) \neq 0$ plays an important role; for otherwise, there exists a functional whose minimum is of C^1 but not of C^2.

Example 13.1   Let $L(t, u, p) = u^2(p - 2t)^2$ and $M = \{u \in C^1([-1, 1]) \mid u(-1) = 0, u(1) = 1\}$, then the functional

$$I(u) = \int_{-1}^1 u^2(\dot{u}(t) - 2t)^2 dt$$

has a minimum

$$u^*(t) = \begin{cases} 0, & t < 0, \\ t^2, & t \geq 0. \end{cases}$$

It is clear $u^* \in C^1 \backslash C^2([-1, 1])$. Moreover, $L_{pp}(t, u^*(t), \dot{u}^*(t)) = 2u^*(t)^2 = 0$, for $t < 0$.

The above lemma elevated the solution from being C^1 to being C^2, based on the Implicit Function Theorem. However, it is not enough to insure the regularity of the solution. In Lecture 9, we pointed out that the space C^1 is not the suitable function space for direct methods. Using direct methods, we can only obtain a solution in some Sobolev space. In order to obtain its regularity, we have to further establish that the generalized solution is indeed a C^1 solution.

Theorem 13.1 *For $1 < r < \infty$, assume L satisfies the following growth condition:*

$$|L(t, u, p)| + |L_u(t, u, p)| + |L_p(t, u, p)| \le C(1 + |p|^r);$$

for $r = \infty$, no growth condition on L is needed.

Furthermore, assume the matrix $(L_{p_i p_j}(t, u, p))$ $\forall (t, u, p) \in \bar{J} \times \mathbb{R}^N \times \mathbb{R}^N$ is positive definite.

If $u^* \in W^{1,r}(J, \mathbb{R}^N)$ is a minimum of the functional $I(u) = \int_\Omega L(t, u(t), \dot{u}(t))dt$, then by changing the values of u^* on a set of measure zero, $u^* \in C^2$.

Proof By Lemma 13.1, it suffices to show that $u^* \in C^1$.

Define the function $\varphi : J \times \mathbb{R}^N \times \mathbb{R}^N \times \mathbb{R}^N \to \mathbb{R}^1$ by

$$\varphi(t, u, p, q) = L_p(t, u, p) - q.$$

By assumption, $\det(L_{p_i p_j}(t, u, p)) \ne 0$, $\forall (t, u, p) \in \bar{J} \times \mathbb{R}^N \times \mathbb{R}^N$. To solve the equation: $\varphi(t, u, p, q) = 0$, we apply the Implicit Function Theorem to see if there exists a unique local C^1-solution

$$p = \lambda(t, u, q). \tag{13.1}$$

On one hand, we show that this solution is globally unique. Suppose p_1, p_2 both satisfy (13.1), then $q = L_p(t, u, p_1) = L_p(t, u, p_2)$. Hence,

$$0 = (L_p(t, u, p_1) - L_p(t, u, p_2), p_1 - p_2) = (B(p_1 - p_2), p_1 - p_2),$$

where

$$B = \int_0^1 L_{pp}(t, u, p_1 + \tau(p_2 - p_1))d\tau.$$

By assumption, B is positive definite, therefore $p_1 = p_2$.

On the other hand, since $u^* \in W^{1,r}(J, \mathbb{R}^N)$, when $1 < r < \infty$, $L_u(t, u^*(t), \dot{u}^*(t)) \in L^1(J)$ and when $r = \infty$, $L_u(t, u^*(t), \dot{u}^*(t)) \in L^\infty(J)$. In any case,

$$q(t) = \int_{t_0}^t L_u(s, u^*(s), \dot{u}^*(s))ds - c \text{ is absolutely continuous.}$$

By the integral form of the E-L equation,

$$\int_{t_0}^{t} L_u(t, u^*(t), \dot{u}^*(t))dt - L_p(t, u^*(t), \dot{u}^*(t)) = c, \quad \text{a.e. } t \in J,$$

it follows that

$$q(t) = L_p(t, u^*(t), \dot{u}^*(t)), \quad \text{a.e. } t \in J$$

and

$$\dot{u}^*(t) = \lambda(t, u^*(t), q(t)), \quad \text{a.e. } t \in J.$$

Since $q(t)$ is absolutely continuous, substituting it into the above expression, by changing the values of u^* on a set of measure zero, we have $\dot{u}^*(t)$ is continuous, i.e. $u^* \in C^1$. Thus, $u^* \in C^2$ follows readily from Lemma 13.1.

In the above proof, the global positive definiteness of the matrix

$$(L_{p_i p_j}(t, u, p)), \qquad \forall \, (t, u, p) \in \bar{J} \times \mathbb{R}^N \times \mathbb{R}^N$$

played a crucial role in establishing both the global uniqueness of the solution as well as its regularity. This is because the derivatives of functions in the Sobolev space $W^{1,r}$ may be discontinuous. Along the graphs of these functions, two timewise nearby points may fall in different image neighborhoods. Since the Implicit Function Theorem only works for neighborhoods in the image space, it is no longer applicable.

The following example demonstrates, by removing the global positive definiteness assumption, the solution becomes non-differentiable.

Example 13.2 Let $L(p) = (p^2 - 1)^2$ and $M = \{u \in \text{Lip}([0, 1]) \,|\, u(0) = u(1) = 0\}$, then the functional

$$I(u) = \int_{0}^{1} (\dot{u}^2(t) - 1)^2 dt$$

has minimal value 0.

Note that $L_{pp}(t, u, p) = 4(3p^2 - 1)$ is not positive definite.

If $u \in C^1$ is a minimum, then $\dot{u}(t) = \pm 1$. However, regardless of $\dot{u}(t) = 1$ or $\dot{u}(t) = -1$, it is impossible to have a solution satisfying the boundary condition $u(0) = u(1) = 0$. In other words, there cannot be a C^1-solution to achieve the minimal value. On the contrary, there are uncountably many solutions of this variational problem, which are Lipschitz sawtooth-like functions satisfying $\dot{u}(t) = \pm 1$ (see Figure 13.1).

Lastly, we give an example to show that without the convexity of the Lagrangian, the weak sequential lower semi-continuity of the functional may not hold.

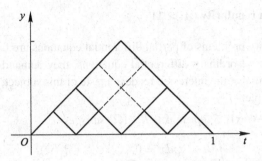

Fig. 13.1

Example 13.3 (Bolza) Let $L(u, p) = u^2 + (p^2 - 1)^2$ and $M = W_0^{1,4}(0, 1)$, the the functional

$$I(u) = \int_0^1 [u^2(t) + (\dot{u}(t)^2 - 1)^2] dt$$

has infimum zero, but I has no minimum in M.

To see this, note on one hand, $I(u) \geq 0$; on the other hand, we define a minimizing sequence of sawtooth-like functions:

$$u_j(t) = \begin{cases} t - \dfrac{k}{j}, & t \in \left[\dfrac{k}{j}, \dfrac{2k+1}{2j}\right], \\ -t + \dfrac{k+1}{j}, & t \in \left[\dfrac{2k+1}{2j}, \dfrac{k+1}{j}\right], \end{cases}$$

where $j = 2, 3, \ldots$ and $0 \leq k \leq j - 1$.

Since $|\dot{u}_j(t)| = 1$ a.e. and $|u_j| \leq \frac{1}{2j}$, we see that

$$I(u_j) \leq \frac{1}{4j^2} \to 0, \quad j \to \infty.$$

Thus, $\inf_{u \in M} I = 0$.

However, I has no minimum in $W_0^{1,4}(0, 1)$. Suppose not, there exists $u_0 \in W_0^{1,4}(0, 1)$ such that $I(u_0) = 0$, hence $\dot{u}_0(t) = 0$ a.e. and $I(u_0) = 1$, a contradiction (see Figure 13.2).

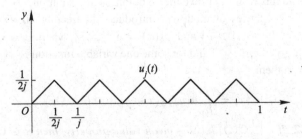

Fig. 13.2

13.2   More on regularity ($n > 1$)

The regularity problems of partial differential equations are far more complicated than those of ordinary differential equations, they demand special knowledge. In order to give the interested readers a taste of this subject, we provide the following example.

Theorem 13.2 (Weyl) *Suppose* $u \in L^1_{\text{loc}}(\Omega)$ *satisfies*

$$\int_\Omega u \cdot \Delta\varphi dx = 0, \quad \forall\varphi \in C_0^\infty(\Omega), \tag{13.2}$$

then after changing its value on a set of measure zero, $u \in C^\infty(\Omega)$.

We call those locally L^1 functions which satisfy Eq. (13.2) *weakly harmonic functions*. The idea of the proof originates from the mean value property of harmonic functions.

If $u \in C^2(\Omega)$ and $\Delta u = 0$ in Ω, then

$$u(x) = \frac{1}{r^{n-1}\varpi_n}\int_{\partial B_r(x)} u(y)d\sigma = \frac{n}{r^n\varpi_n}\int_{B_r(x)} u(y)dy, \quad \forall B_r(x) \subset \Omega, \tag{13.3}$$

where ϖ_n denotes the surface area of the unit sphere $S \subset \mathbb{R}^n$. We call (13.3) the mean value property of u.

To verify (13.3), we assume $B_r(x) \subset \Omega$ and choose $\rho \in (0, r)$, then

$$0 = \int_{B_\rho(x)} \Delta u(y)dy = \int_{\partial B_\rho(x)} \frac{\partial u}{\partial n}d\sigma = \rho^{n-1}\frac{\partial}{\partial\rho}\Big|_{|w|=1}\int u(x + \rho w)dw.$$

This implies

$$\int_{|w|=1} u(x + \rho w)dw = \text{const.} = \varpi_n u(x),$$

hence,

$$u(x) = \frac{1}{r^{n-1}\varpi_n}\int_{\partial B_r(x)} u(y)d\sigma = \frac{n}{r^n\varpi_n}\int_{B_r(x)} u(y)dy.$$

We now extend this result to prove the smoothness of a solution.

Recall the properties of modifiers introduced in Lecture 10: $\forall\delta > 0$, let $\Omega_\delta = \{x \in \Omega \mid d(x, \partial\Omega) > \delta\}$ and $\varphi_\delta(x) = \delta^{-n}\varphi(\frac{x}{\delta})$, where $\varphi \in C_0^\infty(B_1(\theta))$ satisfying $\int_{B_1(\theta)} \varphi(x)dx = 1$ and for some one variable function ρ, $\varphi(rw) = \rho(r)$ for all $|w| = 1$, then

$$u_\delta(x) = \int_\Omega u(y)\varphi_\delta(x - y)dy \in C^\infty(\Omega_\delta).$$

Lemma 13.2 *If* $u \in C(\Omega)$ *has the mean value property, then* $u \in C^\infty(\Omega)$ *and* $u(x) = u_\delta(x), \forall x \in \Omega_\delta$.

Proof $\forall x \in \Omega_\delta$, by the mean value property of u, we have:

$$u_\delta(x) = \delta^{-n} \int_{|y| \le \delta} u(x+y)\varphi(\frac{y}{\delta})dy$$

$$= \int_{|y| \le 1} u(x+\delta y)\varphi(y)dy$$

$$= \int_0^1 r^{n-1}\rho(r) \int_{|w|=1} u(x+\delta rw)dwdr = u(x).$$

Since $u_\delta \in C^\infty(\Omega_\delta)$ and $\delta > 0$ is arbitrary, $u \in C^\infty(\Omega)$. $\qquad \square$

Lemma 13.3 *If $u \in L^1(\Omega)$ is a weakly harmonic function, i.e. it satisfies* (13.2), *then $u \in C^\infty(\Omega)$ and it has the mean value property.*

Proof (1) We first prove $\Delta u_\delta(x) = 0, \forall x \in \Omega_{\delta_0}, 0 < \delta < \delta_0$.

$\forall \psi \in C_0^\infty(\Omega_{\delta_0})$, since supp $\psi_\delta \subset \Omega$, we have

$$\int_{\Omega_{\delta_0}} u_\delta(x)\Delta\psi(x)dx = \int_{\mathbb{R}^n} u_\delta(x)\Delta\psi(x)dx$$

$$= \int_{\mathbb{R}^n} \int_{\mathbb{R}^n} u(y)\varphi_\delta(x-y)\Delta\psi(x)dx$$

$$= \int_\Omega u(y)(\Delta\psi)_\delta(y)dy$$

$$= \int_\Omega u\Delta(\psi_\delta)dy = 0.$$

Thus, $\Delta u_\delta(x) = 0$ in Ω_{δ_0}.

(2) If we can show that $\{u_\delta\}$ is uniformly bounded and equicontinuous on $\bar{\Omega}_{2\delta_0}$, then by the Arzelà–Ascoli theorem, there must be a subsequence which converges uniformly to a continuous function v. Since $u_\delta \to u$ in $L^1(\Omega_{\delta_0})$, $u(x) = v(x)$ a.e. in Ω. Furthermore, since u_δ is harmonic, it has the mean value property. Upon taking limits, v also possesses the mean value property. By Lemma 13.2, it is immediate $v \in C^\infty(\Omega)$.

(3) We now return to prove that $\{u_\delta\}$ is uniformly bounded and equicontinuous on $\bar{\Omega}_{2\delta_0}$. By the mean value property of u_δ,

$$|u_\delta(x)| \le \frac{n}{\delta_0^n \varpi_n} \int_{B_{\delta_0}(x)} |u_\delta(y)|dy \le C_{\delta_0}\|u_\delta\|_1 \le C_{\delta_0}\|u\|_1$$

and

$$|u_\delta(x) - u_\delta(y)| \le \left| \int_\Omega u(\xi) \int_{B_{\delta_0}(\theta)} [\varphi(x+\delta(z-\xi)) - \varphi(y+\delta(z-\xi))]d\xi dz \right|$$

$$\le C_{\delta_0}|x-y|\|u\|_1. \qquad \square$$

Remark 13.1 Theorem 13.2 asserts a generalized solution is smooth in Ω. However, to further examine the smoothness on $\bar{\Omega}$, it becomes exceedingly more difficult, which is beyond the scope of this book. The study of regularity problems of partial differential equations is a specialized area of its own. There are ample textbooks, monographs, and references on this subject; we omit any further discussion and refer the interested readers to the introductory book by D. Gilbarg and N. Trudinger [GT].

For linear strongly elliptic equations, assuming the domain and coefficients are sufficiently smooth, solutions of regular boundary value problems have enough differentiability. The most notable work is due to J. Schauder, S. Agmon, A. Douglise, and L. Nirenberg. For quasi-linear elliptical equations, the most notable work is due to J. Nash, E. De Giorgi, O. A. Ladyzenskaya, and N. N. Uraltzeva. For a system of elliptical equations, generally speaking, its solutions may not possess regularity, cf. M. Giaquinta [Gi].

13.3 The solutions of some variational problems

In this section, we examine the existence and regularity of solutions of variational problems via the following examples.

Example 13.4 (The two-point boundary value problem) Denote the interval $J = (t_0, t_1)$ and the 2-torus $T^2 = \mathbb{R}^2/\mathbb{Z}^2$. Given $a_0, a_1 \in T^2$ and $F \in C^2(J \times T^2)$, find a solution $u \in C^2(J, T^2)$ satisfying the boundary conditions $u(t_i) = a_i$, $i = 0, 1$, and the equation:

$$\ddot{u}(t) = \nabla_u F(t, u(t)). \tag{13.4}$$

Solution In order to turn this into a standard variational problem, we first extend the function F to the entire $\mathbb{R}^2$, namely,

$$F(t, u_1 + 1, u_2) = F(t, u_1, u_2 + 1) = F(t, u_1, u_2), \ \forall t \in J.$$

We still denote it by F. Define the functional and the corresponding space by

$$I(u) = \int_J \left[\frac{1}{2} |\dot{u}(t)|^2 + F(t, u(t)) \right] dt,$$

$$M = u_0 + H_0^1(J, \mathbb{R}^2), \quad u_0 = \frac{a_0(t_1 - t) + a_1(t - t_0)}{t_1 - t_0}.$$

Its E-L equation is exactly (13.4).

Denote $u = u_0 + v$, and we use $\left(\int_J |\dot{v}|^2 \right)^{\frac{1}{2}}$ as the norm on $H_0^1(J, \mathbb{R}^2)$.

Since F is bounded, I is coercive. By the compact embedding $H^1(J, \mathbb{R}^2) \hookrightarrow C(J, \mathbb{R}^2)$ and the continuity of F, I is also weakly sequentially lower semicontinuous. Furthermore, since M is weakly closed, by the existence theorem,

I attains its minimum u on M. Noticing $(L_{p_i p_j}) = \text{Id}$, so it is positive definite. By Theorem 13.1 on regularity, $u \in C^2$ and clearly it satisfies (13.4). □

Example 13.5 (Periodic solutions of forced oscillations) Assume $e \in C[0, T]$ has mean value 0 and period T: $\int_0^T e(t)dt = 0$, and a is a constant. Find the periodic solutions of period $T > 0$ of the following equation:

$$\ddot{u}(t) + a \sin u(t) = e(t).$$

Solution We define $H^1_{\text{per}}(0, T)$ to be the Sobolev space of all $H^1(0, T)$ functions with period $T > 0$, i.e. the closure of T-periodic C^∞ functions in $H^1(0, T)$. Define the functional

$$I(u) = \int_0^T \left(\frac{1}{2} |u'(t)|^2 + a \cos u(t) - E(t)u'(t) \right) dt,$$

where

$$E(t) = \int_0^t e(s)ds.$$

E is also a function of period $T > 0$. Since

$$\int_0^T E(t)u'(t)dt = - \int_0^T e(t)u(t)dt,$$

the E-L equation of I is precisely

$$\ddot{u}(t) + a \sin u(t) = e(t).$$

To verify I attains its minimum in $H^1_{\text{per}}(0, T)$, similar to Poincaré's inequality, we need the following lemma.

Lemma 13.4 (Wirtinger's inequality) *Suppose* $u \in H^1_{\text{per}}(0, T)$ *and* $\bar{u} = \frac{1}{T} \int_0^T u(t)dt = 0$, *then*

$$\int_0^T |u'(t)|^2 dt \geq \frac{4\pi^2}{T^2} \int_0^T |u(t)|^2 dt.$$

Proof On $[0, T]$, we expand the periodic function u as its Fourier series. Since $\bar{u} = 0$,

$$u(t) = \sum_{k=1}^\infty a_k \cos \frac{2\pi kt}{T} + b_k \sin \frac{2\pi kt}{T},$$

hence,

$$u'(t) = \frac{2\pi}{T} \sum_{k=1}^\infty -ka_k \sin \frac{2\pi kt}{T} + kb_k \cos \frac{2\pi kt}{T}.$$

By Parseval's identity,

$$\int_0^T |u'(t)|^2 dt = \frac{(2\pi)^2}{T} \sum_{k=1}^{\infty} k^2 (|a_k|^2 + |b_k|^2)$$

$$\geq \frac{(2\pi)^2}{T} \sum_{k=1}^{\infty} (|a_k|^2 + |b_k|^2)$$

$$= \frac{(2\pi)^2}{T^2} \int_0^T |u(t)|^2 dt. \qquad \square$$

$\forall u \in H^1_{\text{per}}(0, T)$, we decompose

$$u = \tilde{u} + \bar{u},$$

where $\bar{u} = \frac{1}{T} \int_0^T u(t) dt$ is a real number. From Wirtinger's inequality, we see that $\bar{u}$ is not controlled by the values of I. In other words, if we use the H^1-norm directly on $H^1_{\text{per}}(0, T)$, then I is not coercive.

Moreover, since the nonlinear term $\cos u$ in I is 2π-periodic, we have

$$I(u + 2\pi) = I(u).$$

By which, we need not consider I on the entire $H^1_{\text{per}}(0, T)$, but instead, by setting

$$M = \{u = \xi + \eta \mid \xi \in H^1_{\text{per}}(0, T),\ \bar{\xi} = 0, \eta \in [0, 2\pi]\};$$

we then restrict I on M. The advantage is that $\bar{u}$ now only varies on the bounded interval $[0, 2\pi]$.

Noting that M is weakly sequentially closed and

$$I(u) \geq \frac{1}{2} \|\dot{\xi}\|_2^2 - \|E\|_\infty \|\dot{\xi}\|_2 - |a|;$$

by Wirtinger's inequality, $\|\dot{\xi}\|_2$ is an equivalent norm on H^1. I is coercive on M. It is not difficult to verify that I is also weakly lower semi-continuous. According to the existence theorem, there exists a minimum $u \in M \subset H^1_{\text{per}}$. Moreover, since all conditions in Theorem 13.1 on regularity are met, it follows that $u \in C^2$.

$\qquad \square$

Example 13.6 Let $\Omega \subset \mathbb{R}^n$ be a bounded region, $1 < r < p < \infty$, and $f \in L^{\frac{p}{p-r}}(\Omega)$, then

$$I(u) = \int_\Omega \left(\frac{1}{p} |\nabla u(x)|^p - \frac{1}{r} f(x) \cdot |u(x)|^{r-1} u(x) \right) dx, \quad \forall u \in H^1_0(\Omega)$$

has a minimum $u_0 \in W^{1,p}_0(\Omega)$, and it satisfies

$$\int_\Omega [|\nabla u(x)|^{p-2} \nabla u(x) \nabla \varphi(x) - |u|^{r-1} f(x) \varphi(x)] dx = 0, \quad \forall \varphi \in C^\infty_0(\Omega).$$

This is a generalized solution of the equation

$$-\Delta_p u = f|u|^{r-1},$$

where

$$-\Delta_p u = \sum_{i=1}^{n} \partial_i(|\nabla u|^{p-2})\partial_i u$$

is called the p-Laplace operator.

Proof Denote $\|u\| = \left(\int_\Omega |\nabla u(x)|^p dx\right)^{\frac{1}{p}}$, we know that it is an equivalent norm on $W_0^{1,p}(\Omega)$.

(1) We claim I is coercive.

By Poincaré's inequality, Hölder's inequality, and Young's inequality, $\exists C > 0$ such that

$$\int_\Omega |f||u|^r dx \leq \|f\|_{\frac{p}{p-r}} \|u\|_p^r \leq C\|f\|_{\frac{p}{p-r}} \|u\|^r \leq C_\epsilon \|f\|_{\frac{p}{p-r}}^{\frac{p}{p-r}} + r\epsilon\|u\|^p.$$

Thus,

$$I(u) \geq \left(\frac{1}{p} - \epsilon\right)\|u\|^p - \frac{C_\epsilon}{r}\|f\|_{\frac{p}{p-r}}^{\frac{p}{p-r}} \to +\infty, \quad \text{as } \|u\| \to \infty.$$

(2) Since $\int_\Omega |\nabla u(x)|^p dx$ is convex and lower semi-continuous, it is weakly sequentially lower semi-continuous. Using the Rellich–Kondrachov compact embedding theorem, the latter term is also weakly sequentially lower semi-continuous. Thus, I is weakly sequentially lower semi-continuous. $\square$

Remark 13.2 Generally speaking, it is not possible to obtain $u \in C^2$. However, for $p = 2$, $r = 1$, if $f \in C^\gamma(\bar{\Omega})$ $(0 < \gamma < 1)$ is a Hölder function and if $\partial\Omega$ is sufficiently smooth, then using the Schauder's estimate, one can prove $u \in C^{2,\gamma}(\bar{\Omega})$. When $p = 2$, $r = 1$, for $f \in C(\bar{\Omega})$, this is the Poisson equation.

Example 13.7 (Harmonic mappings) Let $\Omega \subset \mathbb{R}^n$ be a bounded open set with smooth boundary $\partial\Omega$. Denote the unit sphere in $\mathbb{R}^{n+1}$ by S^n. Consider the mapping $u = (u^1, \ldots, u^{n+1}) : \Omega \longrightarrow S^n \subset \mathbb{R}^{n+1}$. Let $\varphi = (\varphi^1, \ldots, \varphi^{n+1}) : \partial\Omega \longrightarrow S^n$ be a C^1 mapping defined on the boundary. We define the set

$$M = \{u \in H^1(\Omega, \mathbb{R}^{n+1}) \mid u(x) \in S^n, \text{ a.e. } x \in \Omega, \ u|_{\partial\Omega} = \varphi\},$$

and we want to find

$$\inf\{E(u) \mid u \in M\},$$

where

$$E(u) = \frac{1}{2}\int_\Omega |\nabla u|^2 dx.$$

Since E is convex and lower semi-continuous, it is weakly sequentially lower semi-continuous. Furthermore, E is also coercive.

We now verify M is weakly sequentially closed. If

$$u_j \to u \left(H^1(\Omega, \mathbb{R}^{n+1}) \right),$$

then there exists a subsequence $u_{j'} \to u$ in $L^2(\Omega, \mathbb{R}^{n+1})$ and a further subsequence $u_{j'}(x) \to u(x)$ a.e.

From $u_j(x) \in S$ a.e., it follows that $u(x) \in S^n$ a.e., whence $u \in M$. Consequently, there exists a minimum $u_0 \in M$ satisfying the E-L equation:

$$\int_\Omega \nabla u_0(x) \nabla v(x) dx = 0,$$

$\forall v \in H_0^1(\Omega, \mathbb{R}^{n+1})$ satisfying $v(x) \in T_{u_0(x)} S^n$ a.e. in Ω, where $T_u S^n$ is the tangent space to S^n at $u \in S^n$.

If $u_0 \in C^2(\Omega, \mathbb{R}^{n+1})$, then

$$(\Delta u_0)^T(x) = 0, \quad \text{a.e. in } \Omega,$$

where $(\Delta u_0)^T(x)$ is the tangential projection of $\Delta u_0(x)$ at $u_0(x)$.

The normal projection $(\Delta u_0)^N(x)$ of $\Delta u_0(x)$ at $u_0(x)$ is given by

$$(\Delta u_0)^N(x) = \Delta u_0(x) \cdot u_0(x).$$

Differentiating the equation $|u(x)|^2 = 1$ twice, it yields

$$\Delta u(x) \cdot u(x) = -|\nabla u(x)|^2,$$

thus, u_0 satisfies the equation

$$-\Delta u(x) = u(x)|\nabla u(x)|^2.$$

This is the so-called harmonic mapping equation (cf. Example 7.5 in Lecture 7).
$\square$

Remark 13.3 When $m = 2$, Morrey proved that u_0 is smooth. However, for $m > 2$, F. H. Lin proved that $u_0 = 1/|x|$; in which case, although u_0 is a minimum, it still has singularity.

Example 13.8 (Nonlinear eigenvalue problems) In Lecture 12, we introduced the variational approach to solving eigenvalue problems of linear differential equations. Similarly, for eigenvalue problems in nonlinear differential equations, variational methods are also successful. Consider the following example: given a bounded domain $\Omega \subset \mathbb{R}^n$ and a Carathéodory function $f : \Omega \times \mathbb{R}^1 \to \mathbb{R}^1$ satisfying the growth condition:

$$|f(x, u)| \leq C(1 + |u|^r), \quad 1 \leq r < 2^* - 1,$$

where $2^* = \frac{2n}{n-2}$ $(n > 2)$. Assume

$$f(x,0) = 0,$$

find $u \in H_0^1(\Omega)\backslash\{0\}$ such that

$$-\Delta u(x) = \lambda f(x, u(x)),$$

where λ is a parameter. Furthermore, λ is called an eigenvalue if it corresponds to a nonzero solution $u \in H_0^1(\Omega)$.

Solution Similar to linear equations, we regard this as a constrained variational problem. On $H_0^1(\Omega)$, we consider the functionals

$$I(u) = \frac{1}{2} \int_\Omega |\nabla u(x)|^2 dx,$$

$$N(u) = \int_\Omega F(x, u(x)) dx,$$

where

$$F(x,t) = \int_0^t f(x,s) ds$$

is an anti-derivative of $f(\cdot, t)$. Since the growth of f is restricted, $N(u)$ is well-defined on $H_0^1(\Omega)$.

Likewise, we want to find the minimum of I on $M = N^{-1}(1)$. Since I is coercive and weakly lower semi-continuous, and by the Rellich–Kondrachov compact embedding theorem, M is weakly closed, it follows that I attains its minimum u_0. Moreover, using the Lagrange multipliers, there exists a real number λ_0 such that

$$-\Delta u_0(x) = \lambda_0 f(x, u_0(x)).$$

When $f(x, u) = u$, this is precisely the eigenvalue problem in Lecture 12. However, when $f(x, u) = u^r$ and $1 \le r < 2^* - 1$ with $2^* = \frac{2n}{n-2}$ $(n > 2)$, it generalizes the linear eigenvalue problem. $\square$

Remark 13.4 Recall in Lecture 12, we obtained an increasing sequence of eigenvalues for a linear differential equation. It is natural to ask: for what kind of nonlinear differential equations would we have similar results. It is not a simple task to answer this since for nonlinear problems, we no longer have orthogonality between two eigenvectors with distinct eigenvalues. However, if f is odd in u, that is, $f(x, -u) = -f(x, u)$; similar results are captured by the Liusternik–Schnirelmann theory, which requires more in-depth knowledge of topology.

The following gives an example of a functional which is neither bounded above nor below, so the common variational method does not seem applicable. However, by a certain technique, we can turn it into a minimal value problem.

Example 13.9 (The polarization technique) Find $u \in X = H_0^1(\Omega)$ satisfying the equation:

$$-\Delta u = |u|^{p-2}u,$$

where $2 < p < 2^*$ and 2^* is as defined above.

Notice that this is the E-L equation of the functional

$$I(u) = \int_\Omega \left(\frac{1}{2}|\nabla u|^2 - \frac{1}{p}|u|^p \right) dx.$$

However, I is neither bounded above nor below.

We use polar decomposition on the variable u as follows. Let S be the unit sphere in X, $\forall u \in X \setminus \{\theta\}$, there exists a unique pair $(t, v) \in \mathbb{R}_+^1 \times S$ such that $u = tv$.

Fixing any $v \in S$ and consider the single variable function on the ray

$$t \mapsto I(tv) = \frac{t^2}{2} - \frac{t^p}{p}|v|_p^p,$$

where $|v|_p^p = \int_\Omega |v(x)|^p dx$. It attains its maximum at

$$t = t(v) = \frac{1}{|v|_p^{\frac{p}{p-2}}},$$

which also satisfies

$$\frac{d}{dt}I(tv)|_{t=t(v)} = 0.$$

Substituting into the original functional I, we obtain a functional $\tilde{I}$ on the unit sphere S by

$$\tilde{I}(v) = \left(\frac{1}{2} - \frac{1}{p} \right) \frac{1}{|v|_p^{\frac{2p}{p-2}}}.$$

Using the embedding theorem, there exists a constant $C > 0$ such that

$$|v|_p^p \leq C^p \|v\|^p = C^p.$$

Thus,

$$\frac{1}{|v|_p^p} \geq \frac{1}{C^p}.$$

This means $\tilde{I}$ is indeed a continuous functional on S. Using compact embedding, it is also weakly sequentially continuous. Note that S is itself weakly sequentially compact, hence $\tilde{I}$ must attain its minimum at some v_0.

It remains to show $u_0 = t(v_0)v_0$ is a solution of the original equation. To justify this, on one hand, we have

$$0 = (\tilde{I}'(u_0), w) = (I'(t(v_0)v_0), w) - (I'(t(v_0)v_0), v_0)(v_0, w), \quad \forall w \perp v_0,$$

where $(\cdot, \cdot)$ is the inner product on X. On the other hand,

$$\frac{d}{dt}I(tv)|_{t=t(v)} = (I'(t(v)v), v).$$

Thus,

$$(I'(t(v_0)v_0), w + tv_0) = 0, \quad \forall w \perp v_0, \forall t \in \mathbb{R}^1,$$

i.e.

$$(I'(u_0), \varphi) = 0, \quad \forall \varphi \in X.$$

This affirms that u_0 is a solution of the E-L equation of I. $\qquad\square$

Remark 13.5 In some reference, the polarization technique is also termed the fiberation method.

13.4 The limitations of calculus of variations

At the end of this lecture, we point out, in particular, the solutions of differential equations are not always obtainable by direct methods. Hadamard gave the following counterexample: let

$$g(\vartheta) = \sum_{m=1}^{\infty} \frac{\sin m!\vartheta}{m^2},$$

then

$$u(r, \vartheta) = \sum_{m=1}^{\infty} \frac{r^{m!} \sin m!\vartheta}{m^2}, \quad ((r, \vartheta) \in [0,1] \times [0, 2\pi])$$

converges uniformly to a continuous function u on the unit circle.

The function u is smooth and harmonic in the interior of the unit circle and taking on the boundary value of g. However, the integral of the square of the gradient of u is infinite! In other words, as a harmonic function, u satisfies the E-L equation $\Delta u = 0$ of the Dirichlet integral in some sense, but the corresponding functional (the Dirichlet integral) itself takes on the value of infinity!

Exercises

1. Let $\Omega \subset \mathbb{R}^n$ be a bounded region with smooth boundary and $f \in C(\bar{\Omega})$. On $W_0^{1,4}(\Omega)$, consider the functional

$$I(u) = \int_\Omega \left[\frac{1}{4} |\nabla u|^4 - x^2 |u|^2 - f(x)u \right] dx.$$

   Prove that I is weakly lower semi-continuous and coercive, hence it has a minimal solution.

2. Determine whether each of the following functional is weakly lower semi-continuous, or coercive, and whether the minimal solution exists. Explain why.

   (1)

$$\int_{-1}^{1} t^2 \dot{u}^2 dt, \qquad M = \{ u \in H^1(-1,1) \mid u(\pm 1) = \pm 1 \}.$$

   (2)

$$\int_{0}^{1} (\dot{u}^2 - 1)^2 dt, \qquad M = W_0^{1,4}(0,1).$$

The dual least action principle and the Ekeland variational principle

The main focus of this lecture includes: the dual least action principle and the Ekeland variational principle.

The dual least action principle is mainly applied to Hamiltonian systems and related problems. In general, the functional associated with a Hamiltonian system is neither bounded above nor below, so variational methods are difficult to apply. However, if the Hamiltonian in a Hamiltonian system is a convex function, then by means of convex conjugates, the problem can be transformed into a constrained variational problem. This is the essence of the dual least action principle.

The Ekeland variational principle is a general minimization result with a broad variety of applications. It provides a specific method in choosing a minimizing sequence; consequently, this minimizing sequence along with some other conditions give rise to numerous applications.

14.1 The conjugate function of a convex function

In Lecture 3, we introduced the Legendre transform f^* of a function $f(x)$ on $\mathbb{R}^n$ via

$$f^*(\xi) = \xi \cdot \psi(\xi) - f(\psi(\xi)), \qquad (14.1)$$

where $x = \psi(\xi)$ is the inverse function of $\xi = \nabla f(x)$.

The importance of the Legendre transformation is that it reveals the inverse relation of the gradient ∇f of f and the gradient ∇f^* of f^*, namely,

$$\xi = \nabla f(x) \quad \text{and} \quad x = \nabla f^*(\xi)$$

are inverses of each other. Thus, we can recover f from f^* via the relation

$$f(x) + f^*(\xi) = \langle \xi, x \rangle.$$

However, it is worth noting that the applicability of the Legendre transformation is rather restrictive, since it requires the existence of $(\nabla f(x))^{-1}$ everywhere.

Given a convex function f, whose domain is $D(f) = \{x \in \mathbb{R}^n \mid f(x) < +\infty\}$. The replacement of the idea of the gradient mapping is the sub-differential operator $x \mapsto \partial f(x)$. The sub-differential operator is a set-valued mapping:

$$\xi \in \partial f(x) \Longleftrightarrow f(y) \geq f(x) + \langle \xi, y - x \rangle, \ \forall y \in D(f)$$
$$\Longleftrightarrow f(x) - \langle \xi, x \rangle = \min_{y \in D(f)} \{f(y) - \langle \xi, y \rangle\}.$$

Its inverse is also a set-valued mapping. If we still denote the inverse (set-valued) mapping of $\partial f(x)$ by $\psi(\xi)$, then

$$\psi(\xi) = \{x \in D(f) \mid \xi \in \partial f(x)\},$$

i.e.

$$x \in \psi(\xi) \Longleftrightarrow x \text{ is such that } f(y) - \langle \xi, y \rangle \text{ achieves its minimal value.}$$

Accordingly, we can extend the Legendre transformation to convex functions. Comparing to (14.1), we introduce the following.

Definition 14.1 (Conjugate function) Let $f : \mathbb{R}^n \to \bar{\mathbb{R}} = \mathbb{R}^1 \cup \{+\infty\}$ be a proper function, i.e. $D(f) \neq \emptyset$. We call

$$f^*(\xi) = \sup_{x \in \mathbb{R}^n} \{\langle \xi, x \rangle - f(x)\}$$

the *conjugate function* of f. Sometimes, it is also called the Fenchel transform of f.

A conjugate function has the following properties.

(1) f^* is lower semi-continuous and convex.

This is because f^* is the supremum of a family of affine functions, and since affine functions are convex and lower semi-continuous, the supremum of convex lower semi-continuous functions is itself convex and lower semi-continuous.

(2) If f is a proper, lower semi-continuous, and convex function, then f^* is proper.

Proof Since f is proper, $\exists x_0 \in \mathbb{R}^n$ such that $f(x_0) < +\infty$. This means the epigraph of f $\text{epi}(f) = \{(x, t) \in \mathbb{R}^n \times \mathbb{R}^1 \mid f(x) \leq t\}$ is a closed convex set with non-empty complement. Choose $t_0 < f(x_0)$, then $(t_0, x_0) \notin \text{epi}(f)$. By Ascoli's separation theorem, $\exists (x_0^*, \lambda) \in \mathbb{R}^n \times \mathbb{R}^1$ and $\exists \alpha \in \mathbb{R}^1$ such that

$$\langle x_0^*, x \rangle + \lambda t > \alpha > \langle x_0^*, x_0 \rangle + \lambda t_0 \quad \forall (x, t) \in \text{epi}(f). \tag{14.2}$$

In particular,

$$\langle x_0^*, x_0 \rangle + \lambda f(x_0) > \alpha > \langle x_0^*, x_0 \rangle + \lambda t_0.$$

It follows that $\lambda > 0$ and

$$\left\langle -\frac{1}{\lambda}x_0^*, x \right\rangle - f(x) < -\frac{\alpha}{\lambda}, \quad \forall x \in D(f),$$

i.e.

$$f^*\left(-\frac{x_0^*}{\lambda} \right) < -\frac{\alpha}{\lambda} < +\infty.$$

$\square$

(3) If $f \leq g$, then $g^* \leq f^*$.

(4) (Young's inequality) $\langle x^*, x \rangle \leq f(x) + f^*(x^*)$.

(5) $f(x) + f^*(x^*) = \langle x^*, x \rangle \iff x^* \in \partial f(x)$.

Proof By definition,

$$x^* \in \partial f(x) \iff \langle x^*, y - x \rangle \leq f(y) - f(x), \quad \forall y \in \mathbb{R}^n,$$
$$\iff \langle x^*, y \rangle - f(y) \leq \langle x^*, x \rangle - f(x), \quad \forall y \in \mathbb{R}^n,$$
$$\iff f^*(x^*) \leq \langle x^*, x \rangle - f(x).$$

Combining with Young's inequality, the assertion follows. $\square$

(6) If $g(x) = f(x - x_0) + \langle x_0^*, x \rangle + a$, then

$$g^*(x^*) = f^*(x^* - x_0^*) + \langle x^*, x_0 \rangle - (a + \langle x_0^*, x_0 \rangle).$$

(7) If $g(x) = f(\lambda x)$, $\lambda > 0$, then $g^*(x^*) = f(x^*/\lambda)$.

As long as f^* is a proper function, we can also define its conjugate f^{**}. By property (1), f^{**} is proper, convex, and lower semi-continuous.

Theorem 14.1 (Fenchel–Moreau) *If f is a proper, convex, lower semi-continuous function, then $f^{**} = f$.*

Proof By Young's inequality, we have $f^{**}(x) \leq f(x), \forall x \in \mathbb{R}^n$.

To prove the reversed inequality, we argue by contradiction. Suppose not, there exists $x_0 \in \mathbb{R}^n$ such that $f^{**}(x_0) < f(x_0)$. Following the proof of property (2), we obtain a point $(x_0, f^{**}(x_0)) \notin \mathrm{epi}(f)$, hence the inequality (14.2), where $t_0 = f^{**}(x_0)$.

If $f(x_0) < +\infty$, then $\lambda > 0$ in (14.2) and

$$\langle x_0^*, x \rangle + \lambda f(x) > \alpha, \quad \forall x \in D(f).$$

Likewise, we have

$$f^*\left(-\frac{x_0^*}{\lambda} \right) \leq -\frac{\alpha}{\lambda}.$$

According to the definition of f^{**},

$$f^{**}(x_0) \geq \left\langle -\frac{x_0^*}{\lambda}, x_0 \right\rangle - f^*\left(-\frac{x_0^*}{\lambda} \right),$$

whence

$$\langle x_0^*, x_0 \rangle + \lambda f^{**}(x_0) \geq \alpha. \tag{14.3}$$

This contradicts (14.2).

If $f^{**}(x_0) < +\infty$, then $\lambda > 0$ still holds, which again yields a contradiction.

It remains to consider the case where $f(x_0) = f^{**}(x_0) = +\infty$ and $\lambda = 0$. From (14.2), $\exists \, \varepsilon > 0$ such that

$$\langle x_0^*, x - x_0 \rangle \geq \varepsilon, \quad \forall \, x \in D(f). \tag{14.4}$$

Since f^* is proper, $\exists \, x_1^* \in \mathbb{R}^n$ such that $f^*(x_1^*) < +\infty$ and

$$\langle x_1^*, x \rangle - f(x) - f^*(x_1^*) \leq 0, \quad \forall \, x \in D(f). \tag{14.5}$$

Combining (14.4) and (14.5), $\forall \, n \in \mathbb{N}$, it yields

$$\langle x_1^* - nx_0^*, x \rangle + n\langle x_0^*, x_0 \rangle + n\varepsilon - f(x) - f^*(x_1^*) \leq 0, \quad \forall \, x \in D(f).$$

Consequently,

$$f^*(x_1^* - nx_0^*) + n\langle x_0^*, x_0 \rangle + n\varepsilon - f^*(x_1^*) \leq 0$$

or

$$n\varepsilon + \langle x_1^*, x_0 \rangle - f^*(x_1^*) \leq \langle x_1^* - nx_0^*, x_0 \rangle - f^*(x_1^* - nx_0^*) \leq f^{**}(x_0).$$

Letting $n \to \infty$, it follows that $f^{**}(x_0) = +\infty$, a contradiction. $\qquad\square$

Corollary 14.1 *For a proper, convex, lower semi-continuous function f, we have*

$$\xi \in \partial f(x) \iff x \in \partial f^*(\xi).$$

This directly generalizes the Fenchel–Moreau theorem and property (5).

Corollary 14.2 *If $f : \mathbb{R}^n \to \bar{\mathbb{R}}$ is proper, then*

$$f^{**} = \text{conv}(f) = \sup\{\varphi : \mathbb{R}^n \to \bar{\mathbb{R}} \mid \varphi(x) \leq f(x), \, \forall \, x \in \mathbb{R}^n, \, \varphi \text{ is convex}\}.$$

Proof Suppose $g \leq f$ is convex, then it is proper and convex. By property (3), $f^* \leq g^*$; moreover, $g^{**} \leq f^{**}$. By the Fenchel–Moreau theorem, we see that $g = g^{**} \leq f^{**}$. $\qquad\square$

Example 14.1 Let $f(x) = |x|^p/p$, $1 < p < \infty$, where $|x| = (x_1^2 + x_2^2 + \cdots + x_n^2)^{1/2}$, then

$$f^*(\xi) = \frac{1}{p'}|\xi|^{p'}, \quad \text{where } \frac{1}{p} + \frac{1}{p'} = 1.$$

Example 14.2 Let $f(x) = |x|$, then

$$f^*(\xi) = \begin{cases} 0 & \text{if } |\xi| \leq 1, \\ +\infty & \text{if } |\xi| > 1. \end{cases}$$

Proof Recall

$$f^*(\xi) = \sup_x \{\langle \xi, x \rangle - |x|\}.$$

For $|\xi| > 1$, we let $x = t\xi$ and $t \to +\infty$, then $f^*(\xi) = +\infty$. For $|\xi| \leq 1$, since $\langle \xi, x \rangle - |x| \leq 0$, by choosing $x = 0$, the result follows. $\qquad\square$

14.2 The dual least action principle

Given a convex Hamiltonian $F \in C^1(\mathbb{R}^N \times \mathbb{R}^N)$ satisfying the following growth condition:

$$0 \le H(u, \xi) \le C(|u|^2 + |\xi|^2). \tag{14.6}$$

Find a vector-valued periodic function $(u(t), \xi(t)) \in \mathbb{R}^N \times \mathbb{R}^N$ which satisfies the Hamiltonian system

$$\begin{cases} \dot{\xi}(t) = H_u(t, u(t), \xi(t)), \\ \dot{u}(t) = -H_\xi(t, u(t), \xi(t)). \end{cases} \tag{14.7}$$

As mentioned before, the minimization method cannot be applied directly to the associated functional. We shall use the conjugate function H^* of the convex Hamiltonian H to rephrase the problem.

In the above problem, the period is yet to be determined. Introducing a parameter λ, we wish to find a 2π-periodic solution $(v(t), \eta(t)) \in \mathbb{R}^N \times \mathbb{R}^N$ such that

$$\begin{cases} \dot{\eta}(t) = \lambda H_v(t, v(t), \eta(t)), \\ \dot{v}(t) = -\lambda H_\eta(t, v(t), \eta(t)). \end{cases} \tag{14.8}$$

Suppose we have found a pair $(v(t), \eta(t))$ satisfying (14.8), then by letting

$$u(t) = v(\lambda^{-1}t), \quad \xi(t) = \eta(\lambda^{-1}t),$$

the pair $(u(t), \xi(t))$ satisfies (14.7). If $\lambda > 0$, then both of them have period $2\lambda\pi$.

We now turn (14.8) into a constrained minimization problem. It is the dual problem of the original problem: find $(w, \rho) \in X := H^1_{\text{per}}((0, 2\pi), \mathbb{R}^N)^2 = \{(w, \rho) \in H^1((0, 2\pi), \mathbb{R}^{2N}) \mid w(0) = w(2\pi), \ \rho(0) = \rho(2\pi)\}$ which satisfies

$$I(w, \rho) = \int_0^{2\pi} H^*(\dot{\rho}(t), -\dot{w}(t))dt,$$
$$G(w, \rho) = \int_0^{2\pi} (-\dot{\rho}(t) \cdot w(t) + \dot{w}(t) \cdot \rho(t))dt = -\pi. \tag{14.9}$$

Denoting

$$\langle (w, \rho), (u, \xi) \rangle = \int_0^{2\pi} (w \cdot u + \rho \cdot \xi)dt,$$

a direct calculation shows

$$\langle G'((w, \rho)), (u, \xi) \rangle = 2\langle (-\dot{\rho}, \dot{w}), (u, \xi) \rangle = 2\langle (-\dot{\xi}, \dot{u}), (w, \rho) \rangle$$

and

$$\langle I'((w, \rho)), (u, \xi) \rangle = \int_0^{2\pi} (H^*)'(\dot{\rho}(t), -\dot{w}(t)) \cdot (\dot{\xi}(t), -\dot{u}(t))dt.$$

We have the following conclusions:

(1) $M := G^{-1}(-\pi) \neq \emptyset$.

This is because

$$G(w, \rho) = -2 \int_0^{2\pi} \dot\rho(t) \cdot w(t)dt$$

is linear in both w and ρ.

(2) $G'((w, \rho)) \neq (\theta, \theta), \forall (w, \rho) \in M$.

This is because from (14.9), we see $(w, \rho) \neq (\theta, \theta)$ on M.

(3) I is a convex functional, which is bounded below and continuous on X; therefore, it is also weakly sequentially lower semi-continuous.

It suffices to verify I is bounded below. Combining (14.6), property (3) of a conjugate function, and Example 14.1, we have

$$H^*(w, \rho) \geq \frac{1}{C}(|w|^2 + |\rho|^2)$$

and

$$I(w, \rho) \geq \frac{1}{C} \int_0^{2\pi} (|\dot w|^2 + |\dot\rho|^2)dt.$$

We note that there does not exist any constant vector-valued functions on M, hence

$$\|(w, \rho)\| = \left(\int_0^{2\pi} (|\dot w|^2 + |\dot\rho|^2)dt \right)^{\frac{1}{2}}$$

defines an equivalent norm on X. This simultaneously justifies the following

(4) I is coercive.

(5) M is weakly sequentially closed.

To verify this, suppose $\{(w_j, \rho_j)\} \subset M$ such that $(w_j, \rho_j) \rightharpoonup (w_0, \rho_0)$ in X. Then in $L^2([0, 2\pi], \mathbb{R}^N)$, we have strong convergence $w_j \to w_0$ and $\rho_j \to \rho_0$. Furthermore,

$$-\pi = \lim G(w_j, \rho_j)$$
$$= \lim \int_0^{2\pi} (-\dot\rho_j(t) \cdot w_j(t) + \dot w_j(t) \cdot \rho_j(t))dt$$
$$= \int_0^{2\pi} (-\dot\rho(t) \cdot w(t) + \dot w(t) \cdot \rho(t))dt = G(w_0, \rho_0).$$

Consequently, there exists $(w_0, \rho_0) \in M$, which is the minimum of the constrained problem (14.9). Using Lagrange multipliers, there exists $\lambda \in \mathbb{R}^1$ such that

$$I'(w_0, \rho_0) + \frac{\lambda}{2}G'(w_0, \rho_0) = 0,$$

i.e.

$$(H^*)'(\dot{\rho}_0, -\dot{w}_0) = \lambda(w_0, \rho_0). \tag{14.10}$$

Since the sub-differential of the conjugate function and the sub-differential of the original function are inverses of each other, as stated in Corollary 14.1, (14.10) is equivalent to

$$\begin{cases} \dot{\rho}_0(t) = H_w(t, \lambda w_0(t), \lambda \rho_0(t)), \\ \dot{w}_0(t) = -H_\rho(t, \lambda w_0(t), \lambda \rho_0(t)). \end{cases} \tag{14.11}$$

If we define

$$\begin{cases} \eta = \lambda \rho_0, \\ v = \lambda w_0, \end{cases}$$

by substituting into (14.11), it gives (14.8).

Lastly, we verify $\lambda > 0$. In (14.10), first multiplying both sides by $(\dot{\rho}_0, -\dot{w}_0)$ and then integrating, we obtain

$$\langle (H^*)'(\dot{\rho}_0, -\dot{w}_0), (\dot{\rho}_0, -\dot{w}_0) \rangle = -\lambda G(w_0, \rho_0) = \lambda \pi.$$

It follows from (14.6) that $\nabla H(\theta, \theta) = (\theta, \theta)$. Using property (5) of a conjugate function, we see that

$$H^*(\theta, \theta) = -H(\theta, \theta) = 0.$$

Since H^* is convex, we have

$$H^*(\theta, \theta) - H^*(\dot{\rho}_0, -\dot{w}_0) \geq -\langle (H^*)'(\dot{\rho}_0, -\dot{w}_0), (\dot{\rho}_0, -\dot{w}_0) \rangle,$$

i.e.

$$\langle (H^*)'(\dot{\rho}_0, -\dot{w}_0), (\dot{\rho}_0, -\dot{w}_0) \rangle \geq H^*(\dot{\rho}_0, -\dot{w}_0) \geq 0.$$

Thus, $\lambda \geq 0$.

It remains to show that $\lambda \neq 0$. We argue by contradiction. Suppose $\lambda = 0$, then

$$(H^*)'(\dot{\rho}_0, -\dot{w}_0) = (\theta, \theta).$$

By Corollary 14.1, we see that

$$\begin{cases} \dot{\rho}_0 = -H_u(\theta, \theta), \\ \dot{w}_0 = H_\rho(\theta, \theta). \end{cases}$$

Using condition (14.6), we conclude that $(H_u(\theta, \theta), H_\rho(\theta, \theta)) = (\theta, \theta)$, a contradiction to conclusion (2). $\square$

14.3 The Ekeland variational principle

In view of our previous methods in finding extreme values, except in some special circumstances (mostly linear problems) where the orthogonal projection method is applicable, we almost always rely on the weak convergence (weak sequential lower semi-continuity, weak sequential compactness, etc.). However, the weak topology is in general more complicated, hard to grasp, and often tedious to verify. In the following, we introduce the Ekeland variational principle, a fundamental result proposed by Ekeland in 1970. However, this seemingly simple result can be combined with the Palais–Smale condition, a compactness condition in modern variational calculus, to produce a very useful method in finding extreme values.

Theorem 14.2 (Ekeland) *Let (X, d) be a complete metric space. Let $f : X \to \mathbb{R}^1 \cup \{+\infty\}$ be a proper function, i.e. $f \not\equiv +\infty$. If f is bounded below and lower semi-continuous and if $\exists \, \varepsilon > 0$ and $\exists \, x_\varepsilon \in X$ such that $f(x_\varepsilon) < \inf_X f + \varepsilon$, then $\exists \, y_\varepsilon \in X$ such that*

(1) $f(y_\varepsilon) \leq f(x_\varepsilon)$,
(2) $d(y_\varepsilon, x_\varepsilon) \leq 1$,
(3) $f(x) > f(y_\varepsilon) - \varepsilon d(y_\varepsilon, x), \forall \, x \in X \backslash \{y_\varepsilon\}$.

Proof We point out that the above-mentioned y_ε is itself the minimum of the function $f(x) + \varepsilon d(y_\varepsilon, x)$, which depends on y_ε.

$1°$ We choose a convergent sequence in X recursively.

First, we choose $u_0 = x_\varepsilon$. Suppose u_n has been chosen, we define

$$S_n = \{w \in X \mid f(w) \leq f(u_n) - \varepsilon d(w, u_n)\}.$$

Since $u_n \in S_n, S_n \neq \emptyset$. Choose $u_{n+1} \in S_n$ satisfying

$$f(u_{n+1}) - \inf_{S_n} f \leq \frac{1}{2} \left[f(u_n) - \inf_{S_n} f \right], \quad n = 0, 1, 2, \ldots.$$

$2°$ We show $\{u_n\}$ is a Cauchy sequence. Note

$$\varepsilon d(u_n, u_m) \leq f(u_n) - f(u_m), \quad \forall \, m \geq n. \tag{14.12}$$

Since $\{f(u_n)\}$ is decreasing and f is bounded below, whenever $n.m \to \infty$, $f(u_n) - f(u_m) \to 0$.

Therefore, $\exists \, u^* \in X$ such that $u_n \to u^*$. Since f is lower semi-continuous, we have

$$f(u^*) \leq \lim_{n \to \infty} f(u_n) \leq \lim_{n \to \infty} \inf_{S_n} f. \tag{14.13}$$

$3°$ It remains to verify $y_\varepsilon = u^*$ satisfies (1)–(3).

Since $\{f(u_n)\}$ is decreasing,

$$f(y_\varepsilon) = f(u^*) \leq f(u_n) \leq f(u_0) = f(x_\varepsilon), \qquad \forall n,$$

so (1) holds.

From (14.12),

$$\varepsilon d(x_\varepsilon, y_\varepsilon) = \varepsilon d(u_0, u^*)$$
$$\leq f(x_\varepsilon) - f(u^*)$$
$$\leq f(x_\varepsilon) - \inf_X f < \varepsilon,$$

so (2) holds.

Lastly, we prove (3) by contradiction. Suppose $y_\varepsilon = u^*$ does not satisfy (3), then $\exists w \neq u^*$ such that

$$f(w) \leq f(u^*) - \varepsilon d(u^*, w). \tag{14.14}$$

From (14.12), it follows that

$$\varepsilon d(u_n, u^*) \leq f(u_n) - f(u^*),$$

i.e.

$$f(u^*) \leq f(u_n) - \varepsilon d(u_n, u^*). \tag{14.15}$$

Combining (14.14) and (14.15), we deduce that

$$f(w) \leq f(u_n) - \varepsilon d(u_n, w), \qquad \forall n.$$

Thus, $w \in \bigcap_{n=1}^{\infty} S_n$. Using (14.13), we have $f(u^*) \leq f(w)$, which contradicts (14.14). $\square$

Corollary 14.3 *Let (X, d) be a complete metric space and $f : X \to \mathbb{R}^1 \cup \{+\infty\}$ be a proper function, which is bounded below and lower semi-continuous. Then $\forall \varepsilon > 0, \exists y_\varepsilon \in X$ such that $f(x) > f(y_\varepsilon) - \varepsilon d(x, y_\varepsilon), \forall x \neq y_\varepsilon$.*

It is worth noting that although the Ekeland variational principle only employs the metric topology, which involves neither the weak topology nor the various notions of compactness, but the minimum of the functional f is not reached. However, the significance lies in that by choosing this special minimizing sequence, it produces a special sequence of approximated minima.

14.4 The Fréchet derivative and the Palais–Smale condition

In the following, we connect the Ekeland variational principle to the derivative of a continuously differentiable function.

In Lecture 10, we introduced the Gâteaux derivative of a real-valued function on a Banach space. Let $f : X \to \mathbb{R}^1$, $x_0 \in U \subset X$, where U is an open neighborhood of X. We say that f has Gâteaux derivative at x_0 if $\forall h \in X$, $\exists\, df(x_0, h) \in \mathbb{R}^1$ such that

$$f(x_0 + h) - f(x_0) - t\, df(x_0, h) = o(t), \quad t \to 0, \quad \forall x_0 + h \in U.$$

The Fréchet derivative is closely related to the Gâteaux derivative. If

$$|f(x) - f(x_0) - \langle \xi, x - x_0 \rangle| = o(\|x - x_0\|) \text{ as } x \to x_0,$$

then $\xi \in X^*$ is said to be the Fréchet derivative of f at x_0 and we denote $f'(x_0) = \xi$.

If f has Fréchet derivative $f'(x_0)$, then it has Gâteaux derivative at x_0 and

$$df(x_0, h) = \langle f'(x_0), h \rangle, \quad \forall h \in X.$$

Furthermore,

$$\|f'(x_0)\| = \sup_{h \in X} \frac{df(x_0, h)}{\|h\|}.$$

Conversely, suppose f has Gâteaux derivative $df(x, h)$ everywhere in some neighborhood U of x_0, if $\forall\, x \in U$, $\exists\, \xi(x) \in X^*$ such that

$$\langle \xi(x), h \rangle = df(x, h).$$

If $x \mapsto \xi(x)$ is also continuous, then f has Fréchet derivative $f'(x_0)$ at x_0.

If the Fréchet derivative $f'(x)$ exists everywhere and $x \mapsto f'(x)$ is continuous, then we say that f is continuously differentiable, and we denote $f \in C^1(X, \mathbb{R}^1)$.

The Gâteuax derivative generalizes the directional derivative from $\mathbb{R}^n$ to a Banach space, where the Fréchet derivative generalizes the differential from $\mathbb{R}^n$ to a Banach space.

In Lecture 11, we computed the Gâteaux derivatives of some functionals of integral form. Their appearances are identical to the variational derivatives. As for the Fréchet derivatives, as long as they exist, they should also have the same expression.

Example 14.3 Given $\Omega \subset \mathbb{R}^n$, $F \in C^1(\Omega \times \mathbb{R}^1)$ satisfies

$$|F_s(x, s)| \leq C(1 + |x|)^\mu, \quad \mu \leq 2^* - 1 = \frac{n+2}{n-2} \quad (n > 2).$$

On $H_0^1(\Omega)$, we find the Fréchet derivative of the functional

$$I(u) = \int_\Omega \left[\frac{1}{2} |\nabla u(x)|^2 + F(x, u(x)) \right] dx.$$

We already computed its Gâteaux derivative to be

$$dI(u,v) = \int_\Omega [\nabla u \cdot \nabla v + F_u(x,u)v]dx.$$

On $H_0^1(\Omega)$, using the inner product

$$\langle u,v \rangle = \int_\Omega [\nabla u \cdot \nabla v]dx,$$

we can write it as

$$dI(u,v) = \langle u + KF_u(x,u), v \rangle,$$

where

$$K = (1-\Delta)^{-1} : H_0^1(\Omega)^* \to H_0^1(\Omega).$$

By the embedding

$$H_0^1(\Omega) \hookrightarrow L^{2^*}(\Omega) \tag{14.16}$$

and the growth condition

$$|F_s(x,u(x))| \le C(1 + |u(x)|)^\mu \in L^{\frac{2n}{n+2}} = (L^{2^*}(\Omega))^*,$$

we see the nonlinear mapping $u \mapsto F_s(x,u(x)) : H_0^1(\Omega) \hookrightarrow (L^{2^*(\Omega)})^*$ is continuous.

Moreover, by the continuity of the embedding (14.16), we deduce that its dual mapping

$$(L^{2^*(\Omega)})^* \hookrightarrow (H_0^1(\Omega))^* \tag{14.17}$$

is also continuous. Thus, the Gâteaux derivative $u \mapsto u + KF_u(x,u) : H_0^1(\Omega) \hookrightarrow H_0^1(\Omega)$ is continuous. Consequently, the Fréchet derivative of f exists and is given by

$$I'(u) = u + KF_u(x,u).$$

We call x_0 a critical point of f if $f'(x_0) = \theta$ and the value $f(x_0)$ its critical value.

Subsequently, in variational problems, minima are critical points, and all critical points are solutions of E-L equations.

Definition 14.2 Let X be a Banach space and $f \in C^1(X, \mathbb{R}^1)$. If for any sequence $\{x_j\}_1^\infty \subset X$ satisfying

$$f(x_j) \to c, \quad \|f'(x_j)\| \to 0, \tag{14.18}$$

there is a convergent subsequence, then we say f at c satisfies the Palais–Smale condition, denoted by PS_c. We also call the sequence satisfying (14.18) a Palasi–Smale sequence (or briefly PS-sequence).

If $\forall c \in \mathbb{R}^1$, f satisfies PS_c, then we say f satisfies PS.

The Palais–Smale condition can be extended to a general Banach manifold.

The following corollary is very important, since by combining the Ekeland variational principle and the Palais–Smale condition, we give a criterion for the existence of minimal value.

Corollary 14.4 *Let X be a Banach space (or more generally, a Banach manifold). Let $f \in C^1(X, \mathbb{R}^1)$ be bounded below. Denote*

$$c = \inf_X f.$$

If f satisfies the PS_c condition, then f attains its minimum.

Proof According to the Ekeland variational principle, $\forall n \geq 1$, $\exists\, x_n \in X$ such that

$$\begin{cases} f(x) > f(x_n) - \frac{1}{n}\|x - x_n\|, \ \forall x \neq x_n \\ f(x_n) < c + \frac{1}{n}. \end{cases}$$

The first inequality implies

$$\|f'(x_n)\| = \sup_{\|\varphi\|=1} |df(x_n, \varphi)| \leq \frac{1}{n}.$$

The second inequality implies

$$f(x_n) \to c.$$

According to the PS_c condition, there exists a convergent subsequence $\{x_{n_j}\}$ such that $x_{n_j} \to x^* \in X$. By continuity, it follows that $f(x^*) = c$. $\square$

14.5   The Nehari technique

Since many functionals are neither bounded above nor below, in appearance, it seems difficult to apply variational methods to find their critical points. However, for some particular problems, Nehari provided a special technique, which transforms a critical point problem to an extremal problem.

Let H be a Hilbert space, equipped with inner product $(\cdot, \cdot)$, and a given functional $I \in C^2(H, \mathbb{R}^1)$. We look for the critical points of I, namely, those points for which $I'(u) = 0$.

Define $G(u) = \langle I'(u), u \rangle$. We note that all critical points u of I satisfy:

$$G(u) = 0.$$

If the set $M = \{h \in H \mid G(u) = 0\}$ is a manifold; for example, $G'(u) \neq \theta$, $\forall\, u \in M$, then we can restrict I on M to obtain a new functional $\tilde{I}$. Furthermore, if $\tilde{I}$ has an extreme value, then we can find the extreme value of $\tilde{I}$; in other words, we find the constrained extreme value of I.

Some people may ask: how to handle the Lagrange multiplier associated with $G(u) = 0$? In fact, since

$$\tilde{I}'(u) = I'(u) - \frac{(I'(u), u)}{\|G'(u)\|^2} G'(u),$$

so on the set M, we have $(I'(u), u) = G(u) = 0$. Thus, if $G'(u) \neq \theta$, then

$$\tilde{I}'(u) = 0 \iff I'(u) = 0.$$

We now give a concrete example to demonstrate how to apply this technique.

Example 14.4 Let $\Omega \subset \mathbb{R}^n$ be a bounded domain. Let $a \in C(\bar{\Omega})$, $a(x) \geq \alpha > 0$, $2 < \mu < 2^*$. On $H_0^1(\Omega)$, find the nontrivial critical points of the functional

$$I(u) = \int_\Omega \left[\frac{1}{2} |\nabla u(x)|^2 + a(x) |u(x)|^\mu \right] dx.$$

We compute that

$$G(u) = \langle I'(u), u \rangle = \int_\Omega [|\nabla u|^2 + \mu a(x) |u(x)|^\mu] dx.$$

On $G^{-1}(0)$,

$$\tilde{I}(u) = (\mu - 2) \int_\Omega a(x) |u(x)|^\mu dx = \left(\frac{1}{2} - \frac{1}{\mu} \right) \|u\|^2$$

is nonnegative.

1° Note that

$$\langle G'(u), v \rangle = \int_\Omega [2\nabla u \cdot \nabla v + \mu^2 a(x) |u(x)|^{\mu-2} u(x) v(x)] dx, \quad \forall v \in H_0^1(\Omega),$$

so we have $G(\theta) = 0$ and

$$G'(u) = 2u + (-\Delta)^{-1} a |u|^{\mu-2} u.$$

This means: $\theta \in G^{-1}(0)$ and $G'(\theta) = \theta$.

However, $\forall u \in G^{-1}(0)$, by the embedding theorem, we have

$$\int_\Omega |\nabla u|^2 dx = \mu \left| \int_\Omega a(x) |u(x)|^\mu dx \right| \leq C \left(\int_\Omega |\nabla u|^2 dx \right)^{\frac{\mu}{2}}.$$

It follows that either $u = \theta$ or

$$\left(\int_\Omega |\nabla u|^2 dx \right)^{\frac{\mu}{2} - 1} \geq \frac{1}{C}.$$

This implies θ is the only isolated point in $G^{-1}(0)$. Let $M = G^{-1}(0)\backslash\{\theta\}$, then on M,

$$\langle G'(u), u \rangle = 2\|u\|^2 + \mu^2 \int_\Omega a(x)|u(x)|^\mu]dx > 0,$$

whence $G'(u) \neq 0, \forall u \in M$.

2° It is clear $\tilde{I} \in C^1(H_0^1(\Omega), \mathbb{R}^1)$. We now verify the Palais–Smale condition. Suppose $\{u_j\} \subset M$ is a PS-sequence satisfying $\tilde{I}(u_j) \to 0$ and $|\tilde{I}(u_j)| \leq C$. Since

$$\left(\frac{1}{2} - \frac{1}{\mu} \right) \|u_j\|^2 = \tilde{I}(u_j) \leq C,$$

there exists a subsequence $u_{j'} \rightharpoonup u$. Moreover, since

$$\tilde{I}'(u) = I'(u) = u + (-\Delta)^{-1}\mu a(x)|u(x)|^{\mu-2}u(x)$$

and the embedding

$$H_0^1(\Omega) \hookrightarrow L^\mu(\Omega)$$

is compact, as $\tilde{I}'(u_j) \to 0$, $\{u_j\}$ has a subsequence which converges in $L^\mu(\Omega)$. Composing it with $(-\Delta)^{-1}$, it follows that this subsequence in $H_0^1(\Omega)$ converges to $u_0 \in M$ and

$$I'(u_0) = \tilde{I}'(u_0) = \lim \tilde{I}'(u_j) = 0.$$

Thus, u_0 is the desired nontrivial critical point.   □

Exercises

1. Let $R > a > 0$ and $M = \{u \in C^1([-a, a]) \mid u(\pm a) = \sqrt{R^2 - a^2}\}$,

$$I(u) = \int_{-a}^a \left(\sqrt{1 - \dot{u}^2} - \frac{u}{R} \right) dt.$$

   (1) Compute the first and second variations of I.
   (2) Write the Euler–Lagrange equation.
   (3) Verify that $u_0 = \sqrt{R^2 - t^2}$ is a weak minimal solution.
   (4) Write the Jacobi operator along u_0.

2. Find $\inf\{I(u) \mid u \in M\}$ for each of the following:
   (1)

$$I(u) = \int_1^2 t\sqrt{1 + \dot{u}^2}dt,$$

$$M = \{u \in C^1([1, 2]) \mid u(j) = \cosh^{-1} j, \ j = 1, 2\}.$$

(2)

$$I(u) = \pi \int_a^b u^2 dr,$$

$$M = \left\{ u \in C_0^1([a,b]) \ \middle| \ 2\pi \int_a^b u\sqrt{1 + \dot{u}^2} dt = c \right\}.$$

3. Let $V = C^1(\mathbb{R}^1, \mathbb{R}^1)$. $\forall (u,p) \in \mathbb{R}^n \times \mathbb{R}^N$, let

$$L(u,p) = \frac{1}{2} \sum_{i=1}^N p_i^2 - \sum_{i \neq j} V((u_i - u_j)^2).$$

(1) Write the corresponding functional.

(2) Write the Euler–Lagrange equation.

(3) Write the corresponding Hamiltonian.

(4) Write the corresponding Hamiltonian system.

(5) Suppose $\{u_i(t)\}_1^N$ is a solution of the E-L equation, prove that

$$\frac{1}{2} \sum_{i=1}^N \dot{u}_i^2(t) + \sum_{i \neq j} V((u_i(t) - u_j(t)^2) = \text{const.} \quad \forall t \in \mathbb{R}^1.$$

(6) Write the corresponding Hamilton–Jacobi equation.

4. Let $r \in C([0,1])$, suppose $\exists \, t_0 \in [0,1]$ such that $r(t_0) > 0$. Prove that there exist infinitely many pairs $\{(\lambda_n, u_n)\}_1^\infty$ with $u_n \neq 0$, $\lambda_n \to +\infty$, such that

$$\begin{cases} \ddot{u} = \lambda r u, & \text{in } (0,1), \\ u(0) = u(1) = 0. \end{cases}$$

Lecture 15

The Mountain Pass Theorem, its generalizations, and applications

In this lecture, we introduce the theory of finding critical points other than minima (or maxima). This theory carefully examines the changes of topological structure taking place in the level sets of a functional; subsequently provides criteria to the existence of critical points. Such theory is based on the ideas and machinery from both algebraic and differential topology. Since the 1970s, critical point theory has undergone a rapid development; in particular, it has found profound applications in partial differential equations and dynamical systems with variational structures.

We do not require the reader to possess the needed topological background; instead, we would like to expose the reader to some of the most fundamental and commonly used critical points theorems, such as the Mountain Pass Theorem. To make this more accessible, we will take on a more geometrically intuitive approach.

The standard critical point theory utilizes the gradient flow to accomplish the deformation between the level sets of a functional. However, this treatment is beyond the scope of this book and does not fit well with our current content, we will adopt a more direct approach - introducing these critical point theorems based on the Ekeland variational principle.

15.1 The Mountain Pass Theorem

The following intuitive example illustrates the basic idea used in analyzing saddle points, a special kind of critical point.

Imagine the following scenario: in a valley surrounded by mountains, if a person starting from a point p_1 outside the valley wants to reach a point p_0 in the valley, the optimal path would be along which the highest point is always no higher than the highest point on a nearby path. The highest point on this path is likely to be a saddle point - a critical point which is neither the maximum nor the minimum (see Figure 15.1).

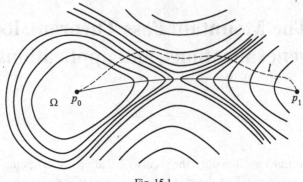

Fig. 15.1

We formulate this mathematically as follows: Let $X = \mathbb{R}^n$ and $\Omega \subset \mathbb{R}^n$ be an open set. Given two points $p_0 \in \Omega$ and $p_1 \notin \bar{\Omega}$. Suppose there exists a function $f \in C^1(X, \mathbb{R}^1)$ such that

$$\alpha = \inf_{x \in \partial\Omega} f(x) > \max\{f(p_0), f(p_1)\}. \tag{15.1}$$

Let

$$\Gamma = \{l \in C([0,1], X) \mid l(i) = p_i, i = 0, 1\} \tag{15.2}$$

be the set of all paths connecting the two points and let

$$c = \inf_{l \in \Gamma} \sup_{t \in [0,1]} f \circ l(t). \tag{15.3}$$

We intend to assert c is a critical value of f, i.e. there exists $x_0 \in X$ such that $f'(x_0) = 0$ and $f(x_0) = c$.

Unfortunately, geometric intuition is not a proof. To validate the assertion, we must impose other conditions on f.

Theorem 15.1 (The Mountain Pass Theorem) *Let X be a Banach space and $f \in C^1(X, \mathbb{R}^1)$. Let $\Omega \subset X$ be an open set. Given two points $p_0 \in \Omega$ and $p_1 \notin \bar{\Omega}$ satisfying* (15.1). *Let c be defined as in* (15.2) *and* (15.3). *If f satisfies the PS_c condition, then $c \geq \alpha$ is a critical value of f.*

Proof We define the metric on Γ by

$$d(l_1, l_2) = \max_{t \in [0,1]} \|l_1(t) - l_2(t)\|,$$

then (Γ, d) is a complete metric space. Let

$$I(l) = \max_{t \in [0,1]} f \circ l(t).$$

From assumption (15.1), $I(l) \geq \alpha$ and I satisfies locally Lipschitzian condition:

$$\begin{aligned}
|I(l_1) - I(l_2)| &\leq \max_{t \in [0,1]} |f \circ l_1(t) - f \circ l_2(t)| \\
&\leq \max_{t, \theta \in [0,1]} \|f'(\theta l_1(t) + (1 - \theta) l_2(t))\| \|l_1(t) - l_2(t)\| \\
&\leq Cd(l_1, l_2),
\end{aligned}$$

where C is a constant, depending only on f', l_1, and l_2. Using the Ekeland variational principle, we obtain a sequence $\{l_n\} \subset \Gamma$ such that

$$c \leq I(l_n) < c + \frac{1}{n}, \tag{15.4}$$

$$I(l) > I(l_n) - \frac{1}{n} d(l, l_n), \quad l \neq l_n, \quad n = 1, 2, \ldots \tag{15.5}$$

Let

$$M(l) = \{t \in [0,1] | \ f \circ l(t) = I(l)\}.$$

Then M is a non-empty compact set and $M \subset (0, 1)$. To see this, suppose $t_0 \in M(l) \cap \{0, 1\}$, then

$$f \circ l(t_0) = \max_{t \in [0,1]} f \circ l(t) \geq \inf_{\partial \Omega} f = \alpha,$$

but

$$f \circ l(t_0) \leq \max\{f(p_0), f(p_1)\} < \alpha,$$

a contradiction.

Denote $\Gamma_0 = \{\psi \in C([0,1], X) | \ \psi(i) = \theta, i = 0, 1\}$, it is a closed linear subspace of $C([0,1], X)$ with norm

$$\|\psi\|_{\Gamma_0} = max_{t \in [0,1]} \|\psi(t)\|.$$

By (15.4), $\forall h \in \Gamma_0$, $\|h\|_{\Gamma_0} = 1$, $\forall \lambda_j \downarrow 0$, $\forall \xi_j \in M(l_n + \lambda_j h)$, we have

$$\lambda_j^{-1} [f \circ (l_n + \lambda_j h)(\xi_j) - f \circ l_n(\xi_j)] \geq -\frac{1}{n}.$$

Since $\{\xi_j\} \subset [0, 1]$, there exists a subsequence, still denoted ξ_j, converging to η_n. The latter depends on l_n, λ_j, and h. Taking the limit, it yields

$$df(l_n(\eta_n), h(\eta_n)) \geq -\frac{1}{n}. \tag{15.6}$$

We want to show $\exists\, \eta_n^* \in M(l_n)$ such that

$$df(l_n(\eta_n^*), \varphi) \geq -\frac{1}{n}, \quad \forall \varphi \in X, \|\varphi\| = 1. \tag{15.7}$$

If this is true, then by letting $x_n = l_n(\eta_n^*)$, it follows that

$$c \leq f(x_n) < c + \frac{1}{n},$$

$$\sup_{\|\varphi\|=1} |df(x_n, \varphi)| \leq \frac{1}{n}.$$

Using the PS_c condition, $\{x_n\}$ has a convergent subsequence $\{x_{n_j}\}$ such that $x_{n_j} \to x^*$, consequently, $f'(x^*) = 0$.

We now prove (15.7) by contradiction. Suppose there does not exist such η_n^* satisfying (15.7), then $\forall \eta \in M(l_n)$, $\exists\, v_\eta \in X$ with $\|v_\eta\| = 1$ such that

$$df(l_n(\eta), v_\eta) < -\frac{1}{n}.$$

So there is a neighborhood of η $O_\eta \subset (0,1)$ such that

$$df(l_n(\xi), v_\eta) < -\frac{1}{n}, \quad \forall \xi \in O_\eta.$$

Since $M(l_n)$ is compact, it has a finite covering, say, $\{O_{\eta_i} \mid i = 1, \ldots, m\}$ and $M(l_n) \subset \bigcup_{i=1}^m O_{\eta_i}$, which corresponds to $\{v_{\eta_i}\}_1^m$, $\|v_{\eta_i}\| = 1$ such that

$$df(l_n(\xi), v_{\eta_i}) < -\frac{1}{n}, \quad \forall \xi \in O_{\eta_i}, i = 1, \ldots, m. \tag{15.8}$$

We construct a partition of unity subordinate to the $\{O_{\eta_i}\}$: $0 \leq \rho_i \leq 1$, supp $\rho_i \subset O_{\eta_i}, 1 \leq i \leq m$ such that

$$\sum_{i=1}^m \rho_i(\xi) = 1, \quad \forall \xi \in M(l_n).$$

Let

$$v = v(\xi) = \sum_{i=1}^m \rho_i(\xi) v_{\eta_i}.$$

Since $M(l_n) \subset (0,1)$, $v \in \Gamma_0$, $\|v\| \leq 1$. In fact, we can choose a finite covering and some $\xi^* \in M(l_n)$ such that there is only one i_0 with $\xi^* \in O_{\eta_{i_0}}$. Hence, $\|v\|_{\Gamma_0} = 1$, and (15.8) implies

$$df(l_n(\xi), v(\xi)) < -\frac{1}{n}, \quad \forall \xi \in M(l_n),$$

contradictory to (15.6). The proof is now complete. $\qquad\square$

Remark 15.1 The Mountain Pass Theorem was stated in the above version by A. Ambrosetti and P. Robinowitz in 1974. Its generalizations as well as variations have since been applied to solving various variational problems. The theorem originated from the Wall Theorem, discovered by M. Morse while studying the multiple-solution problem arisen in minimal surfaces.

The proof of the Mountain Pass Theorem by means of Ekeland variational principle was independently provided by S. Z. Shi (*The Chinese Annal of Mathematics* 1, 1985, 348–355) and J. P. Aubin and I. Ekeland (*Applied Nonlinear Analysis*, John Wiley and Sons, 1984).

Remark 15.2 We comment that in Theorem 15.1, the Palais–Smale condition indeed plays a crucial role. Without it, Brezis and Nirenberg gave the following counterexample:

In $\mathbb{R}^2$, consider the function

$$f(x, y) = x^2 + (1 - x)^3 y^2.$$

Let $c = \inf_{x^2+y^2=\frac{1}{4}} f(x, y) > 0$, it actually has a valley $\Omega = \{(x, y) \in \mathbb{R}^2 \mid f(x, y) \le c\}$, while $f(0, 0) = 0$ and $f(4, 1) = -11$, but by direct computation, f has the only critical point $(0, 0)$.

The geometric structure involved in the Mountain Pass Theorem is a special case of the more general linking structure.

Definition 15.1 Let X be a Banach space, $Q \subset X$ be compact manifold with boundary ∂Q, and $S \subset X$ be a closed subset. We say ∂Q and S link, if

(1) $\partial Q \cap S = \emptyset$,
(2) for any continuous $\varphi : Q \to X$ satisfying $\varphi|_{\partial Q} = \mathrm{id}|_{\partial Q}$, we have $\varphi(Q) \cap S \ne \emptyset$.

Link is a property depicting how two sets intersect with one another under continuous deformations, therefore it is a topological property.

Example 15.1 In the Mountain Pass Theorem, $Q = \{tx_0 + (1-t)x_1 \mid t \in [0, 1]\}$, $S = \partial\Omega$ and $\partial Q = \{x_0, x_1\}$, hence $\partial\Omega$ and S link.

Example 15.2 Let X be a Banach space, X_1 be a finite dimensional linear subspace, X_2 be its complementary space, i.e. $X = X_1 \oplus X_2$. Let

$$S = X_2, \qquad Q = B_R \cap X_1,$$

where B_R is the closed ball in X centered at θ with radius $R > 0$, thus

$$\partial Q = \{x \in X_1 \mid \|x\| = R\}$$

(see Figure 15.2).

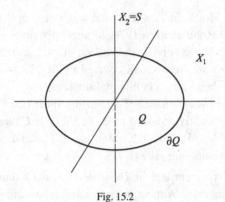

Fig. 15.2

We now show S and ∂Q link. As stated before, since link is a topological property, we must resort to topological machinery. A relatively simple tool from topology is the Brouwer degree.

Given a continuous self mapping f on $\mathbb{R}^n$ and a bounded open set $\Omega \subset \mathbb{R}^n$, fix a point $p \notin f(\partial\Omega)$.

The Brouwer degree $\deg(f, \Omega, p)$ is a function depending on the three variables (f, Ω, p). We list some its important properties:

1. (Kronecker's existence) If $p \notin f(\partial\Omega)$, and $\deg(f, \Omega, p) \neq 0$, then $f^{-1}(p) \cap \Omega \neq \emptyset$.

2. (Homotopy invariance) If $F \in C([0, 1] \times \bar{\Omega}, \mathbb{R}^n)$, $p \notin F([0, 1] \times \partial\Omega)$, then

$$\deg(F(t, \cdot), \Omega, p) = \text{const.}$$

3. (Additivity) Assume $\Omega_1, \Omega_2 \subset \Omega$ are bounded open sets, $\Omega_1 \cap \Omega_2 = \emptyset$, and $p \notin f(\bar{\Omega} \backslash (\Omega_1 \cup \Omega_2))$, then

$$\deg(f, \Omega, p) = \deg(f, \Omega_1, p) + \deg(f, \Omega_2, p).$$

4. (Normality)

$$\deg(f, \Omega, p) = \begin{cases} 1 & \text{if } p \in \Omega, \\ 0 & \text{if } p \notin \Omega. \end{cases}$$

5. Suppose in addition, $f \in C^1(\Omega, \mathbb{R}^n)$ and $p \notin f(\partial\Omega)$ is a regular point of f, i.e.

$$\det(\partial_{x_i} f_j(x)) \neq 0, \qquad \forall\, x \in f^{-1}(p),$$

then

$$\deg(f, \Omega, p) = \sum_{x_k \in f^{-1}(p) \cap \Omega} \text{sgn} \det(\partial_{x_i} f_j(x_k)).$$

We now return to prove $\partial\Omega$ and S link. Clearly, $S \cap \partial Q = \emptyset$, it remains to show $\forall \varphi \in C(Q, X)$ satisfying $\varphi|_{\partial Q} = \text{id}|_{\partial Q}$, we have

$$\varphi(Q) \cap S \neq \emptyset.$$

Equivalently, we must show $\exists \, x_0 \in Q$ such that

$$P \circ \varphi(x_0) = \theta,$$

where $P : X \to X_1$ is a projection operator. For this, we define $F \in C([0,1] \times Q, \mathbb{R}^n)$ as

$$F(t, x) = tP \circ \varphi(x) + (1 - t)x.$$

Since

$$\theta \notin \partial Q = F(t, \partial Q), \qquad \forall t \in [0, 1],$$

by the homotopy invariance property and normality, it follows that

$$\deg(P \circ \varphi, Q, \theta) = \deg(\text{id}, Q, \theta) = 1.$$

Next, by Kronecker's existence property, $\exists \, x_0 \in Q$ such that

$$P \circ \varphi(x_0) = \theta.$$

Thus, S and ∂Q are linked. $\qquad\qquad\qquad\qquad\qquad\qquad\qquad\qquad \square$

Example 15.3 Let X be a Banach space, X_1 be a finite dimensional linear subspace, X_2 be its complementary space, i.e. $X = X_1 \oplus X_2$. Let $e \in X_2$ with $\|e\| = 1$, $R > \rho > 0$. Let

$$S = X_2 \cap \partial B_\rho(\theta)$$

and

$$Q = \{x_1 + te \mid (x_1, t) \in X_1 \times \mathbb{R}_+^1, \|x_1\|^2 + t^2 \leq R^2\},$$

then

$$\partial Q = (B_R(\theta) \cap X_1) \cup (\partial B_R(\theta) \cap (X_1 \oplus \mathbb{R}^1 e))^+,$$

where

$$(\partial B_R(\theta) \cap (X_1 \oplus \mathbb{R}^1 e))^+ = \{x_1 + te \mid (x_1, t) \in X_1 \times \mathbb{R}_+^1, \|x_1\|^2 + t^2 = R^2\},$$

(see Figure 15.3).

We want to show S and ∂Q link. Clearly, $S \cap \partial Q = \emptyset$. It remains to show $\forall \varphi \in C(Q, X)$ satisfying $\varphi|_{\partial Q} = \text{id}|_{\partial Q}$, $\exists \, x_0 \in Q$ such that

$$P \circ \varphi(x_0) = \theta, \quad \|\varphi(x_0)\| = \rho,$$

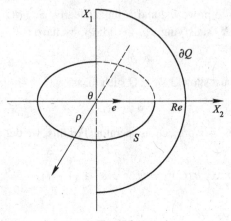

Fig. 15.3

where $P : X \to X_1$ is a projectin operator. This is equivalent to showing

$$P \circ \varphi(x_1 + se) = \theta, \quad \|(I - P) \circ \varphi(x_1 + se)\| = \rho,$$

where $x_1 + se = x_0$.

We define the deformation $F \in C([0, 1] \times Q, X_1 \times \mathbb{R}^1 e)$ via

$$F(t, x_1 + se) = [(1 - t)x_1 + tP \circ \varphi(x_1 + se)]$$
$$+ [(1 - t)s + t\|(I - P) \circ \varphi(x_1 + se)\| - \rho]e,$$

then

$$F(1, x_1 + se) = P \circ \varphi(x_1 + se) + [\|(I - P) \circ \varphi(x_1 + se)\| - \rho]e,$$

$$F(0, x_1 + se) = x_1 + (s - \rho)e.$$

Moreover, when $x_1 + se \in \partial Q$,

$$F(t, x_1 + se) = x_1 + (s - \rho)e \neq \theta.$$

It follows from the homotopy invariance that

$$\deg(F(1, \cdot), Q, \theta) = \deg(F(0, \cdot), Q, \theta).$$

The latter can be computed directly by using property (5):

$$F(0, \cdot)^{-1}(\theta, 0) = (\theta, \rho), \quad \det\left(\frac{\partial(x_1, s - \rho)}{\partial(x_1, s)}\right) = 1.$$

Thus,

$$\deg(F(1, \cdot), Q, \theta) = 1.$$

Next, by Kronecker's existence property, we have

$$F(1, x_1 + se) = (\theta, 0)$$

has a solution. Thus, S and ∂Q are linked.   □

Following the same proof of the Mountain Pass Theorem, we can also prove the following.

Theorem 15.2 *Let X be a Banach space, $Q \subset X$ be compact manifold with boundary ∂Q, and $S \subset X$ be a closed subset linked with ∂Q. Let $f \in C^1(X, \mathbb{R}^1)$. Suppose there exist $\alpha < \beta$ such that*

$$\sup_{x \in \partial Q} f(x) \leq \alpha < \beta \leq \inf_{x \in S} f(x). \tag{15.9}$$

Let

$$\Gamma = \{\varphi \in C(Q, X) \mid \varphi|_{\partial Q} = \mathrm{id}|_{\partial Q}\} \tag{15.10}$$

and

$$c = \inf_{\varphi \in \Gamma} \max_{\xi \in Q} f(\varphi(\xi)). \tag{15.11}$$

If f also satisfies the PS_c condition, then $c \ (\geq \beta)$ is a critical value of f.

15.2 Applications

Both the Mountain Pass Theorem and the linking theorem have numerous applications in variational problems. However, we will only exhibit their usefulness by a few examples to whet the reader's appetite.

Example 15.4 Given a periodic continuous function a with period $T > 0$ on the real line. Define the potential function

$$V(t, x) = -\frac{1}{2}|x|^2 + \frac{a(t)}{p+1}|x|^{p+1}. \tag{15.12}$$

Suppose $p > 1$, $a(t) \geq \alpha > 0$. Find a non-trivial T-periodic solution $x \in C^2([0, T], \mathbb{R}^N)$ of the system

$$\ddot{x} + V_x(t, x) = 0. \tag{15.13}$$

We define, on the space $H^1_{\mathrm{per}}((0, T), \mathbb{R}^N) := \{x \in H^1((0, T), \mathbb{R}^N) \mid x(0) = x(T)\}$, the functional

$$I(x) = \int_0^T \left[\frac{1}{2}(|\dot{x}|^2 + |x|^2) - \frac{a(t)}{p+1}|x|^{p+1} \right] dt. \tag{15.14}$$

It is clear that (15.13) is its E-L equation.

We claim $x = \theta$ is a local minimum of I, therefore it is a trivial solution of (15.13). First, we note

$$I(\theta) = 0.$$

Next, by the embedding theorem, $\|x\|_{p+1} \leq C\|x\|$, where C is a constant and $\|x\| = (\int_0^T [|\dot{u}|^2 + |u|^2] dt)^{\frac{1}{2}}$. Since $\exists\, M > 0$ such that $|a(t)| \leq M, \forall t \in [0,T]$, it follows that

$$(p+1)^{-1} \int_0^T a(t)|x(t)|^{p+1} dt \leq MC^{p+1}\|x\|^{p+1}.$$

Choose $\varepsilon > 0$ so small that whenever $x \in B_\varepsilon(\theta)\backslash\{\theta\}$, it holds that

$$I(x) = \frac{1}{2}\|x\|^2 - (p+1)^{-1} \int_0^T a(t)|x(t)|^{p+1} dt \geq \frac{1}{4}\|x\|^2.$$

Thus, $x = \theta$ is a local minimum.

In the following, we want to obtain a non-trivial critical point by the Mountain Pass Theorem. To do so, we choose a low point outside the "valley": $e = \lambda\xi \sin\frac{2t\pi}{T}, \xi = (1,1,\ldots,1)$. For $\lambda > 0$ sufficiently large, we have

$$I(e) = \lambda^2 n\left(\frac{2\pi^2}{T} + \frac{T}{2}\right) - \lambda^{p+1} n^{\frac{p+1}{2}} \int_0^T (p+1)^{-1} a(t) \left|\sin\left(\frac{2t\pi}{T}\right)\right|^{p+1} dt < 0.$$

Let

$$\Gamma = \{l \in C([0,1], H_0^1((0,T),\mathbb{R}^N)) \mid l(0) = \theta, l(1) = e\}$$

and

$$c = \inf_{l\in\Gamma} \sup_{x\in l} I(x) \geq \frac{1}{4}\varepsilon^2;$$

if we can verify the PS_c condition, then c is a critical value.

Suppose $\{x_n\}$ is a PS sequence with

$$\begin{cases} I(x_n) \to c \\ \|I'(x_n)\| = \sup_{\|\varphi\|=1} dI(x_n,\varphi) \to 0. \end{cases}$$

We want to show that $\{x_n\}$ has a convergent subsequence. Note

$$dI(x_n,\varphi) = \int_0^T \left[\dot{x}_n \cdot \dot{\varphi} + x_n \cdot \varphi - a(t)|x_n|^{p-1}x_n\varphi\right] dt \to 0,$$

$$\forall\, \varphi \in H_{\mathrm{per}}^1((0,T),\mathbb{R}^N). \quad (15.15)$$

In (15.15), by taking $\varphi = x_n$, it yields

$$\int_0^T \left[|\dot{x}_n|^2 + |x_n|^2 - a(t)|x_n|^{p+1}\right] dt = o(\|x_n\|) \quad (15.16)$$

and

$$\int_0^T \left[\frac{|\dot{x}_n|^2 + |x_n|^2}{2} - a(t)\frac{|x_n|^{p+1}}{p+1}\right] dt \to c(\neq 0). \quad (15.17)$$

Comparing (15.16) and (15.17), it yields

$$\left(\frac{1}{2} - \frac{1}{p+1}\right) \int_0^T |\dot{x}_n|^2 + |x_n|^2 dt = C_1 + o(\|x_n\|),$$

where C_1 is a constant. Hence, $\|x_n\|$ is bounded and there exists a subsequence, still denoted by x_n such that

$$x_n \rightharpoonup x_0, \quad \text{in } H^1_{\text{per}}((0,T), \mathbb{R}^N).$$

Lastly, we verify $x_n \to x_0$ in $H^1_{\text{per}}((0,T), \mathbb{R}^N)$ as follows.

By the embedding theorem,

$$x_n \to x_0, \quad \text{in } L^\infty((0,T), \mathbb{R}^N).$$

Furthermore, by the assumption of the PS sequence,

$$x_n - \left(-\frac{d^2}{dt^2} + 1\right)^{-1} (a(t)|x_n|^{p-1} x_n) \to 0, \quad H^1_{\text{per}}((0,T), \mathbb{R}^N).$$

Thus, x_n converges strongly in $H^1_{\text{per}}((0,T), \mathbb{R}^N)$. It follows immediately that $x_n \to x_0$ in $H^1_{\text{per}}((0,T), \mathbb{R}^N)$.

We have now verified all conditions stated in the Mountain Pass Theorem, so this problem has a non-trivial solution $x_0 \in H^1_{\text{per}}((0,T), \mathbb{R}^N)$. Furthermore, by regularity, we see that $x_0 \in C^2_{\text{per}}((0,T), \mathbb{R}^N)$. □

Remark 15.3 The same proof also applies to the case where

$$V(t,x) = \frac{-c}{2}|x|^2 + \frac{a(t)}{p+1}|x|^{p+1},$$

for $c > 0$.

Example 15.5 In Example 15.4, we change the potential to be

$$V(t,x) = \frac{c}{2}|x|^2 + \frac{a(t)}{p+1}|x|^{p+1}, \tag{15.12'}$$

where $c > 0$, and $p > 1$, $a(t) \geq \alpha > 0$. Find a non-trivial T-periodic solution of (15.13).

We still work with the space $H^1_{\text{per}}((0,T), \mathbb{R}^N)$, but we change the functional to be

$$I(x) = \int_0^T \left[\frac{1}{2}(|\dot{x}|^2 - c|x|^2) - \frac{a(t)}{p+1}|x|^{p+1}\right] dt. \tag{15.14'}$$

It is easy to see that $x = \theta$ is still a critical point. However, it is no longer a minimum. In order to find a non-trivial critical point, we must consider how the different level sets of the functional are actually linked.

We linearize Eq. (15.13) and the corresponding linear equation is

$$-\ddot{x}(t) = cx(t), \qquad t \in [0, T],$$

together with periodic boundary conditions.

We first turn our attention to the eigenvalue problem

$$-\ddot{x}(t) = \lambda x(t), \qquad t \in [0, T], \tag{15.18}$$

where $x(0) = x(T)$ and $\dot{x}(0) = \dot{x}(T)$.

In Lecture 12, it has been shown the eigenvalues are

$$\lambda_k = \left(\frac{2k\pi}{T}\right)^2, \qquad k = 0, 1, 2, \ldots$$

with corresponding eigenfunctions

$$\cos \frac{2k\pi t}{T} \otimes e, \quad \sin \frac{2k\pi t}{T} \otimes f,$$

where $e, f \in \mathbb{R}^N$ for $k \geq 1$; $e \in \mathbb{R}^N$ for $k = 0$. For each $k \geq 1$, we denote

$$E_k = \text{span} \left\{ \cos \frac{2k\pi t}{T} \otimes e, \sin \frac{2k\pi t}{T} \otimes f \,\middle|\, e, f \in \mathbb{R}^N \right\},$$

which is a $2N$ dimensional subspace of $H^1_{\text{per}}((0, T), \mathbb{R}^N)$, and $E_0 = \mathbb{R}^N$ is an N dimensional subspace.

We have the direct sum decomposition of the space $X = H^1_{\text{per}}((0, T), \mathbb{R}^N)$: $X = X_1 \oplus X_2$, where

$$X_1 = \bigoplus_{j=0}^{k} E_j \tag{15.19}$$

and $k = \max\{j \in \mathbb{N} \,|\, \lambda_j \leq c\}$.

On the subspace X_2,

$$\int_0^T |\dot{x}(t)|^2 dt \geq \lambda_{k+1} \int_0^T |x(t)|^2 dt.$$

Hence,

$$I(x) \geq \frac{1}{2} \left(1 - \frac{c}{\lambda_{k+1}}\right) \int_0^T |\dot{x}(t)|^2 dt - \frac{1}{p+1} \int_0^T a(t)|x(t)|^{p+1} dt.$$

In Example 15.4, we estimated

$$\int_0^T a(t)|x(t)|^{p+1} dt \leq MC^{p+1} \|x\|^{p+1} = o(\|x\|^2) \quad (x \to \theta).$$

Since $1 - \frac{c}{\lambda_{k+1}} > 0$, there exist $\rho > 0$ and $\beta > 0$ such that

$$I(x) \geq \beta, \quad \text{for } x \in \partial B_\rho(\theta) \cap X_2. \tag{15.20}$$

Let $S = \partial B_\rho(\theta) \cap X_2$.

On X_1, we always have

$$I(x) \leq \int_0^T \left[\frac{1}{2}(|\dot{x}|^2 - c|x|^2) \right] dt \leq \int_0^T \left[\frac{1}{2}(\lambda_k - c)|x|^2 \right] dt \leq 0.$$

We now take $e = \cos \frac{2(k+1)\pi t}{T} \otimes e_1$, where $e_1 = (1, 0, \ldots, 0) \in \mathbb{R}^N$. Since all norms on a finite dimensional space are equivalent, on the space $X_1 \oplus \mathbb{R}^1 e$, as $\|x\| \to \infty$, the following holds uniformly

$$I(x) \leq (\lambda_{k+1} - c) \int_0^T |x|^2 dt - \frac{\alpha}{p+1} \int_0^T |x|^{p+1} dt \to -\infty.$$

Next, we take

$$Q = \{(x_1, t) \in X_1 \times \mathbb{R}_+^1 \mid \|x_1\|^2 + t^2 = R^2\}.$$

For $R > \rho > 0$ sufficiently large,

$$I|_{\partial Q} \leq 0. \tag{15.21}$$

Additionally, according to Example 15.3, we see that S and ∂Q are linked.

Lastly, we verify the PS_c condition. The argument is similar to that of Example 15.5. The only difference is that it is less direct in verifying the boundedness of the PS sequence in $H^1_{\text{per}}((0, T), \mathbb{R}^N)$. Suppose $\{x_j\} \subset H^1_{\text{per}}((0, T), \mathbb{R}^N)$ satisfies $I(x_j) \to c$ and $I'(x_j) \to \theta$, then

$$\int_0^T [|\dot{x}_j(t)|^2 - c|x_j(t)|^2] dt = C_1 + o(\|x_j\|)$$

and

$$\int_0^T a(t)|x_j(t)|^{p+1} dt = C_2 + o(\|x_j\|).$$

By Hölder's inequality, we have

$$\int_0^T |x_j(t)|^2 dt \leq \left(\int_0^T a(t)|x_j(t)|^{p+1} dt \right)^{\frac{2}{p+1}} \left(\int_0^T a^{-\frac{2}{p-1}}(t) dt \right)^{\frac{p-1}{p+1}}.$$

Combining the three inequalities above, it follows that

$$\int_0^T |\dot{x}_j(t)|^2 dt = C_3 + o(\|x_j\|),$$

whence $\|x_j\| \leq C_4$.

According to (15.20), (15.21), and Theorem 15.2, we established that I has a critical value $c \geq \beta > 0$, which corresponds to a non-trivial solution of (15.13).

Similar methods can also be applied to partial differential equations.

Example 15.6 Let $\Omega \subset \mathbb{R}^n$ be a bounded domain. Let $1 < p < 2^* - 1$, $a \in C(\bar{\Omega})$, and $a(x) \geq \alpha > 0$, $\forall x \in \Omega$. Find a weak solution of the equation

$$-\Delta u(x) = a(x)|u(x)|^{p-1}u(x), \tag{15.22}$$

where $u \in H_0^1(\Omega)$. On $H_0^1(\Omega)$, define the functional

$$I(u) = \int_\Omega \left[\frac{1}{2}|\nabla u(x)|^2 - \frac{a(x)}{p+1}|u(x)|^{p+1}\right]dx; \tag{15.23}$$

(15.22) is its E-L equation.

Clearly, $u = \theta$ is a trivial solution. We want to find a non-trivial critical point of I. By Poincaré's inequality,

$$I(u) \geq \int_\Omega \frac{1}{2}|\nabla u(x)|^2 dx - o(\|u\|^2) \geq \frac{1}{4}\int_\Omega |\nabla u(x)|^2 dx \quad \text{as } |u\| \to 0,$$

so for $r > 0$ sufficiently small,

$$I|_{\partial B_r(\theta)} \geq \frac{r^2}{4}.$$

Choose any nonzero function $\varphi \in H_0^1(\Omega)$,

$$I(t\varphi) = t^2 \int_\Omega \frac{1}{2}|\nabla \varphi(x)|^2 dx - t^{p+1} \int_\Omega \frac{a(x)}{p+1}|\varphi(x)|^{p+1}dx \to -\infty.$$

If we take $p_0 = \theta$ and $p_1 = t\varphi$, then for t sufficiently large, we will have the desired geometric structure in the Mountain Pass Theorem.

It remains to verify the PS_c condition. Suppose $\{u_j\}$ satisfies $I(u_j) \to c$ and $I'(u_j) \to \theta$ in $H_0^1(\Omega)$, we want to show that it has a convergent subsequence. In fact, we have

$$\int_\Omega \left[\frac{1}{2}|\nabla u_j(x)|^2 - \frac{a(x)}{p+1}|u_j(x)|^{p+1}\right]dx \to c \tag{15.24}$$

and

$$\int_\Omega [\nabla u_j(x)\varphi(x) - a(x)|u_j(x)|^{p-1}u_j(x)\varphi(x)]dx = o(\|\varphi\|), \quad \forall \varphi \in H_0^1(\Omega). \tag{15.25}$$

Substituting $\varphi = u_j$ into the equation, it yields

$$\int_\Omega [|\nabla u_j(x)|^2 - a(x)|u_j(x)|^{p+1}]dx = o(\|u_j\|). \tag{15.26}$$

Combining (15.24) and (15.26), it follows that

$$\left(\frac{1}{2} - \frac{1}{p+1}\right) \int_\Omega |\nabla u_j(x)|^2 dx = C + o(\|u_i\|).$$

Thus, $\{u_j\}$ is a bounded sequence in $H_0^1(\Omega)$. Consequently, it has a weakly convergent subsequence $u_{j'} \rightharpoonup u_0$.

We prove $u_{j'} \to u_0$ strongly in $H_0^1(\Omega)$. From (15.25), we see that

$$u_{j'} = (-\Delta)^{-1}(a|u_{j'}|^{p-1}u_{j'} + o(\|u_{j'}\|)).$$

We notice

(1) $H_0^1(\Omega) \hookrightarrow L^{p+1}(\Omega)$ is compact,

(2) $u \mapsto a|u|^{p-1}u$ gives a bounded continuous embedding $L^{p+1}(\Omega) \to L^{\frac{p+1}{p}}(\Omega) \hookrightarrow (H_0^1)^*(\Omega)$,

(3) $(-\Delta)^{-1} \in L((H_0^1)^*, H_0^1)$ is continuous;

hence, $u_{j'} \to u_0$ strongly in $H_0^1(\Omega)$.

This completes the verification of the PS_c condition. By the Mountain Pass Theorem, we have the critical value $c \geq \frac{r^2}{4} > 0$, which corresponds to a non-trivial critical point.

When $c < 0$, the equation

$$-\Delta u(x) = cu(x) + a(x)|u(x)|^{p-1}u(x) \tag{15.22'}$$

shares the same conclusion as that of (15.22). The proof is identical, hence omitted.

When $c > 0$, we can apply the linking theorem to prove (15.22') has a non-trivial solution. $\qquad\square$

Remark 15.4 Both the Mountain Pass Theorem and its generalization — the linking theorem are special cases of a more general minimax principle. The original minimax principle can be traced back to G.D. Birkhoff while studying closed geodesics. It then underwent a systematic development by L. Liusternik, L. Schnirelmann, and M. Krasnoselski, and it has since become an important part of critical point theory. In the 1960s, R.S. Palais extended the Liusternik–Schnirelmann theory to infinite dimensional manifolds. Since the work of A. Ambrosetti and P.H. Rabinowitz on the Mountain Pass Theorem, minimax principle has matured rapidly and found numerous applications. At the same time, vast development also took place in the parallel branch of critical point theory — Morse Theory. Together they forge the new area of global variational calculus or topological variational methods. However, due to its broad connection with other subjects, we will omit it from this book and refer the interested reader to references such as [Ch], [MW], [St1], and [Ra1], etc.

Periodic solutions, homoclinic and heteroclinic orbits

16.1 The simple pendulum

We begin this lecture with the motion of a pendulum (see Figure 16.1) to introduce the concepts of periodic solutions, homoclinic and heteroclinic orbits.

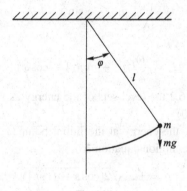

Fig. 16.1

A small ball of mass m is attached to a pendulum of length l and let to swing freely from side to side under gravity. Denote the angle of the pendulum from its stationary position by φ, then the kinetic energy is $\frac{1}{2}m(l\dot{\varphi})^2$ and its potential energy is $mgl(1 - \cos\varphi)$. The Lagrangian is $\frac{1}{2}m(l\dot{\varphi})^2 - mgl(1 - \cos\varphi)$ and its dynamical equation is

$$\ddot{\varphi} + \alpha \sin\varphi = 0,$$

where $\alpha = \omega_0^2 = \frac{g}{l}$.

On the phase plane $(x, y) = (\varphi, \dot{\varphi})$ (see Figure 16.2),

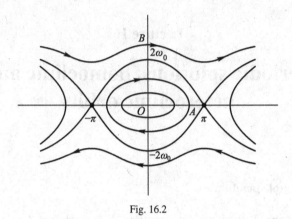

Fig. 16.2

we consider the system

$$\begin{cases} \dot{x} = y \\ \dot{y} = -\alpha \sin x. \end{cases}$$

Its energy is given by

$$E = \frac{m}{2}l^2 y^2 + mgl(1 - \cos x).$$

It is shown in Figure 16.2 the level sets of the energy as well as the equilibrium points $(0, 0)$ and $(\pm\pi, 0)$.

When $A \in (0, \pi)$, the energy at the initial point $(x, y) = (A, 0)$ is $E = mgl(1 - \cos A)$, whose motion equation is

$$y = \pm\omega_0\sqrt{2(\cos x - \cos A)}.$$

The pendulum's motion is periodic with period

$$T(A) = \frac{2\sqrt{2}}{\omega_0} \int_0^A \frac{dx}{\sqrt{\cos x - \cos A}}.$$

Furthermore,

$$\lim_{A \to 0} T(A) = \frac{2\pi}{\omega_0}, \quad \lim_{A \to \pi-0} T(A) = \infty.$$

The orbit: $(x(t), \dot{x}(t)) \to (\pm\pi, 0)$ as $t \to \pm\infty$, connecting $(-\pi, 0)$ and $(\pi, 0)$ through the points $(0, \pm 2\omega_0)$ in the upper or lower half plane is called a *heteroclinic orbit*.

If in addition, we assume $|B| > 2\omega_0$, the solution with energy $E = \frac{ml^2 B^2}{2}$ passing through the initial points $(x, y) = (0, B)$ is given by

$$y = \pm\sqrt{B^2 - 2\omega_0^2(1 - \cos x)}.$$

These are periodic curves in the upper and lower half planes respectively; if we let

$$t(x) = \int_0^x \frac{ds}{\sqrt{B^2 - 2\omega_0^2(1 - \cos s)}}$$

and

$$\tau(B) = t(2\pi),$$

then

$$p(x) = t(x) - \frac{\tau(B)}{2\pi}$$

is a 2π-periodic function. We have

$$\begin{cases} x(t + \tau(B)) = x(t) + 2\pi, \\ y(t + \tau(B)) = y(t). \end{cases}$$

In this sense, we call it a *periodic solution of the second kind*.

In a dynamical system, an orbit $x(t) \to p$ as $t \to \pm\infty$ connecting a saddle equilibrium point p to itself is called a *homoclinic orbit* (see Figure 16.3).

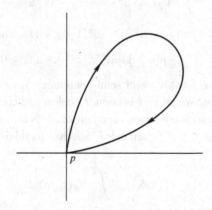

Fig. 16.3

16.2　Periodic solutions

In nonlinear oscillations, we are concerned with the periodic solutions of the following equation

$$\ddot{u} + \nabla_u V(t, u) = 0, \quad u(0) = u(T), \quad \dot{u}(0) = \dot{u}(T). \tag{16.1}$$

For example, the gravitational acceleration g as a result of the moon's gravitational force (tidal force) is a T-periodic function, whose potential energy is

$$V(t, u) = \frac{g(t)}{2\pi} \cos 2\pi u.$$

Generally speaking, we assume $V \in C^1([0, T] \times \mathbb{R}^N, \mathbb{R}^1)$ and introduce the Lagrangian

$$L(t, u, p) = \frac{|p|^2}{2} - V(t, u),$$

whose associated functional is

$$I(u) = \int_0^T \left[\frac{|\dot{u}|^2}{2} - V(t, u) \right] dt.$$

We regard (16.1) as the E-L equation of I and choose the underlying space to be

$$H^1_{\text{per}}([0, T], \mathbb{R}^N) = \{ u \in H^1([0, T], \mathbb{R}^N) \mid u(0) = u(T) \}.$$

We consider the following simple cases separately.

I. Suppose $u \mapsto -V(t, u)$ is continuous and convex, furthermore,

$$F(u) = -\int_0^T V(t, u) \, dt \to +\infty, \quad \text{as } \|u\|_{\mathbb{R}^N} \to \infty. \tag{16.2}$$

For example, let $N = 1$ and $V(t, u) = -|u|^p(1 + \varepsilon \sin t)$, where $p > 1$. Since

$$-V_{uu} = -p(p-1)|u|^{p-2}(1 + \varepsilon \sin t) > 0,$$

I is convex. Since I is weakly lower semi-continuous, it is weakly sequentially lower semi-continuous. We prove I is bounded below and coercive as follows.

Since $u \mapsto -V(t, u)$ is continuous and convex, F is a continuous and convex function on $\mathbb{R}^N$. So $\exists x_0 \in \mathbb{R}^N$ such that F achieves its minimum at x_0. We then have

$$0 = F'(x_0) = -\int_0^T V_u(t, x_0) \, dt. \tag{16.3}$$

By the convexity of $-V$, we have the inequality

$$-V(t, u) \geq -(V(t, x_0) + V_u(t, x_0)(u - x_0)), \tag{16.4}$$

$\forall u \in H^1_T(0,T)$, we have the decomposition $u = \tilde{u} + \bar{u}$, where $\bar{u} = \frac{1}{T}\int_0^T u\, dt$. Combining (16.3) and (16.4), we have

$$-\int_0^T V(t,u(t))dt \geq -\int_0^T (V(t,x_0) + V_u(t,x_0)(u(t) - x_0))dt$$

$$= -\int_0^T (V(t,x_0) + V_u(t,x_0)(u(t) - \bar{u}))dt. \qquad (16.5)$$

Let

$$c_1 = \int_0^T V(t,x_0)\, dt \quad \text{and} \quad c_2 = \left(\int_0^T |V_u(t,x_0)|^2\, dt\right)^{\frac{1}{2}},$$

then

$$I(u) \geq \frac{1}{2}\int_0^T |\dot{\tilde{u}}|^2 - \int_0^T [V(t,x_0) + V_u(t,x_0)\tilde{u}(t)]dt$$

$$\geq \frac{1}{2}\int_0^T |\dot{\tilde{u}}|^2 - c_1 - c_2 \left(\int_0^T |\tilde{u}|^2\right)^{\frac{1}{2}}.$$

By Wirtinger's inequality, $\int_0^T |\tilde{u}|^2 \leq c_3^2 \int_0^T |\dot{\tilde{u}}|^2$, it follows that

$$I(u) \geq \frac{1}{2}\int_0^T |\dot{\tilde{u}}|^2 - c_1 - c_2 c_3 \left(\int |\dot{\tilde{u}}|^2\right)^{\frac{1}{2}}$$

$$\geq \frac{1}{4}\int |\dot{\tilde{u}}|^2 - c_4.$$

Consequently, the coerciveness of I is determined by whether $\bar{u}$ can be bounded by $I(u)$. By convexity,

$$-V(t,\bar{u}/2) = -V\left(t, \frac{u(t) + (-\tilde{u}(t))}{2}\right)$$

$$\leq -\frac{1}{2}(V(t,u(t)) + V(t,-\tilde{u}(t))),$$

it follows that

$$I(u) \geq \frac{1}{2}\int_0^T |\dot{\tilde{u}}|^2 - 2\int_0^T V\left(t, \frac{\bar{u}}{2}\right)dt + \int_0^T V(t,-\tilde{u}(t))dt.$$

Since $\|\tilde{u}\|_\infty \leq C\|\tilde{u}\|_{H^1}$ is bounded by $I(u)$, $-\int_0^T V(t,-\tilde{u}(t))dt$ is also bounded by $I(u)$, whence

$$-\int_0^T V\left(t, \frac{\bar{u}}{2}\right) \leq c_5 I(u) + c_6.$$

By (16.2), $\bar{u}$ is bounded by $I(u)$; this proves that I is coercive.

From this, we see that Eq. (16.1) has a solution $u \in H^1_{\text{per}}([0, T])$. By regularity, $u \in C^2$.

Lastly, we examine the periodic condition. By the embedding theorem and $u \in H^1_{\text{per}}([0, T])$, $u(0) = u(T)$. It remains to show $\dot{u}(0) = \dot{u}(T)$. In the integral form of the E-L equation, we choose a period-T function $\varphi \in C^\infty([0, T])$, then we have

$$0 = \int_0^T [\dot{u}\dot{\varphi} - V_u(t, u)\varphi]dt$$

$$= \int_0^T [-\ddot{u} - V_u(t, u)]\varphi dt + \dot{u}\varphi|_0^T$$

$$= \dot{u}(T)\varphi(T) - \dot{u}(0)\varphi(0).$$

Since $\varphi(0) = \varphi(T)$ is arbitrary, we have shown

$$\dot{u}(T) = \dot{u}(0).$$

Theorem 16.1 *Under the assumption of* (16.2), *Eq.* (16.1) *has a* C^2 *periodic solution.* $\square$

II. V is continuous and periodic. Suppose there exist linearly independent vectors $e_1, \ldots, e_N \in \mathbb{R}^N$ such that

$$V(t, u + e_i) = V(t, u), \quad \forall (t, u) \in [0, T] \times \mathbb{R}^N. \tag{16.6}$$

We must again verify

$$I(u) = \int_0^T \left(\frac{1}{2}|\dot{u}(t)|^2 + V(t, u(t)) \right) dt$$

is weakly sequentially lower semi-continuous and coercive.

According to Morrey's Theorem in Lecture 11, I is for certain weakly sequentially lower semi-continuous.

Since V is continuous and satisfies (16.6), there exists a constant C such that

$$|V(t, u)| \le C.$$

However, we cannot deduce the coerciveness of I directly from

$$I(u) \ge \frac{1}{2}\int_0^T |\dot{u}|^2 - CT.$$

Thus, we choose the decomposition

$$u = \tilde{u} + \bar{u},$$

where $\bar{u} = \frac{1}{T}\int_0^T u(t)\, dt$. Denote

$$X = \{u \in H^1_T([0, T]) \mid \bar{u} = 0\},$$

then X is a closed linear subspace of $H^1_{\text{per}}([0, T])$.

It follows from Wirtinger's inequality that

$$\int_0^T |\dot{u}|^2 = \int_0^T |\dot{\tilde{u}}|^2 \geq \alpha \int_0^T |\tilde{u}|^2;$$

namely, I is coercive on X.

Suppose $\{u_j\}$ is a minimizing sequence of I, we decompose $u_j = \tilde{u}_j + \bar{u}_j$, then $\{\tilde{u}_j\}$ has a weakly convergent subsequence.

Noting that V is periodic: $V(u + \sum \lambda_i e_i) = V(u)$, hence

$$I\left(u + \sum \lambda_i e_i\right) = I(u), \quad \forall (\lambda_1, \ldots, \lambda_n) \in \mathbb{Z}^n.$$

Although $\{\bar{u}_j\}$ could be unbounded, but after removing the integer parts, it becomes bounded, i.e. $\exists (\lambda_1^{(j)}, \ldots, \lambda_n^{(j)}) \in \mathbb{Z}^n$ such that

$$\left\| \bar{u}_j + \sum_{i=1}^N \lambda_i^{(j)} e_i \right\|_{\mathbb{R}^n} \leq \sum_{i=1}^n \|e_i\| \triangleq A.$$

Let $v_j = \tilde{u}_j + (\bar{u}_j + \sum \lambda_i^{(j)} e_i)$, then $I(v_j) = I(u_j)$, whereas $\{v_j\}$ is a bounded minimizing sequence. Consequently, it has a weakly convergent subsequence, still denoted $\{v_j\}$ such that

$$v_j \rightharpoonup u^*.$$

The same argument can be used to show u^* is a minimum. Likewise, the same steps as above can be used to verify u^* is a periodic solution. Lastly, by regularity, $u^* \in C^2(\mathbb{R}^1)$.

Theorem 16.2 *Under the assumption of* (16.6), *Eq.* (16.1) *has a* C^2 *periodic solution.* □

III. Periodic solutions on the torus

Using the same methods as above, we can study the periodic solutions of the E-L equation of a functional defined on the torus $T^N = \mathbb{R}^N / \mathbb{Z}^N$. Given a Lagrangian on the torus

$$L(t, u, p) : T \times T^N \times \mathbb{R}^N \to \mathbb{R}^1$$

which satisfies

$$\dot{L}(t + \mathbb{Z}, u + \mathbb{Z}^n, p) = L(t, u, p)$$

and the following conditions

$$\begin{cases} c^{-1} \leq L_{pp} \leq c, \\ |L_{pt}| + |L_{pu}| \leq c(1 + |p|), \\ |L_u| \leq c(1 + |p|^2), \end{cases}$$

then for the functional

$$I(u) = \int_0^1 L(t, u(t), \dot{u}(t)) dt,$$

its E-L equation

$$\int_0^1 \{L_p(t, u(t), \dot{u}(t))\dot{\phi}(t) + L_u(t, u(t), \dot{u}(t))\phi(t)\} dt = 0,$$

$$\forall \phi(t) \in H^1_{\text{per}}([0, 1], \mathbb{R}^N)$$

has a periodic solution $u \in H^1_{\text{per}}([0, 1], \mathbb{R}^N)$. Moreover, by regularity, we know $u \in C^2([0, 1], \mathbb{R}^N)$. $\square$

IV. $M_{q,p}$-periodic solutions of the second kind on the torus

Under the same assumptions as before, we consider periodic solutions of the second kind.

$\forall (q, p) \in \mathbb{Z}^2$, $q \neq 0$, u is called a periodic solution of type (q, p), if

$$u(t + q) = u(t) + p.$$

A periodic solution of the free oscillation of a pendulum is a type $(1, 0)$ periodic solution, whereas a periodic solution of the second kind is a type $(1, 1)$ periodic solution.

In $M_{q,p} = \{\{\frac{p}{q}t\} \mid t \in \mathbb{R}^1\} + W^{1,2}_{\text{per}}([0, q])$, the minimum of

$$I(u) = \int_0^q L(t, u(t), \dot{u}(t)) \, dt$$

is a desired type (q, p) periodic solution.

Using the same minimization technique, the same argument affirms the existence of the minimum of I. $\square$

16.3 Heteroclinic orbits

By definition, a heteroclinic orbit is an orbit connecting the (non-degenerate) zeros of a vector field. For instance, given a Lagrangian

$$L(t, u, p) = \frac{1}{2}|p|^2 - V(t, u),$$

suppose

1) $V \in C^2(\mathbb{R}^N, \mathbb{R}^1)$;
2) V is T periodic in t;
3) $V(t, u) \leq 0$; there are only two non-degenerate maxima θ and ξ such that

$$V(t, \theta) = V(t, \xi) = 0, \ V_u(t, \theta) = V_u(t, \xi) = 0,$$

where $V_{uu}(t, \theta)$ and $V_{uu}(t, \xi)$ are both negative definite;

4) $\exists\, V_0 < 0$ such that

$$\varlimsup_{|u|\to\infty} V(t,u) \le V_0.$$

Find a solution connecting θ and ξ of the following equation:

$$\begin{cases} \ddot{u} + V_u(t,u) = 0, & t \in \mathbb{R}^1 \\ \dot{u}(-\infty) = \dot{u}(+\infty) = 0,\ u(-\infty) = \theta,\ u(+\infty) = \xi. \end{cases} \tag{16.7}$$

We solve this by variational methods. Choose the space

$$\hat{E} = \left\{ u \in H^1_{\mathrm{loc}}(\mathbb{R}^1, \mathbb{R}^N) \,\middle|\, \int_{\mathbb{R}^1} |\dot{u}|^2 < \infty \right\},$$

with norm

$$\|u\|_{\hat{E}}^2 = \int_{\mathbb{R}^1} |\dot{u}|^2 dt + |u(0)|^2.$$

We define the functional

$$I(u) = \int_{-\infty}^{\infty} \left(\frac{1}{2}|\dot{u}|^2 - V(t,u)dt \right).$$

I may take on value infinity on $\hat{E}$. However, since we are only concerned with a minimizing sequence $\{u_j\}$, it suffices to consider the set on which the value of I is finite. We define

$$\Gamma(\theta,\xi) = \{u \in \hat{E} \,|\, u(-\infty) = \theta, u(+\infty) = \xi\},$$

where $u(\pm\infty) = \lim_{t\to\pm\infty} u(t)$. We intend to find

$$\min_{u \in \Gamma(\theta,\xi)} I(u).$$

Applying Morrey's Theorem on any finite interval and by taking limits, it follows that I is weakly sequentially lower semi-continuous.

Since $I \ge 0$, it is bounded below.

1° We verify I is coercive. Since

$$\frac{1}{2}\int_{\mathbb{R}^1} |\dot{u}|^2 dt \le I(u) \le C,$$

it suffices to bound $|u(0)|$.

When $u \in \Gamma(\theta,\xi)$, we have

$$|u(b) - u(a)| \le \int_a^b |\dot{u}(t)|\,dt$$

$$\le (b-a)^{1/2}\left(\int_a^b |\dot{u}(t)|\,dt \right)^{1/2}.$$

As long as

$$-V(t, u(t)) \geq M_1, \tag{16.8}$$

we have

$$I(u) \geq \int_a^b \left(\frac{1}{2}|\dot{u}|^2 - V(t, u)\right) dt$$

$$\geq \frac{1}{2} \frac{|u(b) - u(a)|^2}{|b - a|} + M_1 |b - a|$$

$$\geq \sqrt{2M_1} |u(b) - u(a)|. \tag{16.9}$$

From assumptions (3) and (4), fixing $\varepsilon > 0$, there exists $M_1 > 0$ such that when

$$u(t) \notin B_\varepsilon(\theta) \cup B_\varepsilon(\xi),$$

(16.8) holds.

Noting that V is T-periodic in t, so by letting

$$(\tau_i u)(t) = u(t - iT), \quad \forall i \in \mathbb{Z},$$

we have $I(\tau_i u) = I(u)$.

In order to bound $|u(0)|$, we use the fact that V is τ_i invariant. $\forall u \in \Gamma(\theta, \xi)$, $\forall \epsilon > 0, \exists i_0$ such that $\tau = \tau_{i_0}$ satisfying

$$\tau u(t) \in B_\varepsilon(\theta), \quad t < 0, \quad \tau u(0) \in \partial B_\varepsilon(\theta).$$

Replacing u by $\tilde{u} = \tau u_j$, it is immediate that $\|\tilde{u}\| = \varepsilon$. This shows I is coercive. Moreover, replacing any $u \in \Gamma(\theta, \xi)$ by $\tilde{u}$, then $I(u) = I(\tilde{u})$.

2° Next, we verify $\Gamma(\theta, \xi)$ is weakly closed with respect to the minimizing sequence. That is, we want to prove: if $\{u_j\} \subset \Gamma(\theta, \xi)$, $u_j \rightharpoonup u$ and $I(u_j) \to \inf_{\Gamma(\theta, \xi)} I$, then $u \in \Gamma(\theta, \xi)$.

Since u is the weak limit of the minimizing sequence and I is weakly lower semi-continuous, it follows that

$$I(u) \leq \inf_{\Gamma(\theta, \xi)} I.$$

By (16.9), $u \in L^\infty(\mathbb{R}^1, \mathbb{R}^n)$. Hence, it has a ω-limit point, i.e. $\exists t_i \to +\infty$, $\exists \alpha \in \mathbb{R}^N$ such that $u(t_i) \to \alpha$.

1) We claim the limit point is unique. Suppose $\exists t_i' \to +\infty$ such that $u(t_i') \to \beta$. Since $I(u) < \infty$, it follows from (16.9) that

$$|u(t_i) - u(t_i')| \leq \frac{1}{\sqrt{2M_1}} \left| \int_{t_i}^{t_i'} \left[\frac{1}{2}|\dot{u}|^2 - V(t, u) dt \right] \right| \to 0,$$

hence $\alpha = \beta$.

2) We claim $\alpha = \theta$ or ξ. We proceed by contradiction. Suppose not, then $\exists\, t_1 > 0$ such that when $t > t_1$, we have

$$u(t) \notin B_\varepsilon(\theta) \cup B_\varepsilon(\xi).$$

Consequently,

$$I(u) \geq -\int_{t_1}^{\infty} V(t, u(t))\mathrm{d}t \geq \int_{t_1}^{\infty} M_1 = \infty,$$

a contradiction.

3) We claim $\alpha = \xi$. Suppose not, then $\alpha = \theta$. $\forall\, \delta > 0$, $\exists\, t_\delta > 0$ such that when $t \geq t_\delta$, we have $u(t) \in B_\delta(\theta)$.

Since $u_j \rightharpoonup u \in \hat{E}$, according to the exposition in the last paragraph of $1°$, we may choose u_j such that $u_j(t) \in B_\varepsilon(\theta)$ for $t \leq 0$ and $u_j(0) \in \partial B_\varepsilon(\theta)$. Choose $\delta < \varepsilon/4$ and $t_1 > t_\delta + 1$ such that $u(t_1) \in B_\delta(\theta)$.

$\exists\, j_0$ such that for $j \geq j_0$, $\|u_j - u\|_{L^\infty}([0, t_1]) < \delta$, hence $u_j(t_1) \in B_{2\delta}(\theta)$. It follows from (16.9) that

$$I(u_j) \geq \int_{t_1}^{\infty} \left[\frac{1}{2}|\dot{u}_j|^2 - V(t, u_j)\right] \mathrm{d}t + \frac{\varepsilon}{2}\sqrt{2M_1}.$$

We construct the sequence

$$v_j(t) = \begin{cases} 0, & t < t_1 - 1, \\ (t - t_1 + 1)u_j(t_1), & t \in [t_1 - 1, t_1], \\ u_j(t), & t > t_1, \end{cases}$$

then $v_j \in \Gamma(\theta, \xi)$ and

$$I(v_j) = \int_{t_1-1}^{t_1} \left[\frac{1}{2}|u_j(t_1)|^2 - V(t, v_j(t))\right] \mathrm{d}t + \int_{t_1}^{\infty} \left[\frac{1}{2}|\dot{u}_j|^2 - V(t, u_j(t))\right] \mathrm{d}t.$$

However,

$$\int_{t_1}^{\infty} \left[\frac{1}{2}|\dot{u}_j|^2 - V(t, u_j(t))\right] \mathrm{d}t \leq I(u_j) - \frac{\varepsilon}{2}\sqrt{2M_1}$$

and

$$\int_{t_1-1}^{t_1} \left[\frac{1}{2}|u_j(t_1)|^2 - V(t, v_j(t))\right] \mathrm{d}t \leq 2\delta^2 + \max_{|u| \leq 2\delta}(-V(t, u)).$$

Choosing $\delta > 0$ such that

$$2\delta^2 + \max_{|u| \leq 2\delta}(-V(t, u)) < \frac{\varepsilon}{4}\sqrt{2M_1},$$

we then obtain

$$I(v_j) \leq I(u_j) - \frac{\varepsilon}{4}\sqrt{2M_1},$$

but $I(u_j) \to \inf_{\Gamma(\theta,\xi)} I$, which contradicts the fact that $\{v_j\} \subset \Gamma(\theta,\xi)$.

We have successfully established $\Gamma(\theta,\xi)$ is weakly closed with respect to the minimizing sequence. Consequently, the functional I achieves its minimum, say u^*. By regularity, $u^* \in C^2(\mathbb{R}^N)$.

3° Lastly, we verify $\dot{u}^*(\pm\infty) = 0$.

We already know when $t > t_1$, $u^*(t) \in B_\varepsilon(\xi)$. Thus, by assumptions (3) and (1), there exist $\beta_1 > 0$ and $\beta_2 > 0$ such that

$$-V(t, u^*(t)) \geq \beta_1 |u^*(t) - \xi|^2,$$
$$|V_u(t, u^*(t))| \leq \beta_2 |u^*(t) - \xi|.$$

It folows that

$$\beta_1 \int_{t_1}^{\infty} |u^*(t) - \xi|^2 \mathrm{d}t \leq -\int_{t_1}^{\infty} V(t, u^*(t))\,\mathrm{d}t \leq I(u^*)$$

and

$$\int_{t_1}^{\infty} |\ddot{u}^*|^2 \mathrm{d}t = \int_{t_1}^{\infty} |V_u(t, u^*(t))|^2 \mathrm{d}t \leq \beta_2^2 \int_{t_1}^{\infty} |u^*(t) - \xi|^2 \mathrm{d}t,$$

from which, we can deduce $\int_{t_1}^{\infty} |\ddot{u}^*|^2 \mathrm{d}t < \infty$. Together with $\dot{u}^* \in L^2(\mathbb{R}^1)$, it follows that $\dot{u}^*(+\infty) = 0$. Likewise, $\dot{u}^*(-\infty) = 0$.

Theorem 16.3 *Under the assumptions* (1)–(4), *Eq.* (16.7) *has a heteroclinic orbit.*

16.4 Homoclinic orbits

Given a continuous periodic function $a \in C^1(\mathbb{R}^1)$ with period $2T > 0$. Suppose $\mu > 2$ and $\exists\, \alpha > 0$ such that $a(t) \geq \alpha$. We define the potential function

$$V(t, x) = -\frac{1}{2}|x|^2 + a(t)|x|^\mu. \tag{16.10}$$

We want to find a homoclinic orbit $x \in H^1(\mathbb{R}^1, \mathbb{R}^1)$ initiating from $x = 0$ which satisfies the equation

$$\ddot{x} + V_x(t, x) = 0. \tag{16.11}$$

We adopt the following method. $\forall k \in \mathbb{N}$, we find a $2kT$-periodic solution x_k of Eq. (16.11). Then by letting $k \to \infty$, we examine whether the sequence of solutions $\{x_k\}$ has a limit. If the limit exists, is it still a solution of (16.11)? Is it a homoclinic orbit?

We begin by defining for $\forall\, k \in \mathbb{N}$, the space $X_k = H^1_{2kT}([-kT, kT])$ with norm

$$\|x\|^2_k = \int_{-kT}^{kT} (|\dot{x}(t)|^2 + |x(t)|^2)dt$$

and the functional

$$I_k(x) = \frac{1}{2}\|x\|^2_k - \int_{-kT}^{kT} a(t)|x(t)|^\mu dt.$$

According to Example 15.5 in Lecture 15, I_k possesses the geometric structure described in the Mountain Pass Theorem, that is,

$$I_k(0) = 0, \quad I_k(x) = \frac{1}{2}\|x\|^2_k + o(\|x\|^2_k), \quad \exists\, \varphi \in X_k\backslash\{0\} \text{ such that } I(t\varphi) \to -\infty.$$

Furthermore, I_k satisfies the Palais–Smale condition. Thus, there exists a critical point $x_k \in X_k$, i.e.

$$\int_{-kT}^{kT} [\dot{x}_k\dot{\varphi} + x_k\varphi - \mu a|x_k|^{\mu-2}x_k\varphi]dt = 0, \quad \forall\, \varphi \in X_k, \qquad (16.11')$$

with

$$c_k = I(x_k) > 0.$$

Noting that a is $2T$-periodic, while $x_k(t)$ is $2kT$-periodic, $x_k(t - jT)|_{[-kT,kT]}$ ($j \in \mathbb{Z}$) are also solutions of (16.10) with the same period and the same critical value. Therefore, we can choose such a x_k that

$$\max_{t\in[0,T]} x_k(t) = \max_{t\in[-kT,kT]} x_k(t). \qquad (16.12)$$

We proceed by the following steps.

$1°$ We claim $\exists\, M > 0, c_k \leq M, \forall\, k \in \mathbb{N}$.

To prove this, we choose $\varphi_1 \in X_1$ such that $\varphi_1(\pm T) = 0$ and $I_1(\varphi_1) \leq 0$. Let

$$M = \max_{t\in[0,1]} I_1(t\varphi_1)$$

and

$$z_k(t) = \begin{cases} \varphi_1(t), & t \in [-T, T], \\ 0, & t \in [-kT, kT]\backslash[-T, T]. \end{cases}$$

Then

$$z_k \in X_k, \quad I_k(z_k) = I_1(\varphi_1) \leq 0.$$

Hence, we have the estimate

$$c_k \leq \max_{t\in[0,1]} I_k(tz_k) = \max_{t\in[0,1]} I_1(t\varphi_1) = M.$$

2° We claim $\exists\, M_1 > 0$ such that $\|x_k\|_k \leq M_1$.

In (16.11), by choosing $\varphi = x_k$, it yields that

$$\int_{-kT}^{kT} [\dot{x}_k^2 + x_k^2 - \mu a(t)|x_k(t)|^\mu]dt = 0.$$

Combining this with the identity

$$c_k = I_k(x_k) = \frac{1}{2}\int_{-kT}^{kT}(\dot{x}_k^2 + x_k^2)dt - \int_{-kT}^{kT}a(t)|x_k(t)|^\mu dt,$$

we obtain that

$$\left(\frac{\mu}{2} - 1\right)\int_{-kT}^{kT}a(t)|x_k(t)|^\mu dt = c_k.$$

Thus,

$$\|x\|_k^2 = 2\left(I_k(z_k) + \int_{-kT}^{kT}a(t)|x_k(t)|^\mu\right)dt \leq \left(2 + \frac{2}{\mu - 2}\right)c_k \leq M_1.$$

3° We want to prove there exists a solution $y \in H^1(\mathbb{R}^1)$. Notice that for any $x \in H^1_{\text{loc}}$, we have the embedding inequality

$$|x(t)|^2 \leq 2\int_{t-\frac{1}{2}}^{t+\frac{1}{2}}(\dot{x}(r)^2 + x(r)^2)dr.$$

This implies there exist constants C and M_2 independent of k such that

$$\|x\|_{L^\infty[-kT,kT]} \leq C\|x\|_k \leq M_2.$$

Substituting into (16.11), we obtain a constant M_3 independent of k such that

$$\|x_k\|_{C^2[-kT,kT]} \leq M_3.$$

Subsequently, on any finite bounded interval $(-R, R)$, x_k as well as its derivative converge uniformly to a continuously differentiable function y, i.e.

$$x_k \to y \quad \text{uniformly in } C^1[-R, R].$$

So $\forall\, k \in \mathbb{N}$,

$$\int_{-kT}^{kT} [\dot{y}^2 + y^2]dt \leq M_1.$$

Since k is arbitrary, we must have

$$\int_{-\infty}^{\infty} [\dot{y}^2 + y^2]dt \leq M_1. \tag{16.13}$$

This means $y \in H^1(\mathbb{R}^1)$ and it satisfies Eq. (16.11) on the entire real line.

4° We want to prove the solution y is a homoclinic orbit. That is, we must show $y(t) \to 0$ as $t \to \pm\infty$. We use the Fourier transform of y:

$$\tilde{y}(\xi) = \frac{1}{2\pi} \int_{-\infty}^{\infty} y(t)e^{i\xi t}dt.$$

By Plancherel's theorem,

$$\int_{-\infty}^{\infty} (1 + |\xi^2|)|\tilde{y}(\xi)|^2 d\xi = \int_{-\infty}^{\infty} [\dot{y}^2 + y^2]dt, \tag{16.14}$$

together with Schwarz's inequality, we have

$$\int_{-\infty}^{\infty} |\tilde{y}(\xi)|d\xi \leq \left(\int_{-\infty}^{\infty} (1 + |\xi^2|)|\tilde{y}(\xi)|^2 d\xi \right)^{\frac{1}{2}} \left(\int_{-\infty}^{\infty} (1 + |\xi|^2)^{-1}d\xi \right)^{\frac{1}{2}} < \infty.$$

Moreover, by the Riemann–Lebesgue theorem, it follows immediately that

$$y(t) \to 0 \quad \text{as } t \to \pm\infty.$$

5° Lastly, we explain why this solution y is non-trivial. The main idea is to show there exists a constant $\delta > 0$ such that

$$\|x_k\|_{L^\infty[-kT,kT]} \geq \delta. \tag{16.15}$$

To carry out this idea, we define the function

$$\phi(r) = \max_{t \in [0,T], |x| \leq r} \frac{a(t)|x|^\mu}{|x|^2}.$$

This is a monotone non-decreasing function which satisfies

$$\begin{cases} \phi(r) > 0, & r > 0, \\ \phi(r) \to \infty, & r \to \infty. \end{cases}$$

Since x_k is a solution of (16.11), we have

$$\|x_k\|_k^2 = \mu \int_{-kT}^{kT} a(t)|x_k(t)|^\mu dt.$$

ϕ and $\|x_k\|_k$ are related via

$$\|x_k\|_k^2 \leq \mu\phi(\|x_k\|_{L^\infty[-kT,kt]}) \int_{-kT}^{kT} |x_k(t)|^2 dt \leq \mu\phi(\|x_k\|_{L^\infty[-kT,kt]})\|x_k\|_k^2.$$

Thus,

$$\phi(\|x_k\|_{L^\infty[-kT,kT]}) \geq \frac{1}{\mu}.$$

By choosing an appropriate δ, (16.15) can be fulfilled.

Consequently, (16.12) implies

$$\max_{t \in [0,T]} |x_k(t)| \geq \delta,$$

whence

$$\max_{t \in \mathbb{R}^1} |y(t)| \geq \delta.$$

Theorem 16.4 *Suppose $\mu > 2$ and $a \in C^1(\mathbb{R}^1)$ is a $2T$-periodic solution $(T > 0)$. If $\exists \, \alpha > 0$ such that $a(t) \geq \alpha$, $\forall t \in \mathbb{R}^1$, then Eq. (16.11) has a non-trivial homoclinic orbit.*

Lecture 17

Geodesics and minimal surfaces

17.1 Geodesics

Let (M, g) be a Riemannian manifold. A curve

$$\gamma : [0, 1] \to M, \quad \dot{\gamma} \neq \theta$$

is called a *geodesic*, if the acceleration vector field

$$\frac{D}{dt} \frac{d\gamma}{dt} = 0.$$

This means, along the curve γ, the velocity vector field is parallel. Hence, along γ,

$$\frac{d}{dt} g\left(\frac{d\gamma}{dt}, \frac{d\gamma}{dt}\right) = 2g\left(\frac{D}{dt}\left(\frac{d\gamma}{dt}\right), \frac{d\gamma}{dt}\right) = 0 \iff \left\|\frac{d\gamma}{dt}\right\| = g(\dot{\gamma}, \dot{\gamma})^{\frac{1}{2}} = \text{const.}$$

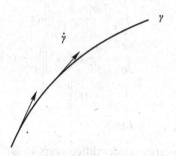

Fig. 17.1

In local coordinates, $\gamma = (u^1, \ldots, u^n)$ and the geodesic equation is given by

$$\frac{d^2 u^k}{dt^2} + \Gamma_{ij}^k(u) \frac{du^i}{dt} \cdot \frac{du^j}{dt} = 0, \ 1 \leq k \leq n,$$

where Γ_{ij}^k are the Christoffel symbols on M, i.e.

$$\Gamma_{ij}^k = \frac{1}{2} g^{kl} \left(\frac{\partial g_{il}}{\partial u^j} + \frac{\partial g_{jl}}{\partial u^i} - \frac{\partial g_{ij}}{\partial u^l} \right),$$

$$g_{ij} = g \left(\frac{\partial}{\partial u^i}, \frac{\partial}{\partial u^j} \right), \quad 1 \leq i, j, k \leq n.$$

$$\sum_{j=1}^n g^{ij} g_{jk} = \delta_k^i.$$

• The arclength functional and the energy functional

For a C^1 curve γ in (M, g), or more generally, $\gamma \in W^{1,1}([0, 1], M)$, we call

$$L(\gamma) = \int_0^1 \left\| \frac{d\gamma}{dt}(t) \right\| dt$$

the arclength functional. For $\gamma \in H^1([0, 1], M)$, we call

$$E(\gamma) = \frac{1}{2} \int_0^1 \left\| \frac{d\gamma}{dt}(t) \right\|^2 dt$$

the energy functional.

By Schwarz's inequality,

$$L(\gamma)^2 \leq 2E(\gamma),$$

and the equality holds if and only if $\| \frac{d\gamma}{dt}(t) \| = \text{const.}$

• L is invariant under parameter diffeomorphisms

$$
\begin{aligned}
L(\gamma \circ s) &= \int_0^T \left\| \frac{d\gamma \circ s}{d\tau}(\tau) \right\| d\tau \\
&= \int_0^T \left\| \frac{d\gamma}{dt}(s(\tau)) \right\| \left\| \frac{ds}{d\tau}(\tau) \right\| d\tau \\
&= \int_0^1 \left\| \frac{d\gamma}{dt}(t) \right\| dt \\
&= L(\gamma).
\end{aligned}
$$

However, E is not invariant under diffeomorphisms!

• The normalized arclength functional

$$l(t) = \frac{1}{L(\gamma)} \int_0^t \left\| \frac{d\gamma}{ds}(s) \right\| ds, \quad t \in [0, 1].$$

Since $l : [0, 1] \to [0, 1]$ is a diffeomorphism, letting $s = l^{-1}$, then $\gamma \circ s : [0, 1] \to M$ satisfies

$$\left\| \frac{d(\gamma \circ s)}{dl}(l) \right\| = \left\| \frac{d\gamma}{dt} \circ s(l) \right\| \frac{ds}{dl}(l) = \left\| \frac{d\gamma}{dt}(t) \right\| \left(\frac{dl}{dt}(t) \right)^{-1} = L(\gamma).$$

If γ is parametrized by its normalized arclength, i.e. it is parametrized by $\gamma \circ s$, then

$$L(\gamma)^2 = 2E(\gamma).$$

Given two points $P_0, P_1 \in M$, let

$$\Gamma = \left\{ \gamma \in H^1([0, 1], M) \mid \gamma(i) = P_i, \ i = 0, 1 \right\},$$

then

$$\boxed{\inf_{\gamma \in \Gamma} L(\gamma) = \inf_{\gamma \in \Gamma} \sqrt{2E(\gamma)}.}$$

Finding the curve with the shortest distance connecting the points P_0 and P_1 is equivalent to finding the curve connecting P_0 and P_1 with minimal energy.

Thus, the curve γ with the shortest arclength also minimizes the energy.

Consequently, γ satisfies the E-L equation of E:

$$\frac{d}{dt} L_p(t, u(t), \dot{u}(t)) = L_u(t, u(t), \dot{u}(t)),$$

where $L(t, u, p) = \frac{1}{2} \sum_{i,j=1}^{n} g_{ij}(u) p_i p_j$. Namely,

$$\sum_{j=1}^{n} \left[\frac{d}{dt}(g_{ij}(u(t))\dot{u}(t)) - \frac{1}{2} \frac{\partial g_{kj}}{\partial u^i}(u(t)) \dot{u}^k(t)\dot{u}^j(t) \right] = 0$$

$$\iff \sum_{j=1}^{n} \left[2g_{ij}\ddot{u}^j + 2\sum_{k=1}^{n} \frac{\partial g_{ij}}{\partial u^k}\dot{u}^k\dot{u}^j - \sum_{k=1}^{n} \frac{\partial g_{kj}}{\partial u^i}\dot{u}^k\dot{u}^j \right] = 0$$

$$\iff \ddot{u}^i + \frac{1}{2}\sum_{l=1}^{n}\sum_{k=1}^{n}\sum_{j=1}^{n} g^{il}(2g_{lj,k}\dot{u}^k\dot{u}^j - g_{kj,l}\dot{u}^k\dot{u}^j) = 0$$

$$\iff \ddot{u}^i + \frac{1}{2}\sum_{l=1}^{n}\sum_{k=1}^{n}\sum_{j=1}^{n} g^{jl}(g_{lj,k} + g_{kl,j} - g_{jk,l})\dot{u}^k\dot{u}^j = 0$$

$$\iff \ddot{u}^i + \sum_{k,j=1}^{n} \Gamma^i_{jk}\dot{u}^k\dot{u}^j = 0.$$

This is precisely the geodesic equation. So the curves of the shortest arclengths are themselves geodesics!

• The existence of geodesics

Assume (M, g) is a compact Riemannian manifold, given two points $P_0, P_1 \in M$, does there exist a geodesic connecting P_0 and P_1?

By The Nash Embedding Theorem, we can isometrically embed (M, g) into some $\mathbb{R}^N$ for N sufficiently large. For the functional

$$E(u) = \frac{1}{2} \int_0^1 \|\dot{u}(t)\|^2_{\mathbb{R}^N} dt,$$

consider the set

$$\begin{aligned} S &= \{ u \in H^1([0,1], \mathbb{R}^N) \mid u(0) = P_0,\ u(1) = P_1,\ u(t) \in M,\ \forall t \in [0,1] \} \\ &= C_*([0,T], M) \cap H^1([0,1], \mathbb{R}^N), \end{aligned}$$

where

$$C_*([0,T], M) = \{ u \in C([0,T], M) \mid u(i) = P_i,\ i = 0, 1 \}.$$

This is because the embedding $i : H^1 \hookrightarrow C$ is continuous.

Moreover, we have the estimate

$$\begin{aligned} \|u(t_1) - u(t_2)\|_{\mathbb{R}^N} &= \left\| \int_{t_1}^{t_2} \dot{u}(t) dt \right\| \\ &\leq \int_{t_1}^{t_2} \|\dot{u}(t)\| dt \\ &\leq |t_1 - t_2|^{\frac{1}{2}} \left(\int_{t_1}^{t_2} \|\dot{u}(t)\|^2 dt \right)^{\frac{1}{2}} \\ &\leq (2E(u))^{\frac{1}{2}} |t_1 - t_2|^{\frac{1}{2}}. \end{aligned}$$

This implies S is a weakly closed subset of $H^1([0,T], \mathbb{R}^N)$.

Using the direct method, E attains its minimum, which is also the minimum of L.

We arrive at the following.

Theorem 17.1 *Let (M, g) be a compact Riemannian manifold and let $P_0, P_1 \in M$ be given. Then there exists a curve joining P_0 and P_1 with the shortest arc length. Furthermore, this curve satisfies the geodesic equation.*

Remark 17.1 In the above approach, we did not seek the minimum of $L(\gamma)$ directly, this is because $L(\gamma)$ is invariant under any diffeomorphism on $[0, 1]$. However, the diffeomorphism group itself is vast and without compactness.

On the contrary, $E(\gamma)$ is not invariant under the diffeomorphism group action. Compactness is also available. In the original proof due to Hilbert, the use of "parametrizing by the arclength" was intended to avoid the diffeomorphism group. So in essence, it plays the same role as minimizing the energy functional.

17.2 Minimal surfaces

In geometry, a minimal surface is defined to be a surface whose mean curvature is zero.

For a surface X over the domain $\Omega \subset \mathbb{R}^2$, we adopt the parametric equation

$$X : \Omega \longrightarrow \mathbb{R}^n, \quad (x, y) \mapsto X = (X^1, \dots, X^n).$$

In isothermal coordinates, the parametrization X of a minimal surface satisfies

$$\begin{cases} \Delta X^i = 0, \ 1 \le i \le n & \text{(harmonic equation)} \\[2mm] \displaystyle\sum_{i=1}^{n} [(X_x^i)^2 - (X_y^i)^2] = \sum_{i=1}^{n} X_x^i \cdot X_y^i = 0, & \text{(weak conformal condition)}. \end{cases} \tag{17.1}$$

Example 17.1 The catenoid

$$\begin{cases} x = \cosh u \sin v, \\ y = \cosh u \cos v, \\ z = u. \end{cases}$$

The origin of its name is related to the following problem.

• The Plateau problem of minimal surfaces

Given a Jordan curve Γ in $\mathbb{R}^n$ (see Figure 17.2).

Find a (disklike) surface $X : \bar{D} \to \mathbb{R}^n$ bounded by Γ such that its enclosed area is a minimum. Denote

$$\bar{D} = \{ w = (x, y) \in \mathbb{R}^2 \,|\, |w| \le 1 \},$$
$$X = (X^1(w), \dots, X^n(w)).$$

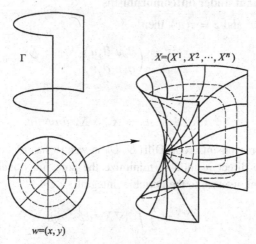

Fig. 17.2

Recall the square of the area determined by the two vectors X_x and X_y is

$$|X_x \wedge X_y|^2 = |X_x|^2 |X_y|^2 - (X_x \cdot X_y)^2,$$

hence the area of X is

$$A(X) = \int_D |X_x \wedge X_y| dx \wedge dy.$$

• **Boundary conditions** Let $\hat{X} = X|_{\partial D}$. Suppose $\hat{X} : \partial D \longrightarrow \Gamma$ is an oriented parametrization; namely, it is an orientation preserving homeomorphism. Denote

$$C(\Gamma) = \{ X \in H^1(\bar{D}, \mathbb{R}^n) \mid \hat{X} = X|_{\partial D} \in C(\partial D, \Gamma)$$
$$\text{is a weakly monotone parametrization} \}.$$

Before applying variational methods to study minimal surfaces, let us first insert certain condition on Γ such that

$$\inf_{X \in C(\Gamma)} A(X) < \infty,$$

which ensures $C(\Gamma) \neq \emptyset$.

For example, if we assume Γ is rectifiable, then we prove there exists $X_0 \in C(\Gamma)$ such that

$$A(X_0) = \inf_{X \in C(\Gamma)} A(X).$$

• A **is invariant under diffeomorphisms**

Let $z = (u, v)$ and $z = z(w)$, then

$$\frac{dz}{dw} = \begin{pmatrix} \partial_x u & \partial_y u \\ \partial_x v & \partial_y v \end{pmatrix}.$$

Thus,

$$X_x \wedge X_y \, dx \wedge dy = X_u \wedge X_v \, du \wedge dv.$$

Since the diffeomorphsim group $\text{Diff}(\bar{D}, \bar{D})$ is vast, it has no compactness. We will employ the direct method to minimize the area functional A. As in the geodesic problem, we turn to the Dirichlet integral

$$D(X) = \frac{1}{2} \int_D |\nabla X|^2 dx \wedge dy,$$

where $\nabla X = (X_x^i, X_y^i), 1 \leq i \leq n$.

• **Conformal mappings** A diffeomorphism

$$g : \mathbb{R}^2 \to \mathbb{R}^2, \quad w \mapsto z$$

is conformal, if it satisfies the following weakly conformal condition:

$$|g_x|^2 - |g_y|^2 = g_x \cdot g_y = 0.$$

A direct calculation shows that although the Diriihlet integral is not diffeomorphically invariant, it is conformally invariant!

$$\int_D |\nabla_w X|^2 dx \wedge dy = \int_D |\nabla_z X|^2 du \wedge dv.$$

We also note the following relations:

$$|\alpha \wedge \beta| \leq \frac{1}{2}(|\alpha|^2 + |\beta|^2), \quad "=" \iff |\alpha|^2 - |\beta|^2 = \alpha \cdot \beta = 0;$$

$$A(X) \leq D(X), \quad "=" \iff X \text{ is weakly conformal.}$$

We have the following important result.

Theorem 17.2 (Morrey–Lichtenstein) *Let* $\Gamma \in C^2$ *be a Jordan curve, then* $C(\Gamma) \neq \emptyset$, *and*

$$\inf_{X \in \mathbf{C}(\Gamma)} A(X) = \inf_{X \in \mathbf{C}(\Gamma)} D(X).$$

In order to avoid disrupting the proof of existence, we shall postpone the proof of the above theorem until the end of this lecture.

The advantage of minimizing the Dirichlet integral instead of the area functional is that the conformal group is far smaller than the diffeomorphism group.

The conformal group is the set of all conformal transformations forming a group. It is generated by three real parameters:

$$G = \left\{ g(w) = e^{i\phi} \frac{w + a}{1 + \bar{a}w} \,\middle|\, |a| < 1, \phi \in \mathbb{R}^1 \right\}.$$

It is worth noting when $a = a_j \to 1$, $g_j(w)$ is concentrated at one point on the unit circle. This means $\forall X \in C(\Gamma)$, the closure of the orbit $\{X \circ g \mid g \in G\}$ in the weak topology on $H^1(D, \mathbb{R}^n)$ contains constant functions. However, the latter cannot be a weakly monotone parametrization of Γ, so $C(\Gamma)$ cannot be a weakly closed subset in H^1. Even replacing the functional by the Dirichlet integral, it is still impossible to get around the weak closedness issue of $C(\Gamma)$.

To rectify this, we further reduce the influence of the conformal group G: by imposing additional restriction on $C(\Gamma)$, we hope to "mod out" the action of G.

Choose three arbitrary points P_0, P_1, and P_2 on Γ and let

$$C^*(\Gamma) = \left\{ X \in C(\Gamma) \,\middle|\, X\left(e^{\frac{2k\pi}{3}i} \right) = P_k, \, 0 \le k \le 2 \right\}.$$

We call $X\left(e^{\frac{2k\pi}{3}i} \right) = P_k$ for $0 \le k \le 2$ the three-point condition. Since

$$\inf_{X \in C(\Gamma)} A(X) = \inf_{X \in C(\Gamma)} D(X) = \inf_{X \in C^*(\Gamma)} D(X),$$

we instead find

$$\min\{D(X), \, X \in C^*(\Gamma)\}.$$

If X is the minimizing function, then it satisfies

(1)

$$\frac{d}{d\varepsilon} D(X + \varepsilon\varphi) \bigg|_{\varepsilon=0} = 0,$$

(2)

$$\frac{d}{d\varepsilon} D(X \circ g_\varepsilon^{-1}, g_\varepsilon(D)) \bigg|_{\varepsilon=0} = \frac{d}{d\varepsilon} \frac{1}{2} \int_{g_\varepsilon(D)} \left| \nabla_z (X \circ g_\varepsilon^{-1}(z)) \right|^2 du \wedge dv \big|_{\varepsilon=0} = 0,$$

where $g_\varepsilon : D \to g_\varepsilon(D)$ is a diffeomorphism.

The rationale is as follows. Although $C^*(\Gamma)$ is not a differentiable manifold, for any $\varphi \in C_0^\infty(D, \mathbb{R}^n)$, $X + \varphi \in C^*(\Gamma)$, so (1) holds. From this, we can deduce the E-L equation

$$\Delta X = 0 \quad \text{in } D,$$

which is the first equation in (17.1).

As for (2), we argue by contradiction. Suppose $\frac{d}{d\varepsilon} D(X \circ g_\varepsilon^{-1}, g_\varepsilon(D))|_{\varepsilon=0} \neq 0$, then there must exist a diffeomorphism $\bar{g}_\epsilon \in C^1(\bar{D}, \mathbb{R}^2)$ such that $\bar{X}_\varepsilon = X \circ \bar{g}_\varepsilon^{-1}$ satisfying

$$D(\bar{X}_\varepsilon, \bar{g}_\varepsilon(D)) < D(X).$$

However, $\bar{g}_\varepsilon(D)$ is simply connected, so by the Riemann Mapping Theorem, there exists a conformal mapping $h_\varepsilon : D \to \bar{g}_\varepsilon(D)$. Let $\tilde{X}_\varepsilon = \bar{X}_\varepsilon \circ h_\varepsilon$, then $\tilde{X}_\varepsilon \in C(\Gamma)$. Furthermore, by the conformal invariance of the Dirichlet integral, we have

$$D(\tilde{X}_\varepsilon) = D(\bar{X}_\varepsilon, \bar{g}_\varepsilon(D)) < D(X).$$

This is a contradiction!

From (2) and a similar argument used in Example 8.4 in Lecture 8, it follows that on D,

$$\begin{cases} |X_x|^2 = |X_y|^2, \\ X_x \cdot X_y = 0, \end{cases}$$

which is the second equation in (17.1).

Suppose $\frac{dg_\varepsilon}{d\varepsilon}|_{\varepsilon=0} = \tau$, $\tau = (\tau^1, \tau^2)$. Applying Noether's formula in Lecture 8 to $L(p) = \frac{1}{2}(\sum_{i=1}^n (p_1^i)^2 + (p_2^i)^2)$, it yields

$$\int_D \operatorname{div}(L\tau + L_{p^i} \cdot \varphi^i) \, dxdy = 0,$$

where $\varphi^i = -(X_x^i \tau^1 + X_y^i \tau^2)$.

Since

$$L\tau + L_{p^i} \cdot \varphi^i$$
$$= \frac{1}{2}(|X_x|^2 + |X_y|^2)(\tau^1, \tau^2) - \Sigma_i (X_x^i \tau^1 + X_y^i \tau^2)(X_x^i, X_y^i)$$
$$= \left(-\frac{1}{2}(|X_x|^2 - |X_y|^2)\tau^1 - X_x \cdot X_y \tau^2, \ -X_x \cdot X_y \tau^1 + \frac{1}{2}(|X_x|^2 - |X_y|^2)\tau^2 \right),$$

it follows that

$$\operatorname{div}(L\tau + L_{p^i} \cdot \varphi^i) = \frac{1}{2}(|X_x|^2 - |X_y|^2)(\tau_y^2 - \tau_x^1) - X_x \cdot X_y (\tau_x^1 + \tau_y^2).$$

We have thus proved

$$\frac{d}{d\varepsilon} D(X \circ g_\varepsilon^{-1}, g_\varepsilon(D))|_{\varepsilon=0}$$

$$= \frac{d}{d\varepsilon} \frac{1}{2} \int_{g_\varepsilon(D)} \left| \nabla_z (X \circ g_\varepsilon^{-1}) \right|^2 du \wedge dv$$

$$= -\frac{1}{2} \int_D [(|X_x|^2 - |X_y|^2)(\tau_x^1 - \tau_y^2) + 2X_x \cdot X_y (\tau_y^1 + \tau_x^2)] \, dxdy.$$

Since $\lambda = \tau_x^1 - \tau_y^2$ and $\mu = \tau_y^1 + \tau_x^2$ are arbitrary, we conclude

$$|X_x|^2 - |X_y|^2 = X_x \cdot X_y = 0 \quad \text{a.e. in } D. \qquad \square$$

We have shown the minimizing function of the energy functional is indeed a generalized solution of the minimal surface equation.

• Replacing $\mathbb{R}^n$ by a more general Riemannian manifold (N^n, h), while Γ is an embedded Jordan curve in N. The corresponding equation becomes

$$\begin{cases} \operatorname{trace}(\nabla dX) = 0, \\ h(X_x, X_x) - h(X_y, X_y) = h(X_x, X_y) = 0. \end{cases}$$

Via the isometric embedding $N \hookrightarrow \mathbb{R}^N$, it can also be written as

$$\begin{cases} \Delta X = A(X)(\nabla X, \nabla X), \\ |X_x|^2 - |X_y|^2 = X_x \cdot X_y = 0, \end{cases}$$

where $A(X)(\cdot, \cdot)$ denotes the second fundamental form of X.

The existence of solution for the Plateau problem of a minimal surface was proved by Douglas (1931), Rado (1933), Courant (1945), and Struwe (1988), etc. The regularity of solution was proved by Hilbrandt (1969, 1971), which states: If $\Gamma \in C^{2,\alpha}$, then $X \in C^{2,\alpha}(\bar{\Omega}, \mathbb{R}^N)$ for $0 < \alpha < 1$.

Proof of existence It is of particular note that in order to minimize $D(X)$, it suffices to only consider its boundary value $\hat{X}$ instead of the whole function X. The reason is, from E-L equation, we know a harmonic function achieves the minimal value of the Dirichlet integral $D(X)$, and a harmonic function is determined by its boundary values. In fact, consider the Fourier expansion of $\hat{X}$

$$\hat{X}(e^{i\theta}) = a_0 + \sum_{k=1}^{\infty}(a_k \cos k\theta + b_k \sin k\theta), \tag{17.2}$$

where

$$a_0 = \frac{1}{2\pi}\int_0^{2\pi} \hat{X}(e^{i\theta})d\theta, \quad a_k = \frac{1}{\pi}\int_0^{2\pi} \hat{X}(e^{i\theta})\cos k\theta d\theta,$$

$$b_k = \frac{1}{\pi}\int_0^{2\pi} \hat{X}(e^{i\theta})\sin k\theta d\theta, \quad k = 1, 2, \ldots,$$

$$\sum_{k=1}^{\infty} k(|a_k|^2 + |b_k|^2) < \infty,$$

then

$$X(re^{i\theta}) = \frac{1}{2}a_0 + \sum_{k=1}^{\infty}(a_k \cos k\theta + b_k \sin k\theta)r^k. \tag{17.3}$$

It is evident that we may regard

$$D(X) = \pi \sum_{k=1}^{\infty} k(|a_k|^2 + |b_k|^2)$$

as a functional defined on the set $C^*(\Gamma) \subset \mathcal{X} = \{\hat{X} = X|_{\partial D} \mid D(X) < \infty\}$; on $\mathcal{X}$, if we define the norm to be

$$\|\hat{X}\| = \left(\frac{1}{4}a_0^2 + \sum_{k=1}^{\infty} k(|a_k|^2 + |b_k|^2)\right)^{\frac{1}{2}},$$

then it is a Hilbert space. D is clearly weakly sequentially lower semi-continuous. Since the range of $\hat{X}$ falls on Γ, $|a_0|$ is bounded, hence $D(X)$ is coercive.

However, in order to apply the direct method, the notion of "compactness" is of particular concern.

For a given Γ, $\hat{X}$ is merely a parametrization of Γ.

For a family of minimizing functions, their continuous moduli are in fact continuous moduli of Γ under diffeomorphic transformations. The crucial step in proving the equicontinuity of a sequence of minimizing functions is that for a family of parametrizations of finite energy, intervals cannot be concentrated to a single point. The following lemma plays a key role.

Lemma 17.1 (Courant–Lebesgue) *Let* $X \in H^1(D, \mathbb{R}^n)$, *then* $\forall w \in \bar{D}$, $\forall \delta \in (0, 1)$, $\exists \rho \in [\delta, \sqrt{\delta}]$ *such that*

$$\int_{C'_\rho} |\partial_s X|^2 ds \leq 4 \frac{D(X)}{\rho |\ln \delta|},$$

where ∂_s *denotes the tangential derivative,* C'_ρ *is the arc whose center lies on the unit circle and whose radius is* ρ *(see Figure 17.3).*

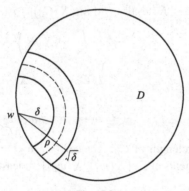

Fig. 17.3

Proof By Fubini's Theorem, for almost all $\rho \in (0, 1)$, $|\partial_s X| \in L^2(C_\rho)$ and

$$2D(X) \geq \int_{(B_{\sqrt{\delta}}(w) \setminus B_\delta(w))} |\nabla X|^2 dz$$

$$\geq \int_\delta^{\sqrt{\delta}} \int_{C_\rho} |\partial_s X|^2 ds d\rho$$

$$\geq \operatorname*{ess\ inf}_{\delta < \rho < \sqrt{\delta}} \left(\rho \int_{C_\rho} |\partial_s X|^2 ds \right) \int_\delta^{\sqrt{\delta}} \frac{d\rho}{\rho}.$$

Moreover, from

$$\int_\delta^{\sqrt\delta} \frac{d\rho}{\rho} = \frac{1}{2}|\ln\delta|,$$

the assertion now follows. □

As a consequence, we have the following.

Lemma 17.2 $\forall M > 0$, *the set*

$$C_M = \left\{ \hat{X} : \partial D \longrightarrow \Gamma \,\middle|\, \hat{X} = X|_{\partial D},\ X \in C^*(\Gamma),\ D(X) \le M \right\}$$

is equicontinuous.

Proof Since Γ is a Jordan curve, $\forall \varepsilon > 0,\ \exists d > 0,\ \forall P \ne P'$,

$$\boxed{|P - P'| < d}$$

$$\implies \boxed{\text{Among the two components of } \Gamma\backslash\{P, P'\}, \text{ at least one has diam} < \varepsilon.}$$

$1°.$ $\forall \varepsilon > 0$, assume $\varepsilon < \min_{i \ne j} |P_i - P_j|$. Choose $d > 0$ as above, choose $\delta > 0$ such that $\frac{8\pi M}{|\ln\delta|} < d^2$. $\forall X \in C_M,\ \forall w \in \partial D$, set $C_\rho = D \cap \partial B_\rho(w)$. By the Courant–Lebesgue lemma,

$$\begin{aligned}
|X(w_1) - X(w_2)|^2 &\le [\operatorname{diam} X(C_\rho)]^2 \\
&\le \left(\int_{C_\rho} |\partial_s X| ds \right)^2 \\
&\le L(C_\rho) \int_{C_\rho} |\partial_s X|^2 ds \\
&\le \frac{8\pi M}{|\ln\delta|}, \qquad \forall w_1, w_2 \in C_\rho,
\end{aligned}$$

where $L(C_\rho)$ is the arclength of C_ρ. Hence, $\{w_1, w_2\} = \partial D \cap \partial B_\rho(w) \implies$ among the two components of $\Gamma\backslash\{X(w_1), X(w_2)\}$, at least one has diameter less than ε.

$2°$ Choose $\delta > 0$ such that $\forall z \in \partial D$, at least two values of k satisfy

$$\left| z - e^{\frac{2\pi k}{3} i} \right| \ge \sqrt{\delta}.$$

Suppose $\partial D\backslash\{w_1, w_2\} = C_1 \cup C_2,\ C_1 \cap C_2 = \{w_1, w_2\}$, then at least one of C_1, C_2, say C_1, contains at most one point of $\left\{ e^{\frac{2k\pi}{3} i} \,\middle|\, k = 1, 2, 3 \right\}$. Hence,

$$\operatorname{diam} \hat{X}(C_1) < \varepsilon.$$

$\forall z, z' \in \partial D$, when $|z - z'| < \delta$, choose $w_0 \in \partial D$ and $\rho > 0$ such that $z, z' \in C_1$. Thus,

$$|\hat{X}(z) - \hat{X}(z')| \le \operatorname{diam} \hat{X}(C_1) < \varepsilon \implies C_M \text{ is equicontinuous.} \qquad \square$$

Theorem 17.3 *If* $\Gamma \in C^2$ *is a Jordan curve, then* $\exists X_0 \in C^*(\Gamma)$ *such that*

$$D(X_0) = \inf_{X \in \mathbf{C}^*(\Gamma)} D(X).$$

Proof We consider the functional $D(X)$ on $C^*(\Gamma)$

$$\hat{X} \leftrightarrow X \mapsto D(X),$$

where the relation of X and $\hat{X}$ is determined by (17.2) and (17.3). We already know that it is both weakly sequentially lower semi-continuous and coercive, it remains to show $C^*(\Gamma)$ is weakly closed. Suppose $\{\hat{X}_j\}$ converges to $\hat{X}_0$ weakly, then their norms are bounded. By Lemma 17.2, the sequence is equicontinuous. Furthermore, since the range of $\hat{X}$ lies on Γ, $\{\hat{X}_j\}$ is uniformly bounded. By the Arzelà–Ascoli theorem, it has a subsequence $\{\hat{X}'_j\}$ which converges uniformly to a continuous function. Consequently, $X_0 : \partial D \to \Gamma$ is continuous and weakly monotone. Furthermore, it also satisfies the three-point condition, so $X_0 \in C^*(\Gamma)$.

In summary, the Plateau problem of a minimal surface has a solution in $C(\Gamma)$. By Weyl's theorem, the solution is C^∞ in D. As for the regularity of the solution on $\bar{D}$, it is beyond the scope of this book, hence omitted. $\qquad\square$

We now return to the proof of Theorem 17.2, that is, $A(X)$ and $D(X)$ share the same infimum. We have shown previously $A(X) \leq D(X)$ and the equality holds if and only if X is weakly conformal.

Lemma 17.3 *Let* $\Gamma \in C^2$ *be a Jordan curve and* $X \in C(\Gamma) \cap C^2(\bar{D}, \mathbb{R}^n)$, *then* $\forall \varepsilon > 0$, $\exists g \in H^1 \cap C(\bar{D}, \mathbb{R}^2)$ *with* $g : \bar{D} \to \bar{D}$ *surjective and* $g : \partial D \to \partial D$ *monotone, such that* $X_\varepsilon \circ g$ *is weakly conformal, where* $X_\varepsilon(x, y) = (X(x,y), \varepsilon x, \varepsilon y) \in C^2(\bar{D}, \mathbb{R}^{n+2})$.

Proof Let

$$S = \{g \in C^1(\bar{D}, \mathbb{R}^2) \mid g : \bar{D} \to \bar{D} \text{ is a diffeomorphism,}$$
$$g(e^{\frac{2k\pi}{3} i}) = e^{\frac{2k\pi}{3} i} \text{ for } 1 \leq k \leq 3\}$$

and $\bar{S}$ be the weak closure of S in $H^1(D, \mathbb{R}^2)$. We also define

$$E(g) = D(X_\varepsilon \circ g) = \frac{1}{2} \int_D |(\nabla X_\varepsilon) \circ g) \cdot \nabla g|^2 \, dx dy.$$

Since

$$E(g) \geq \varepsilon D(g),$$

E is a weakly sequentially lower semi-continuous and coercive functional on $\bar{S}$. According to the proof of Theorem 17.3, E has a minimum g_0 on $\bar{S}$, i.e.

$$D(X_\varepsilon \circ g_0) = E(g_0) \leq D(X_\varepsilon \circ g), \quad \forall g \in \bar{S}.$$

From (2), it follows that $X_\varepsilon \circ g_0$ is weakly conformal.

At this point, we have already obtained

$$A(X_\varepsilon \circ g_0) = D(X_\varepsilon \circ g_0).$$

Although $A(X_\varepsilon \circ g_0) = D(X_\varepsilon \circ g_0)$ is true for any diffeomorphism g, but $g_0 \in \bar{S}$ may not be a diffeomorphism. In the following, we endeavor to extend the diffeomorphic invariance of the area to $\bar{S}$. If we are successful, then the proof of Theorem 17.2 is complete. But first, let us study $\bar{S}$.

Lemma 17.4 $\bar{S} \subset C(\bar{D}, \bar{D}) \cap C(\partial D, \partial D)$. $\forall g \in \bar{S}$, $g : \bar{D} \to \bar{D}$ *is surjective and it is weakly monotone on* ∂D; *furthermore, it also satisfies the three-point condition.*

Proof 1. For boundary points, by the Courant–Lebesgue lemma, or for $\Gamma = \partial D$, directly by Lemma 17.2, we have $\forall \varepsilon > 0$, $\exists \rho > 0$ such that

$$\sup_{w, w' \in C_\rho} |g(w) - g(w')| < \varepsilon,$$

where $C_\rho = \partial B_\rho(w_0) \cap D$.

2. For interior points, since $g \in S$ is a diffeomorphism, it maps $B_\rho(w_0) \cap D$ into a small neighborhood contained in $g(C_\rho)$,

$$\sup_{|w - w'| < \delta} |g(w) - g(w')| \le \sup_{w_0 \in \bar{D}} \left\{ \sup_{w, w' \in B_\rho(w_0) \cap D} |g(w) - g(w')| \right\}$$

$$= \sup_{w_0 \in \bar{D}} \left\{ \sup_{w, w' \in C_\rho} |g(w) - g(w')| \right\} < \varepsilon.$$

This implies the H^1 bounded subset of S is equicontinuous.

3. $\forall g \in \bar{S}$, $\exists \{g_j\}_1^\infty \subset S$ such that $g_j \rightharpoonup g$ in $H^1(D)$. We have proved earlier that g_j is equicontinuous on $\bar{D}$. According to the Arzelà–Ascoli theorem, $g_j \to g$ uniformly on $\bar{D}$, so $g \in C(\bar{D}, \bar{D})$ is surjective and $g \in C^*(\partial D)$. □

By adding differentiability to X, we can further extend this invariance.

Lemma 17.5 *Let* $\Gamma \in C^2$ *be a Jordan curve, then* $\forall g \in \bar{S}$, $\forall X \in C^*(\Gamma) \cap C^2(\bar{D}, \mathbb{R}^n)$, *we have* $X \circ g \in C^*(\Gamma)$ *and* $A(X \circ g) = A(X)$.

Proof From $X \in C^*(\Gamma)$ and Lemma 17.4, $g \in C^*(\partial D)$, hence $X \circ g \in C^*(\Gamma)$.

We already knew the area functional is invariant under diffeomorphisms, we now extend this property to $\bar{S}$ via weak limits. To do so, we divide into two steps.

1. Fix $w = (h(z)) \in \bar{S}$, $\forall g \in S$, consider the integral

$$\int_D |X_u \times X_v|_{w=(h(z))} \det(\nabla g(z)) dz.$$

Noting

$$\det(\nabla g) = \partial_x(g_1 \partial_y g_2) - \partial_y(g_1 \partial_x g_2),$$

and using integration by parts, it yields

$$2 \int_D |X_u \times X_v|_{w=(h(z))} \det(\nabla g) \, dxdy$$

$$= - \int_D [G_h(g_1 \partial_y g_2 - g_2 \partial_y g_1) - H_h(g_1 \partial_x g_2 - g_2 \partial_x g_1)] \, dxdy$$

$$+ \int_{\partial D} |X_u \times X_v|_{w=(h(e^{i\theta}))} \psi'(\theta) \, d\theta,$$

where

$$\psi'(\theta) = g_1 \partial_\theta g_2 - g_2 \partial_\theta g_1,$$
$$G_h = |X_u \times X_v|_u \partial_x h_1 + |X_u \times X_v|_v \partial_x h_2,$$
$$H_h = |X_u \times X_v|_u \partial_y h_1 + |X_u \times X_v|_v \partial_y h_2.$$

Since $G_h, H_h \in L^2(D)$, the area integral on the right-hand side of the equation can be extended to $g \in \bar{S}$.

In the line integral on the right-hand side of the equation, we notice

$$g_1^2 + g_2^2 = 1,$$

whence

$$\psi(\theta) = \arctan \frac{g_2(e^{i\theta})}{g_1(e^{i\theta})}.$$

When $g \in \bar{S}$, $g \in C(\bar{D}, \bar{D})$ and $g : \partial D \to \partial D$ is monotone, i.e. ψ is monotone. So $\psi'(\theta)$ exists almost everywhere. In addition,

$$\int_{\partial D} |\psi'(\theta)| d\theta = 2\pi.$$

From which, we conclude $\psi \in W^{1,1}(\partial D)$. This means the line integral on the right-hand side of the equation can also be extended to $g \in \bar{S}$. Namely, if $\{g^j\} \subset S$, $g \in \bar{S}$ such that $g^j \rightharpoonup g$, then

$$2 \int_D |X_u \times X_v|_{w=(g(z))} \det(\nabla g) \, dxdy$$

$$= \lim_{j \to \infty} [- \int_D [G_g(g_1^j \partial_y g_2^j - g_2^j \partial_y g_1^j) - H_g(g_1^j \partial_x g_2^j - g_2^j \partial_x g_1^j)] \, dxdy$$

$$+ \int_{\partial D} |X_u \times X_v|_{w=(g(e^{i\theta}))} \psi_j'(\theta) d\theta]$$

$$= \lim_{j \to \infty} 2 \int_D |X_u \times X_v|_{w=(g(z))} \det(\nabla g^j) \, dxdy,$$

where $\psi'(\theta) = g_1^j \partial_\theta g_2^j - g_2^j \partial_\theta g_1^j$.

2. We now prove: if $\{g_j\} \subset S,\, g_j \rightharpoonup g$ in $H^1 \cap C(\bar{D}, \bar{D})$, then

$$\int_D \left| |X_u \times X_v|_{w=(g_j(z))} - |X_u \times X_v|_{w=(g(z))} \right| \det(\nabla g_j)\, dxdy \to 0.$$

On one hand, Lemma 17.4 implies $g_j \to g$ uniformly on $\bar{D}$, hence

$$\left| |X_u \times X_v|_{w=(g_j(z))} - |X_u \times X_v|_{w=(g(z))} \right| \to 0 \text{ uniformly on } \bar{D}.$$

On the other hand, we have

$$\int_D |\det(\nabla g_j)|\, dxdy \le \int_D |\nabla g_j|^2\, dxdy \le M,$$

the assertion follows.

3. From

$$A(X \circ g) = A(X), \quad \forall g \in S,$$

it follows readily that

$$A(X \circ g) = A(X), \quad \forall g \in \bar{S}. \qquad \square$$

Proof of Theorem 17.2 For a Jordan curve Γ with differentiability assumption, by using it as boundary value to solve the harmonic equation, we see that $C(\Gamma) \ne \emptyset$.

From our earlier analysis, it suffices to show $\inf D(X) \le \inf A(X)$. Combining Lemmas 17.3 and 17.5, for $X \in C^*(\Gamma) \cap C^2(\bar{D}, \mathbb{R}^n)$, $\exists\, g \in \bar{S}$ such that

$$A(X_\varepsilon) = A(X_\varepsilon \circ g) = D(X_\varepsilon \circ g).$$

Since

$$D(X \circ g) \le D(X_\varepsilon \circ g)$$

and

$$\lim_{\varepsilon \to 0} A(X_\epsilon) = A(X),$$

it follows that

$$\inf_{X \in C(\Gamma)} D(X) \le \inf_{X \in C(\Gamma) \cap C^2(\bar{D}, \mathbb{R}^n)} A(X).$$

Moreover, since $\Gamma \in C^2$, $C^2(\bar{D}, \mathbb{R}^n) \cap C(\Gamma)$ is dense in $C(\Gamma)$, this completes the proof. $\qquad \square$

Numerical methods for variational problems

18.1 The Ritz method

The direct method for solving a variational problem consists of two main parts:

(1) the construction of a minimizing sequence,

(2) taking the limit of the minimizing sequence in order to obtain a solution.

L. Rayleigh and W. Ritz proposed a numerical method for finding a minimizing sequence, known as the Rayleigh–Ritz method. Let M_0 be a linear subspace of the functions space X over some domain Ω, I is the integral functional $\int_\Omega L(x, u(x), \nabla u(x))\, dx$ on Ω, whose domain $M = \varphi_0 + M_0$, where φ_0 is a function on Ω (determined by its inhomogeneous boundary value).

The idea of the Rayleigh–Ritz method is to choose a complete basis $\{e_1, e_2, \ldots\}$ for M_0 and let

$$M_n = \varphi_0 + \operatorname{span}\{e_1, e_2, \ldots, e_n\} \subset M.$$

We then search, on M_n, for the minimum $\varphi_n \in M_n$ of the functional I. Since any function u in M_n can be expressed as

$$u = \varphi_0 + \sum_{i=1}^n \xi_i e_i \qquad \xi = (\xi_1, \ldots, \xi_n) \in \mathbb{R}^n \tag{18.1}$$

$I|_{M_n}$ is in fact a function in the n real variables $\xi_1, \ldots, \xi_n$:

$$I(u) = I\left(\varphi_0 + \sum_{i=1}^n \xi_i e_i\right). \tag{18.2}$$

Consequently, we can either directly apply optimization technique to find the minimum φ_n of $I|_{M_n}$, or by solving the system of n homogeneous equations:

$$\frac{\partial}{\partial \xi_j} I\left(\varphi_0 + \sum_{i=1}^n \xi_i e_i\right) = 0, \qquad j = 1, 2, \ldots, n. \tag{18.3}$$

Closely related to this idea is the Ritz–Galerkin method. However, this method is not targeted at the functional I, rather at its E-L equation

$$\delta I(u, \varphi) = \int_{\Omega} (L_u(\tau)\varphi(x) + L_p(\tau)\nabla\varphi(x)) \, dx = 0, \quad \forall \varphi \in M_0,$$

where $\tau = (x, u(x), \nabla u(x))$. On M, we write u in the form of (18.1), then by choosing $\varphi = e_j, j = 1, \ldots, n$, successively, we obtain

$$\int_{\Omega} (L_u(\tau_n(x))e_j(x) + L_p(\tau_n(x))\nabla e_j(x)) \, dx = 0, \quad j = 1, \ldots, n, \qquad (18.4)$$

where

$$\tau_n(x) = \left(x, \varphi_0(x) + \sum_{i=1}^{n} \xi_i e_i(x), \nabla\varphi_0(x) + \sum_{i=1}^{n} \xi_i \nabla e_i(x) \right).$$

Note that (18.4) is also a system of n equations in the n variables $\xi = (\xi_1, \ldots, \xi_n)$.

Example 18.1 Let $f \in L^2(\Omega)$ and define the fucntional

$$I(u) = \int_{\Omega} \left[\frac{1}{2}|\nabla u|^2 - f \cdot u \right] dx, \quad u \in H_0^1(\Omega). \qquad (18.5)$$

Let $V_n = \text{span}\{e_1, \ldots, e_n\}$ be an n dimensional linear subspace of $H_0^1(\Omega)$. According to the Rayleigh–Ritz method, we convert it to be the problem of finding the minimal value of

$$J_n(\xi_1, \ldots, \xi_n) = \int_{\Omega} \left[\frac{1}{2} \sum_{i=1}^{n} |\xi_i \nabla e_i(x)|^2 + f(x) \sum_{i=1}^{n} \xi_i e_i(x) \right] dx$$

$$= \frac{1}{2} \sum_{i,j=1}^{n} a_{ij} \xi_i \xi_j + \sum_{j=1}^{n} b_j \xi_j,$$

where $a_{ij} = \int_{\Omega} \nabla e_i(x) \cdot \nabla e_j(x) \, dx$ for $i, j = 1, \ldots, n$, and $b_j = \int_{\Omega} f(x)e_j(x) \, dx$ for $j = 1, \ldots, n$.

This is equivalent to solving the following system of linear equations:

$$\sum_{j=1}^{n} a_{ij} \xi_j = b_i, \quad i = 1, 2, \ldots, n. \qquad (18.6)$$

Since the corresponding E-L equation is

$$\int_{\Omega} [\nabla u \cdot \nabla \varphi - f \cdot \varphi] \, dx = 0, \quad \forall \varphi \in H_0^1(\Omega),$$

according to the Ritz–Galerkin method, we must solve the system of equations

$$\sum_{i=1}^{n}\xi_i\int_{\Omega}\nabla e_i(x)\nabla e_j(x)\,dx - \int_{\Omega}f(x)e_j(x)\,dx = 0, \qquad j=1,\dots,n,$$

namely,

$$\sum_{i=1}^{n}a_{ij}\xi_i - b_j = 0, \qquad j=1,\dots,n. \tag{18.7}$$

Since $a_{ij}=a_{ji}$ for all i and j, (18.6) and (18.7) are identical to each other.

Remark 18.1 The Ritz–Galerkin method is not only applicable to the E-L equation of a functional, but also applicable to all differential equations whose weak solutions are expressed in terms of integrals. For example, suppose a_{ij}, b_j, and c are continuous functions, we define a weak solution $u \in H_0^1(\Omega)$ of the equation

$$\begin{cases} \displaystyle\sum_{i,j=1}^{n}\frac{\partial}{\partial x_j}\left(a_{ij}(x)\frac{\partial}{\partial x_j}u(x)\right) + \sum_{j=1}^{n}b_j(x)\frac{\partial}{\partial x_j}u(x) + c(x)u(x) = f(x), & x\in\Omega \\ u\big|_{\partial\Omega} = 0 \end{cases}$$

$$\tag{18.5$'$}$$

to be such that $\forall\,\varphi \in H_0^1(\Omega)$, it satisfies the integral equation

$$\int_{\Omega}\left[-\sum_{i,j=1}^{n}a_{ij}(x)\frac{\partial u(x)}{\partial x_j}\frac{\partial\varphi(x)}{\partial x_i} + \left(\sum_{j=1}^{n}b_j(x)\frac{\partial u(x)}{\partial x_j} + c(x)u(x) - f(x)\right)\varphi(x)\right]dx = 0.$$

By substituting $u = \sum_{k=1}^{n}\xi_k e_k(x)$ and $\varphi(x)=e_l(x)$, $l=1,2,\dots,n$, we can again obtain a system of n linear equations in n variables.

18.2 The finite element method

In view of the previous discussion, we are confronted with the challenge of how to construct such finite dimensional approximating subspaces. We propose a criterion, according to which, a sequence of finite dimensional subspaces approximating any given space can be constructed. As examples, in variational problems, we frequently encounter the space $H_0^1(\Omega)$ (the Dirichlet problem with zero boundary value) and the space $H^1(\Omega)$ (the zero Neumann boundary value), we follow the steps outline below to construct function spaces on Ω as finite dimensional subspaces of the above mentioned spaces.

For simplicity, we assume the domain Ω is a bounded polyhedron. In other words, $\partial\Omega$ consists of piecewise hyperplanes. We select spaces consisting of piecewise linear functions to approximate $H_0^1(\Omega)$ (or $H^1(\Omega)$).

We first introduce some terminology. We call $K \subset \mathbb{R}^n$ an n-simplex, denoted $K = \{P_0, P_1, \dots, P_n\}$, if it is the closure of the convex hull of the $n+1$ points

$\{P_0, P_1, \ldots, P_n\}$, where $P_i \in \mathbb{R}^n$ for $i = 0, 1, \ldots, n$ such that

$$\{P_i - P_0 \mid 1 \le i \le n\}$$

are linearly independent.

It is clear a 0-simplex is a point, a 1-simplex is a line segment, and a 2-simplex is a triangle, etc.

We call $J = \{K_1, \ldots, K_m\}$ a triangulation of Ω if

(1) $K_i \cap K_j = \emptyset$ or a p-simplex for $0 \le p \le n - 1$,

(2) $\Omega = \bigcup_{K_i \in J} K_i$.

Denote $h_J = \max\{\text{diam}(K_i) \mid 1 \le i \le m\}$ and we call it the mesh of J.

For the space $H^1(\Omega)$, the vertices of $\{K_i\}_1^m$ are called nodes, they are denoted by $N = \{N_j \mid 1 \le j \le M\}$.

For the space $H_0^1(\Omega)$, we ignore all vertices on $\partial\Omega$, but only take the vertices in the interior of Ω, they are denoted by $N = \{N_j \mid 1 \le j \le M_0\}$.

We correlate the nodes with the dual basis for the finite dimensional approximating subspace V_h as follows.

Choose piecewise affine functions as elements of V_h with basis functions $\{e_i\}_1^M$, they are determined by the node set via

$$e_i(N_j) = \delta_{ij} \qquad i, j = 1, 2, \ldots, M.$$

In fact, for any simplex $K \in J$, the value of the basis function e_i on K is completely determined by its value at each vertex (if the vertex lies on $\partial\Omega$, then set the value be zero for homogeneous Dirichlet boundary condition and set the value be a fixed constant for inhomogeneous Dirichlet boundary condition).

In particular, taking $K = (P_0, \ldots, P_n)$, $P_i = (p_i^1, \ldots, p_i^n)$ for $i = 0, 1, \ldots, n$, we define $n + 1$ functions via

$$\begin{cases} v_j(x) = x^j & j = 1, 2, \ldots, n \\ v_0(x) = 1, \end{cases}$$

where $x = (x^1, \ldots, x^n)$, it follows that

$$\begin{cases} x^j = \displaystyle\sum_{i=0}^n p_i^j e_i(x), \\ 1 = \displaystyle\sum_{i=0}^n e_i(x). \end{cases} \tag{18.8}$$

Thus, $\forall i$, the support of e_i is $\bigcup_{K \in J}\{K \mid N_i \in K\}$; namely, the union of the simplices which have N_i as a vertex.

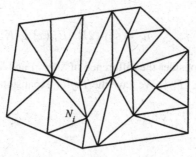

Fig. 18.1

An arbitrary function $v \in V_h$ can be expressed as

$$v(x) = \sum_{j=1}^{M} v(N_j)e_j(x).$$

We call (Ω, V_h, N) a finite element (see Figure 18.1).

For a given function $u \in C(\bar{\Omega})$, how do we construct an element in V_h to approximate u? A natural yet simple way is to use interpolation.

Denote Πu the corresponding element of u in V_h. $\forall K \in J$, denote $\Pi_K u = \Pi u|_K$. Since the functions in V_h are piecewise affine, $\Pi_K u$ is completely determined by the values of u at the vertices of K. Let

$$\Pi_K u(x) = \sum_{j=0}^{n} u(P_j)e_j(x),$$

where $K = \{P_0, \ldots, P_n\}$.

Consequently, $\Pi : C(\bar{\Omega}) \to H_0^1(\Omega)$ is a bounded linear operator with

$$\|\Pi u\|_{H^1} \leq C\|u\|_{C(\bar{\Omega})},$$

where $C \doteq n \sup_{1 \leq j \leq M} \|e_j\|_{H^1}$.

Lemma 18.1 *Let* $u \in C^2(\bar{\Omega})$, *then*

$$\|\nabla u - \nabla(\Pi u)\|_{1,\infty} \leq \frac{n^2(n+1)}{2} \frac{h_J^2}{\rho_J} \|\nabla^2 u\|_\infty, \tag{18.9}$$

$$\|u - \Pi u\|_\infty \leq \frac{n^2(n+1)}{2} h_J^2 \|\nabla^2 u\|_\infty, \tag{18.10}$$

where $\|\cdot\|_{1,\infty}$ *is the norm on* $W^{1,\infty}$, $\|\cdot\|_\infty$ *is the norm on* L^∞,

$$h_J = \sup_{K \in J} h_K, \quad h_K = \text{diam}(K),$$

and

$$\rho_J = \inf_{K \in J} \rho_K, \quad \rho_K = \sup\{2R \mid B_R(x) \subset K, \ x \in K\}.$$

Proof It suffices to consider the difference of $\nabla u(x)$ and $\nabla(\Pi_K u)$ on each simplex $K = \{P_0, P_1, \ldots, P_n\}$. We have the Taylor expansion of $\nabla(\Pi_K u)$:

$$\nabla(\Pi_K u)(x) = \sum_{j=0}^{n} u(P_j)\nabla e_j(x)$$
$$= \sum_{j=0}^{n} \left[u(x) + \nabla u(x)(P_j - x) + \frac{1}{2}\nabla^2 u(\xi_x)(P_j - x)^2 \right] \nabla e_j(x),$$

$$(18.11)$$

where $\xi_x \in k$. Differentiating (18.8), we obtain

$$\begin{cases} \delta_k^j = \sum_{i=0}^{n} p_i^j \dfrac{\partial e_i}{\partial x_k}, & j, k = 1, 2, \ldots, n \\ 0 = \sum_{i=0}^{n} \dfrac{\partial e_i}{\partial x_k}. \end{cases}$$

Substituting into (18.11), it yields

$$\nabla(\Pi_K u)(x) = \nabla u(x) + \sum_{j=0}^{n} R_j(x)\nabla e_j(x),$$

where

$$R_j(x) = \frac{1}{2}\nabla^2 u(\xi_x)(P_j - x)^2.$$

Since

$$\|P_j - x\| \leq \mathrm{diam}K \leq h_J, \quad \left|\frac{\partial e_i}{\partial x_k}\right| \leq \frac{1}{\rho_K} \leq \frac{1}{\rho_J},$$

this validates (18.9):

$$\|\nabla u - \nabla(\Pi u)\|_{1,\infty} \leq \frac{n^2(n+1)}{2}\frac{h_J^2}{\rho_J}\|\nabla^2 u\|_{\infty}.$$

Likewise, we can prove (18.10). $\qquad\square$

Theorem 18.1 *Let $u \in C(\bar{\Omega}) \cap H^2(\Omega)$. Suppose for a simplicial subdivision J, there exists $\beta > 0$ such that*

$$\frac{\rho_K}{h_K} \geq \beta, \quad \forall K \in J, \tag{18.12}$$

then

$$\|u - \Pi u\|_{L^2} \leq Ch_J^2 |u|_{H^2}$$

and

$$|u - \Pi u|_{H^1} \leq C_\beta h_J |u|_{H^2},$$

with the semi-norm

$$|v|_{H^r} = \left(\sum_{|\alpha|=r} \int_\Omega |D^\alpha v|^2 \, dx \right)^{\frac{1}{2}}, \quad r = 1, 2.$$

Proof According to Lemma 18.1, for any simplex $K \in J$ we have

$$\|u - \Pi u\|_{L^2(K)} \leq C h_K^2 |u|_{H^2(K)}$$

and

$$|u - \Pi u|_{H^1(K)} \leq C \frac{h_K^2}{\rho_K} |u|_{H^2(K)}.$$

Thus,

$$
\begin{aligned}
\|u - \Pi u\|_{L^2(\Omega)}^2 &= \sum_{K \in J} \|u - \Pi_K u\|_{L^2(K)}^2 \\
&\leq \sum_{K \in J} C^2 h_K^4 |u|_{H^2(K)}^2 \\
&\leq C^2 h_J^4 |u|_{H^2}^2
\end{aligned}
$$

and

$$
\begin{aligned}
|u - \Pi u|_{H^1(\Omega)}^2 &= \sum_{K \in J} |u - \Pi_K u|_{H^1(K)}^2 \\
&\leq \sum_{K \in J} C^2 \frac{h_K^4}{\rho_K^2} |u|_{H^2(K)}^2 \\
&\leq C^2 h_J^2 \frac{1}{\beta^2} |u|_{H^2(K)}^2. \qquad \square
\end{aligned}
$$

Remark 18.2 The geometric meaning of condition (18.12) is that in a simplicial subdivision, the angle of each simplex cannot be arbitrarily small.

Remark 18.3 The advantage of adopting the interpolation method in constructing approximating functions is evident not only in theoretical research but also in practical computations. The reason is that the interpolation method only requires the solution to be continuous, i.e. $u \in C(\bar{\Omega})$, however, the resulting approximation is in the sense of H^1. In particular, for linear second order elliptic equations, $u \in H^2(\Omega)$. When $n \leq 3$, by the embedding theorem, $H^2(\Omega) \hookrightarrow C(\bar{\Omega})$.

18.3   Cea's theorem

We begin this section with an abstract theorem. Let H be a Hilbert space equipped with an inner product $(\cdot, \cdot)$ (whose induced norm is $\| \cdot \|$). Let V be a closed linear subspace of H. Given $F \in V^*$.

Suppose

(1) $a(u, v)$ is a bounded bilinear form on H, i.e.

$$|a(u, v)| \le C\|u\|\|v\|,$$

$$a(\alpha_1 u_1 + \alpha_2 u_2, v) = \alpha_1 a(u_1, v) + \alpha_2 a(u_2, v), \quad \forall \alpha_1, \alpha_2 \in \mathbb{R}^1,$$

$$a(u, \beta_1 v_1 + \beta_2 v_2) = \beta_1 a(u, v_1) + \beta_2 a(u, v_2), \quad \forall \beta_1, \beta_2 \in \mathbb{R}^1.$$

(2) a is coercive on V, i.e. $\exists \alpha > 0$ such that

$$a(u, v) \ge \alpha\|v\|^2, \quad \forall v \in V.$$

(3) V_h is a finite dimensional closed subspace of V.

According to the Lax–Milgram theorem, under assumptions (1) and (2), there exists a unique $u \in V$ such that

$$a(u, v) = F(v), \quad \forall v \in V.$$

Using the Ritz–Galerkin method, we can obtain a unique approximation solution u_h such that

$$a(u_h, v) = F(v), \quad \forall v \in V_h.$$

We ask in what sense u_h approximates u.

Theorem (Cea) *Under the assumptions (1)–(3), if $u \in V$ is the unique solution of*

$$a(u, v) = F(v), \quad \forall v \in V$$

and if $u_h \in V_h$ is the solution of the approximated equation

$$a(u_h, v) = F(v), \quad \forall v \in V_h,$$

then

$$\|u - u_h\| \le \frac{C}{\alpha} \min_{v \in V_h} \|u - v\|.$$

Proof Since

$$a(u, v) = F(v), \quad \forall v \in V,$$
$$a(u_h, v) = F(v), \quad \forall v \in V_h,$$

it follows that

$$a(u - u_h, v) = 0, \quad \forall v \in V_h.$$

In particular,

$$\begin{aligned}
\alpha \|u - u_h\|^2 &\le a(u - u_h, u - u_h) \\
&= a(u - u_h, u - v) + a(u - u_h, v - v_h) \\
&\le C \|u - u_h\| \, \|u - v\|,
\end{aligned}$$

where $v \in V_h$. This means

$$\|u - u_h\| \le \frac{C}{\alpha} \|u - v\|, \quad \forall v \in V_h,$$

i.e.

$$\|u - u_h\| \le \frac{C}{\alpha} \min_{v \in V_h} \|u - v\|. \qquad \square$$

We now return to address the linear second order elliptic equation (18.5′) mentioned in Remark 18.1. Suppose $\alpha, \beta, \gamma > 0$ satisfy $\beta^2 < \alpha\gamma$ and

(1) $\sum_{i,j=1}^{n} a_{ij}(x)\xi_i\xi_j \ge \alpha|\xi|^2, \forall \xi = (\xi_1, \ldots, \xi_n) \in \mathbb{R}^n, |a_{ij}(x)| \le C$ for $1 \le i, j \le n, \forall x \in \Omega$,

(2) $c(x) \ge \gamma, \forall x \in \Omega$,

(3) $\max_{1 \le i \le n} \|b_i\|_\infty < 2\beta$.

Let $\delta = \alpha - \beta^2/\gamma$, then $\delta > 0$. We also let

$$a(u, v) = \int_\Omega \left(\sum_{i,j}^{n} a_{ij}(x)\partial_j u \, \partial_i v + \sum_{i=1}^{n} b_i(x)\partial_i u \cdot v + c(x)u \cdot v \right) \, \mathrm{d}x,$$

for $u, v \in H_0^1(\Omega)$.

The so-defined $a(u, v)$ is now a bounded bilinear form on $H_0^1(\Omega)$. Furthermore, from the inequality

$$\left| \int_\Omega \sum_{i=1}^{n} b_i(x)\partial_i u(x) \cdot u(x) \, \mathrm{d}x \right| \le \max_{1 \le i \le n} \|b_i\|_\infty \|\nabla u\|_2 \|u\|_2,$$

it follows that $a(u, v)$ is coercive on $H_0^1(\Omega)$ as affirmed by

$$a(v, v) \ge \alpha \|\nabla v\|^2 - 2\beta \|\nabla v\| \, \|v\|_2 + \gamma \|v\|_2^2 \ge \delta \|\nabla v\|^2 \to \infty.$$

Now, $\forall f \in L^2(\Omega)$, by the Lax–Milgram theorem, we obtain a unique solution $u \in H_0^1(\Omega)$ which satisfies

$$a(u, v) = \int_\Omega f \cdot v \, dx, \quad \forall v \in H_0^1(\Omega).$$

Moreover, by the regularity of elliptic equations, we know $u \in H^2(\Omega)$ and there exists a constant C such that

$$|u|_{H^2(\Omega)} \leq C\|f\|_{L^2(\Omega)}.$$

If we apply the Ritz–Galerkin method to find the approximated solution, adopting the previously discussed finite element (Ω, V_h, N), we obtain $u_h \in V_h$. Lastly, according to Cea's theorem and Theorem 18.1, we have:

$$\|u - u_h\|_{H_0^1(\Omega)} \leq \frac{C}{\delta} \min_{v \in V_h} \|u - v\|_{H_0^1(\Omega)}$$

$$\leq \frac{C}{\delta} h_J |u|_{H^2(\Omega)}$$

$$\leq \frac{C'}{\delta^2} h_J \|f\|_{L^2(\Omega)}.$$

In summary, using the finite element method, the interpolated function is not only an approximation of the original function in the sense of H_0^1-norm, but as an approximated solution, it is also an approximation of the real solution u in the sense of H_0^1-norm. The accuracy of the approximation is proportional to the mesh size h_J.

18.4 An optimization method — the conjugate gradient method

Optimization technique is a common technique used in finding the extremal values of a function numerically. Given a real-valued function $f : \mathbb{R}^n \to \mathbb{R}^1$, we intend to design an iterative algorithm in order to produce a minimizing sequence $x^0, x^1, \ldots, x^i, \ldots$ which will lead to the minimal value of f.

Starting from a point x^i, we want to choose x^{i+1} carefully so that the function value decreases in each step of the iteration, i.e. $f(x^{i+1}) < f(x^i)$, if it is possible.

It is worth noting, from the point x^i to another point x^{i+1}, we must take into account two important factors — the direction as well as the step length,

$$x^{i+1} - x^i = \lambda^i \sigma^i, \ \lambda^i > 0, \ \|\sigma^i\| = 1, \tag{18.13}$$

where σ^i denotes the direction of descent and λ^i its step length (see Figure 18.2).

We linearize f near x^i by

$$f(x) = f(x^i) + (\nabla f(x^i), x - x^i) + o(\|x - x^i\|).$$

Locally speaking, if we choose the direction of the negative gradient

$$\sigma^i = -\nabla f(x^i)/\|\nabla f(x^i)\|, \tag{18.14}$$

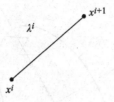

Fig. 18.2

then it gives the steepest descent. The step length λ^i can then be computed by minimizing the single variable function

$$\phi(t) = f(x^i + t\sigma^i)$$

in the direction of σ^i. This iterative algorithm is known as the gradient method, which is outlined as follows.

1. Given an initial point x^0.
2. Starting from x^i, use (18.14) to compute σ^i; then use

$$\begin{cases} \phi'(\lambda^i) = 0, \\ \phi(\lambda^i) < \phi(\lambda), \ \forall \lambda \in [0, \lambda^i), \end{cases}$$

to compute $\lambda^i > 0$, where $\phi(t)$ is defined as in (18.15).

3. Use (18.13) to compute x^{i+1} based on x^i, σ^i, and λ^i.

If $\sigma^{i+1} = 0$, then stop and x^{i+1} is the desired point. Otherwise, continue.

It is not difficult to show, if f is coercive with only one minimum, then starting from any initial point x^0, the gradient method will either reach the minimum in finite steps, or it will produce a minimizing sequence whose limit is the minimum.

However, the gradient method is not ideal, since the rate of convergence could be painfully slow. The following example clearly portrays such weakness. Let

$$f(x_1, x_2) = x_1^2 + \frac{1}{\varepsilon} x_2^2, \quad \varepsilon > 0.$$

This is a quadratic function whose level curves are all ellipses and whose minimum is at $(0,0)$. Starting from $x^i = (x_1^i, x_2^i)$, by direct calculation, we have

$$x^{i+1} = \left((1 + \lambda^i) x_1^i, \left(1 + \frac{\lambda^i}{\varepsilon} \right) x_2^i \right),$$

where

$$\lambda^i = \frac{-\left((x_1^i)^2 + \frac{1}{\varepsilon^2} (x_2^i)^2 \right)}{(x_1^i)^2 + \frac{1}{\varepsilon^3} (x_2^i)^2}.$$

So for $\varepsilon > 0$ very small,

$$x^{i+1} \approx \left((1 - \varepsilon) x_1^i, -\varepsilon^2 \frac{(x_1^i)^2}{x_2^i} \right).$$

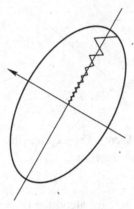

Fig. 18.3

This is why the descending path zigzags slowly approaching $(0,0)$ (see Figure 18.3).

The conjugate gradient method emerged to compensate for this weakness. Its main idea is in a small neighborhood of the minimum, the function f can be approximated by its second order Taylor polynomial:

$$f(x) = f(x_0)+(\nabla f(x_0), x-x_0)+\frac{1}{2}\left(\nabla^2 f(x_0)(x-x_0), x-x_0\right)+o(\|x-x_0\|^2).$$

For the quadratic function

$$c + (b, x) + \frac{1}{2}(Ax, x), \quad c \in \mathbb{R}^1, \ b \in \mathbb{R}^n, \ A \in \mathbb{R},$$

where A is positive definite, it has a unique minimum. When $n = 2$, its level curves are ellipses. We will take advantage of the special feature of this quadratic function to avoid the zigzagging descending path.

From analytic geometry, we know a given vector $w = (w_1, w_2)$ determines an axis of a family of ellipses, while its conjugate axis is the trace of the midpoints lying on the chords parallel to this vector. We call the direction u of the conjugate axis the conjugate direction of w (see Figure 18.4).

We are inspired by a geometric fact: for a positive definite quadratic function on the plane, regardless of the initial point x^0, given any direction w, the minimum along w must be the midpoint of a chord in the ellipse parallel to w. The next step is, if we can find the minimum in the conjugate direction u of w, then we can reach the center of the family of ellipses in merely two steps, which coincides with the minimum of the original function.

In order to extend this geometric fact to higher dimensions, we must be able to analytically express the concept of "conjugate" direction. For simplicity, we take $b = 0$.

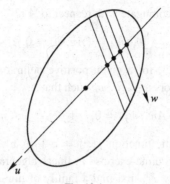

Fig. 18.4

Given a vector w, $\forall v \in \mathbb{R}^2$, the parametric equation of the line passing through v parallel to w is $x = v + tw$. It intersects the ellipse

$$c + \frac{1}{2}(Ax, x) = 0$$

at the points

$$x_i = v + t_i w, \quad i = 1, 2,$$

where t_1 and t_2 are two real roots of the equation

$$2c + t^2(Aw, w) + 2t(Av, w) + (Av, v) = 0.$$

If v lies on the conjugate axis, then $v = \frac{x_1 + x_2}{2}$. Thus,

$$t_1 + t_2 = 0,$$

namely,

$$(Av, w) = 0.$$

In other words, if we redefine an inner product on $\mathbb{R}^2$ using the positive definite matrix A by

$$[v, w]_A = (Av, w),$$

then

v is conjugate to w $\iff$ v is orthogonal to w with respect to $[\cdot, \cdot]_A$.

This can easily be extended to $\mathbb{R}^n$. Let A be an $n \times n$ positive definite matrix, $\forall v, w \in \mathbb{R}^n$, define

$$[v, w]_A = (Av, w).$$

It is straightforward to verify $[\cdot, \cdot]_A$ is in fact an inner product on $\mathbb{R}^n$.

We say v and w are conjugate with respect to A if

$$[v, w]_A = (Av, w) = 0.$$

Consequently, on $\mathbb{R}^n$, for a given positive definite matrix A, there exist n mutually conjugate vectors $w_1, \ldots, w_n$ such that

$$(Aw_i, w_j) = 0, \quad 1 \leq i < j \leq n.$$

Consider the quadratic function $f(x) = c + (b, x) + \frac{1}{2}(Ax, x)$, where A is positive definite. As a routine exercise in linear algebra, it is straightforward to verify: starting from any x^0, first find a family of mutually conjugate unit vectors $w_1, \ldots, w_n$, and then find the one dimensional minimum along these vectors iteratively; in n steps, we can reach the minimum of f.

Based on this idea, we outline the conjugate gradient method as follows:

1. Given an initial point $x^0 \in \mathbb{R}^n$ and an initial direction $\sigma^0 = -\nabla f(x^0)$.

2. Suppose we already found x^i and σ^i set

$$r^i = \nabla f(x^i) \ (= Ax^i + b),$$
$$\lambda^i = -(r^i, \sigma^i)/(A\sigma^i, \sigma^i),$$
$$\sigma^{i+1} = -r^{i+1} + \beta^i \sigma^i,$$
$$\beta^i = (Ar^{i+1}, \sigma^i)/(A\sigma^i, \sigma^i).$$

Set

$$x^{i+1} = x^i + \lambda^i \sigma^i.$$

3. If $r^{i+1} = 0$, then stop and x^{i+1} is the desired minimum; otherwise, continue. For a quadratic function, the above iteration will terminate in at most n steps.

In fact, it is not difficult to verify by mathematical induction: for $m \leq n$, we have

(1) if $\phi(t) = f(x^i + t\sigma^i)$, then $\phi'(\lambda^i) = 0$;

(2) $\mathrm{span}\{\sigma^0, \ldots, \sigma^m\} = \mathrm{span}\{r^0, \ldots, r^m\} = \mathrm{span}\{r^0, Ar^0, \ldots, A^m r^0\}$;

(3) $[\sigma^i, \sigma^j]_A = 0$ for $i \neq j$ and $1 \leq i < j \leq m$.

Remark 18.4 Since the optimization problem of a quadratic function can be reduced to solving a system of linear equations, the conjugate gradient method is also a very effective iterative algorithm in solving systems of linear equations.

The systems of linear equations generated by the finite element method in general correspond to sparse matrices, where the conjugate gradient method is proven to be particularly effective.

Remark 18.5 The conjugate gradient method is not only applicable to optimization problems of quadratic functions, but is also applicable to convex functions. Despite the fact that it may not reach the minimum in only n steps, the rate of convergence is nevertheless notably improved.

Remark 18.6 Of particular note, in calculations of quantum mechanics and mathematical physics, it is often required to compute the spectrum of a positive definite self-adjoint operator. Under certain compactness assumption, the spectrum consists only of eigenvalues. In which case, based on the minimax description of the eigenvalues (see Lecture 12), we can apply the Ritz method to obtain approximated solutions. Both the finite element method as well as the conjugate gradient method are numerical implementations of the Ritz method.

Lecture 19

Optimal control problems

Optimal control problems are a special kind of variational problems, which have profound applications in engineering, economics, and other areas. In general, since the function realizing the extremal value of a functional need not be continuous (it usually involves jump discontinuity, such as switching), it cannot be obtained by solving the traditional Euler–Lagrange equation. Instead, the necessary condition is replaced by the Pontryagin Maximal Principle. The foundation of optimal control theory has been developed into a subject of its own. In this lecture, we only give a brief introduction.

19.1 The formulation of problems

In order to assist the reader with an intuitive understanding of the formulation and possible solution of an optimal control problem, we begin with a rather simple example.

Example 19.1 (A rolling cart problem) A small rolling cart with mass $m = 1$ is placed on the x-axis. Starting from the stationary point $x = 0$, the cart is to reach the point $x = L$ at exactly time T. Without friction and air resistance, we seek how to apply a nonnegative external force $u \geq 0$ on the cart such that the total work done W is minimum.

Denote the position of the cart at time t by $x(t)$. Let $v(t)$ be the velocity of the cart at time t and $u(t)$ be the external force put on the cart at time t. Their relations are given by

$$\begin{cases} \dfrac{dx}{dt} = v, \\[2mm] \dfrac{dv}{dt} = u, \end{cases} \tag{19.1}$$

$$\begin{pmatrix} x(0) \\ v(0) \end{pmatrix} = \begin{pmatrix} 0 \\ 0 \end{pmatrix}, \tag{19.2}$$

$$x(T) = L.$$

The controlled external force belongs to the set

$$U = \{u \in L^\infty([0,T]) | u \geq 0\}.$$

The work done by the external force is

$$W = \int_0^L u\,dx = \int_0^T uv\,dt = \int_0^T \dot{v}v\,dt = \frac{v^2(T)}{2}.$$

We are asked to solve the constrained minimization problem

$$\min\left\{ W = \int_0^T uv\,dt \,\middle|\, (19.1),\ (19.2),\ \int_0^T v\,dt = L,\ u \in U \right\}.$$

Analysis $1°$ Under the constraint, we claim $W \geq \frac{L^2}{2T^2}$.

To prove this, since v is continuous, by the Mean Value Theorem of integrals, $\exists\, t_0 \in (0,T)$ such that $v(t_0) = L/T$. Since $u \geq 0$, v is non-decreasing, hence

$$W = \frac{v^2(T)}{2} \geq \frac{v^2(t_0)}{2} = \frac{L^2}{2T^2}.$$

$2°$ We claim $W > \frac{L^2}{2T^2}$.

To prove this, suppose there exists a control function u such that $W = \frac{L^2}{2T^2}$, then $\exists\, t_0 \in (0,T)$ which satisfies

$$\begin{cases} v(t) = v(t_0) = \dfrac{L}{T}, \quad \forall t \in [t_0, T], \\[2mm] v(t_0) = \max_{t \in [0,t_0]} v(t). \end{cases}$$

It follows that

$$L = \int_0^{t_0} v(t)\,dt + \frac{L}{T}(T - t_0),$$

i.e.

$$\frac{L}{T}t_0 = \int_0^{t_0} v(t)\,dt.$$

Thus,

$$v(t) \equiv \frac{L}{T}.$$

This contradicts (19.2).

3° We consider a discontinuous u. Let

$$u(t) = \begin{cases} a, & t \in [0, T_0], \\ 0, & t \in (T_0, T], \end{cases}$$

where a and T_0 are parameters. The solution is

$$v(t) = \begin{cases} at, & t \in [0, T_0], \\ aT_0, & t \in (T_0, T]. \end{cases}$$

Now we have

$$W = \int_0^{T_0} a^2 t \, dt = \frac{a^2 T_0^2}{2},$$

$$L = \int_0^T v \, dt = \frac{aT_0^2}{2} + aT_0(T - T_0) = aTT_0 - \frac{aT_0^2}{2}.$$

When $aT_0 = \frac{L}{T}$, as $T_0 \to 0$, $W \to \frac{L^2}{2T^2}$, the controlled external force $u(t) = \frac{L}{T}\delta(t)$, where $\delta(t)$ is a pulse function.

Therefore, the pulse force $u(t) = \frac{L}{T}\delta(t)$ produces constant velocity $v = \frac{L}{T}$, which realizes the minimal work $W = \frac{L^2}{2T^2}$. However, pulse force is not physical.

4° If u is allowed to change signs, then the control set becomes $U = L^\infty([0, T])$.

If we set $u = a(\frac{T}{2} - t)$, then

$$v = a\left(\frac{T}{2}t - \frac{t^2}{2}\right), \quad v(T) = 0,$$

hence $W = 0$. However, there are many solutions for $W = 0$. For example, if we take

$$u(t) = \begin{cases} a, & t \in [0, T_0], \\ 0, & t \in (T_0, T - T_0), \\ -a, & t \in [T - T_0, T], \end{cases}$$

then

$$v(t) = \begin{cases} at, & t \in [0, T_0], \\ aT_0, & t \in (T_0, T - T_0), \\ a(T - t), & t \in [T - T_0, T], \end{cases}$$

is also a solution of $W = 0$. In particular, if we choose $T_0 = T/2$, then the control function u fluctuates in between the positive and negative constants $\pm a$, which is why it is also called a Bang-Bang control.

From this example, we discover in the formulation of an optimal control problem, variables can be divided into two groups: state variables and control variables.

A state variable describes the state of a system, which is typically a vector (for instance the variable x in our previous example). Its range is a subset Y of the vector space $\mathbb{R}^n$.

A control variable is controllable (for instance the variable u in our previous example). All permissible control variables constitute a set U. It is a subset of $L^1([0, T])$; it is often comprised of piecewise continuous functions.

In an optimal control problem, it is required to prescribe a state equation which dictates the system's changes:

$$\dot{x} = f(t, x, u), \quad t \in (0, T), \tag{19.3}$$

It is also required to prescribe the initial state $x(0) = x_0$ and the terminal set $x(T) \in B$. Using the Lagrangian $L : [0, T] \times \mathbb{R}^n \times \mathbb{R}^N \to \mathbb{R}^1$, we can define the price functional

$$J(u) = \int_0^T L(t, x(t), u(t)) \, dt,$$

the goal is to find $u \in U$ such that

$$\min_{u \in U} J(u).$$

Example 19.1 (continued) In the above example, the state equations are

$$\begin{cases} \dot{x} = v, \\ \dot{v} = u, \end{cases} \quad t \in [0, T],$$

the initial state and the terminal set are respectively given by

$$\begin{pmatrix} x(0) \\ v(0) \end{pmatrix} = \begin{pmatrix} 0 \\ 0 \end{pmatrix}, \quad B = \begin{pmatrix} L \\ * \end{pmatrix}.$$

The set of permissible control functions is

$$U = \{u \in L^\infty([0, T], \mathbb{R}^1) \mid u(t) \geq 0\}.$$

The Lagrangian $L = u \cdot v$, which defines the price functional

$$J(u) = W = \int_0^T u \cdot v \, dt = \int_0^T L(v, u) \, dt.$$

Example 19.2 We find how to reinvest the products in order to maximize the overall production of the goods.

Suppose the production rate of a certain goods (the unit time production) is $q = q(t)$. Suppose its rate of increase $\dot{q}$ is proportional to the percentage of reinvestment u,

$$\dot{q} = \alpha u q, \quad t \in [0, T],$$

where $\alpha > 0$ is given. The initial state is also given,

$$q(0) = q_0.$$

However, the terminal state has no restriction whatsoever, i.e. $B = \mathbb{R}^1$.

The overall production is given by

$$I = \int_0^T (1 - u) q \, dt.$$

The controlled reinvestment percentage

$$u \in U = \{v \in L^\infty(0, T) \mid v(t) \in [0, 100]\}.$$

The goal is to maximize the overall production, i.e.

$$\max\{I \mid u \in U\}.$$

Example 19.3 (The Ramsey economic growth model) In 1928, Ramsey proposed the following economic growth model.

Suppose the investment per worker is i, the consumption per worker is c, and the output per worker is $x = i + c$.

Suppose x is a function $f(k)$ of the capital intensity (capital per worker) k. Suppose the population growth rate is α, then the labor L, as a function of time t, follows the exponential growth $L(t) = L_0 e^{\alpha t}$.

Since the total capital is $K = kL$, the total investment $I = \frac{dK}{dt}$. They are related via

$$i = \frac{I}{L} = \frac{1}{L}\frac{dK}{dt} = \frac{dk}{dt} + \frac{k}{L}\frac{dL}{dt} = \frac{dk}{dt} + \alpha k.$$

The target function measuring the consumption of the society is

$$W(c) = \int_0^\infty e^{-pt} u(c(t)) dt,$$

where $p > 0$ is called the depreciation rate of capital, while $u(c)$ is called the consumer's utility function. Ramsey proposed to maximize $W(c)$ under the constraint

$$\frac{dk}{dt} = f(k) - \alpha k - c, \quad k(0) = k_0.$$

19.2 The Pontryagin Maximal Principle

Consider the following optimal control problem.

Assume the state variable $x \in W^{1,\infty}((0,T), \mathbb{R}^n)$, the control variable $u \in U = \{v$ is a piecewise continuous function taking values in $\mathbb{R}^n \mid v(t) \in C\}$, where $C \subset \mathbb{R}^N$ is compact. The state equation is $\dot{x} = F(x, u)$, where $F \in C^1(\mathbb{R}^n \times \mathbb{R}^N, \mathbb{R}^n)$. Both the initial state $x(0) = x_0$ and the terminal state $x(T) = x_1$ are given.

We also assume $L \in C(\mathbb{R}^n \times \mathbb{R}^N, \mathbb{R}^1)$ is a given Lagrangian. Define the target functional

$$J(x, u) = \int_0^T L(x(t), u(t))dt.$$

We have the following maximal principle.

Theorem 19.1 *Suppose $u = u^*$ is a solution of the above optimal control problem which corresponds to the orbit $x = x^*(t)$ of the optimal state variable, i.e. $(x^*(t), u^*(t))$ attains the minimum of*

$$J(x, u) = \int_0^T L(x(t), u(t))\, dt.$$

Then there exists a piecewise continuously differentiable function $\lambda = \lambda(t) \in \mathbb{R}^n$ such that

$$\dot{\lambda} = -H_x(x^*, \lambda^*, u^*),$$

where

$$H(x, \lambda, u) = -L(x, u) + \lambda F(x, u). \tag{19.4}$$

Furthermore, along the optimal orbit $\{(x^*(t), \lambda(t), u^*(t)) \mid t \in (0, T)\}$, H is the constant

$$E = \sup_{v \in C} H(x^*(t), \lambda(t), v).$$

The function $H(x, \lambda, u)$ is also called the Hamiltonian, whereas λ is called the conjugate variable.

Proof (The proof is rather lengthy and it is beyond the scope of this book. We hereby adopt the outline of the proof given by Macki and Struass [MS], we also refer to MacCluer [Mac] for details.)

$1°$ Suppose for any initial state x^0, $\exists\, u = u(x^0, t) \in U$ minimizes the functional. This means there exists an optimal path, $y = x(x^0, t)$ for $t \in [0, T_0]$ such that $\dot{y} = F(y, u)$, $y(0) = x^0$, $y(T_0) = x_1$, and that

$$I(x^0) := \min_{v \in U} \int_0^{T_0} L(y(t), v(t)) \, dt$$

$$= \int_0^{T_0} L(y(t), u(t)) \, dt.$$

According to the local minimizing principle, $(u(x^0, t), x(x^0, t))$ is also optimal in $(t, T_0]$, hence

$$I(x(x^0, t)) = \int_t^{T_0} L(x(x^0, s), u(x^0, s)) \, ds.$$

2° Suppose I is differentiable and u is piecewise continuous. Then except for finitely many points, for all t, we have

$$0 = L(x(x^0, t), u(x^0, t)) + \nabla I(x(x^0, t)) \cdot \dot{x}(x^0, t)$$
$$= L(x(x^0, t), u(x^0, t)) + \nabla I(x(x^0, t)) F(x(x^0, t), u(x^0, t)).$$

When $x^0 = x_0$, by setting $u = u^*(x_0, t)$, we have $x = x^*(x_0, t)$ and

$$0 = L(x^*(x_0, t), u^*(x_0, t)) - \lambda F(x^*(x_0, t), u^*(x_0, t))$$
$$= -H(x^*(x_0, t), \lambda(t), u^*(x_0, t)),$$

where $\lambda(t) = -\nabla I(x^*(x_0, t))$. Since $x^*(x_0, t)$ is piecewise differentiable, λ is piecewise continuous and along the optimal orbit (x^*, λ, u^*),

$$H(x^*(x_0, t), \lambda(t), u^*(x_0, t)) = 0.$$

3° We now prove

$$H(x^*(t), \lambda(t), v) \le 0. \quad \forall v \in C, \ \forall t \in (0, T).$$

In fact, $\forall v \in C$, the system

$$\begin{cases} \dot{x} = F(x, v), \\ x(0) = x_0, \quad \forall t \in [0, T_0) \end{cases}$$

has a local solution $x = \tilde{x}(t)$. Generally speaking, $\forall v \in C$, $\tilde{x}$ is not the optimal orbit from x_0 to $\tilde{x}(t)$, we only have

$$I(x_0) \le \int_0^t L(\tilde{x}, v) \, ds + I(\tilde{x}(t)).$$

So for $t > 0$ sufficiently small,

$$-\frac{I(\tilde{x}(t)) - I(x_0)}{t} \le \frac{1}{t} \int_0^t L(\tilde{x}, v) \, ds.$$

From this, we can deduce that

$$-\frac{d}{dt}I(\tilde{x}(t))|_{t=0} \leq L(x_0, v),$$

while the differential of the left-hand side is equal to

$$-\nabla I(x_0) \cdot \dot{\tilde{x}}(0) = -\nabla I(x_0) \cdot F(x_0, v).$$

Since we can choose any state $x^*(x_0, t)$ along the orbit to be an initial value, it follows that

$$H(x^*(x_0, t), \lambda(t), v) \leq 0, \quad \forall v \in C, \quad \forall t \in [0, T].$$

4° Lastly, we prove λ satisfies the conjugate equation. Let

$$h(x, u) = -L(x, u) - \nabla I(x)F(x, u).$$

From our earlier discussion, we know that $\forall t \in (0, T)$, the function h, except for finitely many points, achieves its maximum when $x = x^*(t)$ and $u = u^*(t)$.

Suppose $I \in C^2$. By direct computation, it yields

$$
\begin{aligned}
0 &= \frac{\partial h}{\partial x_i} \\
&= -\frac{\partial L}{\partial x_i} - \frac{\partial(\nabla I \cdot F)}{\partial x_i} \\
&= -\frac{\partial L}{\partial x_i} - \frac{\partial}{\partial x_i}\left(\sum_{j=1}^{N} \frac{\partial I}{\partial x_j} F_j\right) \\
&= -\frac{\partial L}{\partial x_i} - \sum_{j=1}^{N} \frac{\partial^2 I}{\partial x_i \partial x_j} F_j - \sum_{j=1}^{N} \frac{\partial F_j}{\partial x_i} \frac{\partial I}{\partial x_j} \\
&= -\frac{\partial L}{\partial x_i} + \sum_{j=1}^{N} \frac{\partial \lambda_i}{\partial x_j} \dot{x}_j + \sum_{j=1}^{N} \lambda_j \frac{\partial F_j}{\partial x_i},
\end{aligned}
$$

whence

$$\dot{\lambda}_i = \sum_{j=1}^{N} \frac{\partial \lambda_i}{\partial x_j} \dot{x}_j = \frac{\partial L}{\partial x_i} - \sum_{j=1}^{N} \lambda_j \frac{\partial F_j}{\partial x_i} = -\frac{\partial H}{\partial x_i}.$$

That is,

$$\dot{\lambda} = -H_x(x^*(x_0, t), \lambda(t), u^*(x_0, t)). \qquad \square$$

Remark 19.1 We call the quantity $H(x, \lambda, u)$ defined in (19.4) the Hamiltonian, since the orbit $(x^*(t), \lambda(t), u^*(t))$ satisfies the Hamiltonian system

$$
\begin{cases}
\dot{x} = F(t, x, u) = H_\lambda(x, \lambda, u), \\
\dot{\lambda} = -H_x(x, \lambda, u).
\end{cases}
$$

Remark 19.2 In this theorem, if the terminal vector $x(T) = x_1$ has some components which are not pre-determined, then the corresponding components in the conjugate variable $\lambda(T)$ must be zero.

We now revisit the previous two examples using the maximal principle.

Example 19.1 (further continuation) Suppose the control variable has a constraint $0 \leq u(t) \leq a$, where $a > 0$ is a prescribed constant. By introducing the conjugate variable $\lambda(t) = (\alpha(t), \beta(t))$, the Hamiltonian is

$$H = -L + \lambda F = -uv + (\alpha, \beta)\begin{pmatrix} v \\ u \end{pmatrix} = \alpha v + u(\beta - v).$$

Suppose $(x^*(t), v^*(t))$ is the optimal orbit with conjugate orbit (α, β) such that it achieves the maximum of the Hamiltonian. Since $v^*(t)$ is fixed, $u(\beta - v^*)$ must attain its maximum. Thus,

$$u = \begin{cases} a, & \beta > v^*, \\ 0, & \beta < v^*. \end{cases}$$

Since $\dot{v} = u$ and $v(0) = 0$, it follows that there exists a constant v_0 such that

$$v^* = \begin{cases} at, & \beta > v^*, \\ v_0, & \beta < v^*. \end{cases}$$

In addition, the conjugate variable satisfies the equation

$$\dot{\lambda} = (\dot{\alpha}, \dot{\beta}) = -H_{(x,v)} = (0, u - \alpha).$$

From $\lambda(t) = (\alpha(t), \beta(t))$, we deduce $\alpha = \alpha_0$ must be a constant, and

$$\beta = \begin{cases} (a - \alpha_0)t + c, & u = a, \\ -\alpha_0 t + d, & u = 0, \end{cases}$$

where c and d are constants. Since both v^* and β are continuous, it follows that

$$\beta - v^* = -\alpha_0 t + c, \quad d = c + v_0.$$

Noting the terminal $v(T)$ is indeterminate, according to Remark 19.2, $\beta(T) = 0$. However, since

$$-\alpha_0 t + c = 0$$

has only one real root $T_0 = c/\alpha_0$, this implies $c \leq \alpha_0 T$.

$$v^*(t) = \begin{cases} at, & t \in [0, T_0), \\ aT_0, & t \in [T_0, T]. \end{cases}$$

$$u(t) = \begin{cases} a, & t \in [0, T_0), \\ 0, & t \in [T_0, T]. \end{cases}$$

Moreover, by the terminal condition $x(T) = L$, we also have

$$L = \int_0^T v\,dt = aTT_0 - \frac{aT_0^2}{2},$$

$$T_0 = \begin{cases} T - \sqrt{T^2 - \dfrac{2}{a}L}, & a \geq \dfrac{2L}{T^2}, \\ \text{no solution}, & a < \dfrac{2L}{T^2}. \end{cases}$$

Example 19.2 (continued) We return to analyze how to maximize the overall production.

Introducing the conjugate variable λ, the Hamiltonian is

$$H = -L + \lambda F = (1 - u)q + \lambda(\alpha uq) = (1 - u + \lambda\alpha u)q.$$

1° Since $q \geq 0$, maximizing H is equivalent to maximizing $-u + \lambda\alpha u = (\lambda\alpha - 1)u$. Hence, the optimal control variable should be

$$u(t) = \begin{cases} 1, & \alpha\lambda(t) > 1, \\ 0, & \alpha\lambda(t) < 1. \end{cases}$$

2° The conjugate variable λ satisfies the equation

$$\dot\lambda = -H_q = -(1 - u + \lambda\alpha u),$$

namely, $\dot\lambda + \lambda\alpha u = u - 1$, therefore,

$$\begin{cases} \dot\lambda + \alpha\lambda = 0, & u = 1, \\ \dot\lambda = -1, & u = 0. \end{cases}$$

It yields the solution

$$\lambda(t) = \begin{cases} ce^{-\alpha t}, & u = 1 \\ -t + d, & u = 0, \end{cases}$$

where c and d are constants.

3° The terminal condition. Since $q(T)$ is indeterminate, $\lambda(T) = 0$. This implies $u(T) \neq 1$, which is only possible if

$$u = 0, \quad d = T.$$

This conclusion is reasonable, since at the end of a production cycle, there is no need to reinvest.

4° The change of the control variable. From 1°, we already know, the values of the control variable u should change at $\alpha\lambda(t) = 1$, i.e. when $t_s = T - \frac{1}{\alpha}$. Thus, the final solution is given by

$$u(t) = \begin{cases} 1, & t \in (0, t_s), \\ 0, & t \in (t_s, T), \end{cases}$$

$$q(t) = \begin{cases} q_0 e^{\alpha t}, & 0 < t < t_s \\ q_0 e^{\alpha t_s}, & t_s < t < T, \end{cases}$$

$$I = \int_{t_s}^{T} q_0 e^{\alpha t_s} \, dt = q_0 e^{\alpha t_s}(T - t_s) = \frac{q_0}{\alpha} e^{\alpha(T - \frac{1}{\alpha})}.$$

Example 19.3 (continued) Returning to the Ramsey's economic growth model, to maximize $W(c)$ is equivalent to minimizing $-W(c)$. By introducing the conjugate variable λ and the Hamiltonian

$$H(k, \lambda, c) = e^{-pt}u(c) + \lambda(f(k) - \alpha k - c),$$

we obtain the optimal conditions

$$\frac{\partial H}{\partial c} = u'(c)e^{-pt} - \lambda = 0,$$

and

$$\begin{cases} k'(t) = f(k) - \alpha k - c, \\ \lambda'(t) = -\dfrac{\partial H}{\partial k} = -\lambda(f'(k) - \alpha) \end{cases}$$

together with the initial condition $k(0) = k_0$. Furthermore, since the problem is proposed on the positive half line, we also have the terminal condition

$$\lim_{t \to \infty} \lambda k(t) = 0.$$

In 1928, Ramsey applied variational methods to discuss this problem; however, its importance was neglected. It was not until 1965 that Ramsey's original idea was rediscovered and improved by economists Cass and Koopmans respectively, and it has since been termed the Ramsey–Cass–Koopmans' model.

19.3 The Bang-Bang principle

A specific control model is called a Bang-Bang control, if each component of the control variable $u \in \mathbb{R}^N$ takes on at most two different values in the entire process.

Given an $n \times n$ matrix A and an $n \times N$ matrix B, let $x \in \mathbb{R}^n$ be the state variable and $u \in \mathbb{R}^N$ be the control variable. Consider the following linear system

$$\begin{cases} \dot{x} = Ax + Bu, \\ x(0) = x^0. \end{cases} \tag{19.5}$$

Theorem 19.2 *In the linear system* (19.5), *if there exists a control variable* $u^0 \in L^\infty(0, T)$ *to move from the initial state* $x(0) = x^0$ *to the terminal state* $x(T) = x^1$, *then it must be realized by a bang-bang control* u.

This theorem is in fact a corollary of the Krein–Milman theorem in functional analysis. The Krein–Milman theorem concerns the extreme points of a compact convex subset in a locally convex topological vector space. It is an equivalent statement to the Hahn–Banach theorem.

Let X be a locally convex topological vector space and $E \subset X$ be a convex subset. A point $x \in E$ is called an extreme point, if there are no $y, z \in E$ such that $x = (y + z)/2$. The set of all extreme points is called the extremal set of E, denoted by $D(E)$.

The Krein–Milman theorem *Let X be a locally convex topological vector space and $E \subset X$ be a compact convex subset, then E is the closure of the extremal set $D(E)$.*

Proof of Theorem 19.2 Without loss of generality, we may assume $\|u^0(t)\|_\infty \le 1$. Let

$$U = \left\{ v \in L^\infty(0, T) \mid \|v\|_\infty \le 1 \ x(T) = e^A x^0 + \int_0^t e^{A(t-s)} Bv(s) ds \right\}.$$

Clearly, U is a weak-* closed convex set and $u^0 \in U$.

According to the Banach–Alaoglu theorem, the unit ball in $L^\infty(0, T)$ is weak-* compact, so must be U itself. According to the Krein–Milmann theorem, it must have extreme points.

In the following, we prove: if $u \in U$ is an extreme point of U, then u must be a bang-bang control. For simplicity, we only assume $N = 1$. Choose any $\varepsilon \in (0, 1)$, consider the set

$$S = \{ t \in (0, T) \mid u(t) \in (-1 + \epsilon, 1 - \epsilon) \}.$$

Note that the solution of the linear system can be expressed as

$$x(t) = e^A x^0 + \int_0^t e^{A(t-s)} Bu(s) ds.$$

Choose a function $v \in L^\infty(0, T)$ such that

$$\int_S e^{-As} Bv(s)ds = 0.$$

We define v to be zero outside S, by rescaling if necessary, we have $\|v\|_\infty = 1$. Thus, $u \pm \varepsilon v \in U$ and $u = \frac{1}{2}[(u + \epsilon v) + (u - \epsilon v)]$, which contradicts u is an extreme point. This implies $S = \emptyset$. However, since $\varepsilon > 0$ is arbitrary, u must be a bang-bang control as claimed. $\qquad\square$

Functions of bounded variations and image processing

In many modern applied problems (such as problems arising from image restoration, phase transition, etc.), the solutions of variational problems are not always continuous; more specifically, these kinds of discontinuity often occur to rectifiable curves with obvious breaking (or measurable higher dimensional hypersurfaces). This phenomenon suggests that the desired solutions do not belong to any appropriate Sobolev spaces, instead, the focus should be turned to the space of functions of bounded variations.

Starting from the mid-20th·century, De Giorgi *et al.* adopted methods from measure theory to study Plateau problems of minimal surfaces. Functions of bounded variations and measures are naturally related, the theoretical framework revealing their intertwined connection has been systematically developed into the area of geometric measure theory. However, these topics are far beyond the scope of this book, we can only demonstrate the essence by some concrete examples with applications.

We first recall the definition as well as some chief properties of functions of bounded variations in one variable, and then we will extend this to the case of several variables.

20.1 Functions of bounded variations in one variable — a review

Definition 20.1 A function $u : J = (a, b) \to \mathbb{R}^1$ is said to be of bounded variation, denoted BV, if there is a constant $M > 0$ such that

$$S_\pi(u) = \sum_{i=1}^n |u(t_{i+1}) - u(x_i)| \leq M$$

for every partition $\pi = \{a < t_1 < \cdots < t_{n+1} < b\}$ of (a, b). We call

$$V_a^b(u) = \sup_\pi S_\pi(u)$$

the total variation of u on J.

The set of BV functions on J is denoted by $BV(J)$, which is itself a linear space. The following simple properties are straightforward to verify:

(1) A bounded monotone function on J is a BV function.

(2) $\text{Lip}(J, \mathbb{R}^1) \subset BV(J)$.

(3) $\forall u \in BV(J)$, u can be decomposed as the difference of two monotone increasing functions:

$$u(x) = \frac{1}{2}(T_u(x) + u(x)) - \frac{1}{2}(T_u(x) - u(x)),$$

where

$$T_u(x) = \sup \left\{ \sum_{i=1}^m |u(t_{i+1}) - u(t_i)| \,\big|\, \sigma : a < t_1 < \cdots < t_{m+1} < x \text{ is any partition of } J \right\}.$$

(4) $\forall u \in BV(J)$, the set of jumping points $D_u = \{x \in J | u(x - 0) \neq u(x + 0)\}$ is at most countably infinite.

Since the value of a BV function at a jump discontinuity is not determined, in order to avoid unnecessary fuss caused by this, we choose to normalize

$$u(x) = u(x - 0), \quad \forall x \in J.$$

To further signify the variation part of u on J, we also normalize $u(a + 0) = 0$. The set of all normalized functions of bounded variations on J is still a linear space, denoted by $NBV(J)$. On which, we define the norm

$$\|u\| = V_a^b(u).$$

Then $NBV(J)$ is a Banach space.

In the following, we establish the one-to-one correspondence between the NBV functions and the σ-additive Borel measurable functions.

Let $u : J \to \mathbb{R}^1$ be a monotone increasing normalized BV function, define

$$\mu_u((a, x)) = u(x).$$

We claim that μ_u can be extended to a measure on J, i.e. it is a σ-additive non-negative set function. In fact, $\forall [\alpha, \beta) \subset J$, let

$$\mu_u([\alpha, \beta)) = u(\beta) - u(\alpha),$$

then $\mu_u(\{\alpha\}) = u(\alpha + 0) - u(\alpha)$. Hence, for any Borel set $E \subset J$, we have

$$\mu_u(E) = \mathcal{L}^1 \left(\bigcup_{x \in E} [u(x), u(x + 0)] \right),$$

where $\mathcal{L}^1$ denotes the Lebesgue measure on $\mathbb{R}^1$.

Conversely, for a given Borel measure μ on J, $u(x) = \mu((a, x))$ is monotonically increasing and left continuous; moreover, it satisfies $u(a + 0) = 0$.

From property (3), $\forall u \in \mathrm{BV}(J)$, there exist two monotone increasing, left continuous functions u_1 and u_2 such that $u = u_1 - u_2$. By the above statement, they correspond to two respective Borel measures $\mu_i = \mu_{u_i}$ for $i = 1, 2$. By defining

$$\mu(E) = \mu_1(E) - \mu_2(E),$$

it follows that μ is a σ-additive Borel measurable set function, which satisfies

$$\mu((a, x)) = \mu_1((a, x)) - \mu_2((a, x)) = u(x), \quad \forall x \in J.$$

Subsequently, for any σ-additive Borel measurable set function μ,

$$u(x) = \mu((a, x)) \tag{20.1}$$

defines a unique normalized BV function.

As a consequence, for any σ-additive Borel measurable set function μ and any Borel measurable set $E \in \mathcal{B}$, we can also define a "total variation":

$$|\mu|(E) = \sup \left\{ \sum_{i=1}^{\infty} |\mu(B_i)| \ \middle|\ B_i \in \mathcal{B}, B_i \cap B_j = \emptyset, i \neq j, \bigcup_{i=1}^{\infty} B_i = E \right\}.$$

It is not difficult to verify:

(1) $|\mu|$ is a Borel measure,
(2) $|\mu(E)| \leq |\mu|(E), \forall E \in \mathcal{B}$,
(3) $|\mu|(J) = V_a^b(u)$.

We denote the space of all σ-additive Borel measurable set functions on J by $\mathcal{M}(J, \mathbb{R}^1)$. On $\mathcal{M}(J, \mathbb{R}^1)$, define the norm to be

$$\|\mu\|_{\mathcal{M}} = |\mu|(J),$$

then it is also a Banach space.

The above one-to-one correspondence $u \mapsto \mu$ is an isomorphism between the Banach spaces $\mathrm{NBV}(J)$ and $\mathcal{M}(J, \mathbb{R}^1)$.

Recall the Riesz representation theorem for the space of continuous functions on J:

$$C_0(J)^* \cong \mathcal{M}(J, \mathbb{R}^1),$$

where $C_0(J) = \{\varphi \in C(J) | \varphi(a) = \varphi(b) = 0\}$.

Since $C_0(J)$ is separable, $\mathcal{M}(J, \mathbb{R}^1)$ is the dual space of a separable Banach space. This means: $\forall F \in C_0(J)^*$, there exists a unique $\mu \in \mathcal{M}(J, \mathbb{R}^1)$ such that

$$F(\phi) = \int_J \phi \, d\mu, \quad \forall \phi \in C_0(J),$$

with $\|F\| = \|\mu\|_{\mathcal{M}}$, and the map $F \mapsto \mu$ is surjective.

By the Lebesgue–Nikodym decomposition theorem, every BV function has the following decomposition:

$$u(x) = v(x) + r(x) + s(x),$$

where $v(x)$ is the absolutely continuous part of u, $r(x)$ is the Cantor part of u, and $s(x)$ is the jumping part of u. In particular, s is itself a jump function, while r is a non-constant function of bounded variation, but whose derivative equals zero almost everywhere.

Since a BV function u is itself a bounded measurable function, $u \in L^1(J)$. As a function in $L^1(J)$, it has a generalized derivative (in the sense of distribution) Du, which satisfies

$$\langle Du, \varphi \rangle = -\int_J u \cdot \varphi' \, dx, \quad \forall \varphi \in C_0^\infty(J). \tag{20.2}$$

It is worth noting, contrasting to Sobolev spaces, for a BV function u, Du and the almost everywhere derivative u' are not identical to each other! Only when $r = s = 0$, $u \in AC(J)$, which implies $Du = u'$.

If $s \neq 0$, then Du may be a measure. For example, let $J = (-1, 1)$ and

$$u(x) = \begin{cases} 0, & x \le 0, \\ 1, & x > 0, \end{cases}$$

then $Du = \delta(x)$. That is, if we define $\mu = Du$, then $\mu(\{0\}) = 1$, $\mu(B) = 0$, $\forall B \in \mathcal{B}$ and $0 \notin B$.

We now reveal the explicit relation between a BV function u and its corresponding measure μ. Notice as the dual space of $C_0(J)$, we have a natural dual pairing:

$$\langle \mu, \varphi \rangle = \int_J \varphi \, d\mu, \quad \forall (\mu, \varphi) \in \mathcal{M}(J, \mathbb{R}^1) \times C_0(J).$$

Thus,

$$\|\mu\|_{\mathcal{M}} = \sup \left\{ \int_J \varphi \, d\mu \mid \|\varphi\|_{C_0} \le 1 \right\}.$$

Furthermore, according to the one-to-one correspondence of $\mathcal{M}(J, \mathbb{R}^1)$ and $\mathrm{NBV}(J)$, we also have

$$\langle \mu, \varphi \rangle = \int_J \varphi \, du(x) = -\int_J u(x) \varphi'(x) \, dx.$$

Combining with (20.2), it follows immediately that

$$\boxed{Du = \mu.}$$

Since there is no notion of end (left and right) points in the domain of functions of several variables, the idea of normalizing the function value seems impractical. However, the above formula sheds new light on how to generalize functions of bounded variations from a single variable to several variables. More precisely, we are led to the following definition:

Definition 20.2 On BV(J), we define the norm

$$\|u\|_{\mathrm{BV}} = \|u\|_{L^1} + \|Du\|_{\mathcal{M}}.$$

It is not difficult to verify that BV(J) is a Banach space.

20.2 Functions of bounded variations in several variables

Let $\Omega \subset \mathbb{R}^n$ be a bounded open subset. Denote $C_0^1(\Omega, \mathbb{R}^n)$ the function space of all continuously differentiable functions on Ω with compact support in $\mathbb{R}^n$. Analogous to the single variable situation, we define

Defintion 20.3 A function $u \in L^1(\Omega)$ is said to be of bounded variation, if

$$\|Du\|(\Omega) = \sup\left\{ \int_\Omega u \operatorname{div} \varphi \, dx \,\middle|\, \varphi \in C_0^1(\Omega, \mathbb{R}^n), \quad \|\varphi\|_\infty \le 1 \right\} < \infty.$$

We call $\|Du\|(\Omega)$ the total variation of u.

Denote the space of all BV functions on Ω by BV(Ω) with the norm

$$\|u\|_{BV} = \|u\|_{L^1} + \|Du\|(\Omega).$$

Just like in the single variable setting, for any BV function u, we consider the relation between the generalized gradient Du:

$$\langle Du, \varphi \rangle = \langle u, \operatorname{div} \varphi \rangle := \int_\Omega u \operatorname{div} \varphi dx, \quad \forall \varphi \in C_0^\infty(\Omega, \mathbb{R}^n)$$

and the corresponding measure.

From the fact that total variation of u is bounded, we can affirm Du is a vector-valued Radon measure.

This is because $\forall \varphi_0 \in C_0(\Omega, \mathbb{R}^n)$, in the Banach space consisting of all continuous functions on $\bar{\Omega}$ with zero boundary values, there exists a sequence $\{\varphi_k\} \subset C_0^1(\Omega, \mathbb{R}^n)$ such that $\varphi_k \to \varphi_0$ uniformly on $\bar{\Omega}$, and $|\varphi_k|_\infty \le |\varphi_0|_\infty$, $\forall k$. Define the linear functional

$$L(\varphi) = \int_\Omega u \operatorname{div} \varphi \, dx, \quad \forall \varphi \in C_0^1(\Omega, \mathbb{R}^n),$$

it satisfies

$$|L(\varphi)| \le C\|\varphi\|_\infty, \quad \forall \varphi \in C_0^1(\Omega, \mathbb{R}^n).$$

By letting

$$\overline{L}(\varphi_0) = \lim_{k \to \infty} L(\varphi_k),$$

we see that not only the limit $\overline{L}(\varphi_0)$ exists, but it is independent of the particular choices of φ_k as well. This means L has a unique linear continuous extension

$$\overline{L} : C_0(\Omega, \mathbb{R}^n) \to \mathbb{R}^1.$$

Namely, $\overline{L}$ is a linear continuous functional on $C_0(\Omega, \mathbb{R}^n)$. By the Riesz representation theorem, there exists a vector-valued Radon measure $\mu = (\mu_1, \dots, \mu_n)$ such that

$$\int_\Omega u \operatorname{div} \varphi \, dx = -\int_\Omega \sum_{i=1}^n \varphi_i d\mu_i = -\int_\Omega \varphi \, d\mu, \forall \varphi = (\varphi_1, \dots, \varphi_n) \in C_0^1(\Omega, \mathbb{R}^n).$$

This means

$$\boxed{Du = \mathrm{d}\mu.}$$

We henceforth denote the space of all Radon measures on Ω taking values in $\mathbb{R}^n$ by $\mathcal{M}(\Omega, \mathbb{R}^{n+1})$, on which, by defining the norm

$$\|\mu\|_{\mathcal{M}} = \sup \left\{ \int_\Omega \varphi \, \mathrm{d}\mu \;\middle|\; \|\varphi\|_{C_0(\Omega, \mathbb{R}^{n+1})} \le 1 \right\},$$

it becomes a Banach space.

Example 20.1 Let $u \in W^{1,1}(\Omega)$, then $u \in \mathrm{BV}(\Omega)$ and $\|u\|_{W^{1,1}} = \|u\|_{\mathrm{BV}}$. To prove this, note that on one hand,

$$\left| \int_\Omega u \operatorname{div} \phi \right| = \left| \int_\Omega Du \cdot \phi \right| \le \|Du\|_1, \quad \forall \phi \in C_0^1(\Omega, \mathbb{R}^n), \|\phi\|_\infty \le 1.$$

On the other hand, $\forall \varepsilon > 0, \exists \varphi_\varepsilon \in C_0^1(\Omega, \mathbb{R}^n)$ with $\|\varphi_\varepsilon\|_\infty \le 1$, such that

$$\int_\Omega |Du| \le \int_\Omega Du \cdot \varphi_\varepsilon \, dx + \varepsilon = \int_\Omega u \operatorname{div} \varphi_\varepsilon \, dx + \varepsilon \le \|Du\|(\Omega) + \varepsilon.$$

Consequently, $\|Du\|(\Omega) = \|Du\|_1$. $\square$

Example 20.2 Let $S \subset \mathbb{R}^n$ be a C^∞ compact closed hypersurface of dimension $n - 1$. We equip S the induce metric from $\mathbb{R}^n$ and denote its $(n-1)$ dimensional Hausdorff measure by $\mathcal{H}^{n-1}(S)$.

Assume Ω is the region bounded by S with characteristic function χ_Ω. Then $\chi_\Omega \in \mathrm{BV}(\mathbb{R}^n)$ and

$$\|D\chi_\Omega\|(\mathbb{R}^n) = \mathcal{H}^{n-1}(S).$$

Proof $1°$ $\forall \phi \in C_0^1(\mathbb{R}^n, \mathbb{R}^n)$, by Gauss's theorem,

$$\int_{\mathbb{R}^n} \chi_\Omega \cdot \operatorname{div} \phi \, dx = \int_\Omega \operatorname{div} \phi \, dx = \int_S n(x) \cdot \phi(x) \, d\mathcal{H}^{n-1},$$

where $n(x)$ is the unit normal vector. Thus,

$$\|D\chi_\Omega\|(\mathbb{R}^n) \leq \mathcal{H}^{n-1}(S).$$

$2°$ On the other hand, we can extend $n(x)$ to be a C^∞ vector field V on $\mathbb{R}^n$ using the standard partition of unity argument. Furthermore, we can make sure $\|V(x)\| \leq 1, \forall x \in \mathbb{R}^n$.

$\forall \rho \in C_0^\infty(\mathbb{R}^n, \mathbb{R}^1)$, $|\rho(x)| \leq 1, \forall x \in \mathbb{R}^n$, let $\phi = \rho V$, then

$$\int_{\mathbb{R}^n} \chi_\Omega \operatorname{div} \phi \, dx = \int_S \rho \, d\mathcal{H}^{n-1}.$$

Thus,

$$\|D\chi_\Omega\|(\mathbb{R}^n) \geq \sup \left\{ \int_\Omega \chi_\Omega \operatorname{div} \phi \, dx \mid \phi \in C_0^\infty(\mathbb{R}^n, \mathbb{R}^n), \|\phi\|_\infty \leq 1 \right\}$$

$$\geq \sup \left\{ \int_S \rho \, d\mathcal{H}^{n-1} \mid \rho \in C_0^\infty(\mathbb{R}^n, \mathbb{R}^1), |\rho(x)| \leq 1, \forall x \in \mathbb{R}^n \right\}$$

$$= \mathcal{H}^{n-1}(S). \qquad \square$$

From the above two examples, we see that $W^{1,1}(\Omega) \subset BV(\Omega)$, but they are not equal.

BV(Ω) has the following properties.

(1) (Lower semi-continuity) Suppose $\{u_j\} \subset BV(\Omega)$ with $u_j \to u$ in $L^1(\Omega)$, then for any open subset $U \subset \Omega$,

$$\|Du\|(U) \leq \varliminf \|Du_j\|(U).$$

Moreover, from $\sup\{\|Du_j\|(\Omega) \mid j = 1, 2, \ldots\} < \infty$, we can deduce that $\|Du\|(\Omega) < \infty$.

Proof $\forall \phi \in C_0^1(\Omega, \mathbb{R}^n)$, $\|\phi\|_\infty \leq 1$, we have

$$\int_\Omega u \operatorname{div} \phi \, dx = \lim_{j \to \infty} \int_\Omega u_j \operatorname{div} \phi \, dx \leq \varliminf \|Du_j\|(\Omega).$$

The assertion now follows. $\qquad \square$

(2) BV(Ω) is complete.

Proof Let $\{u_j\} \subset BV(\Omega)$ be a Cauchy sequence. By definition, this is also a Cauchy sequence in $L^1(\Omega)$. Hence, $u_j \to u$ in $L^1(\Omega)$. By Property (1), $u \in BV(\Omega)$. It remains to show

$$\|D(u_j - u)\|(\Omega) \to 0.$$

Again by property (1), $\forall \varepsilon > 0$, $\exists j_0 > 0$ such that

$$\|D(u_j - u)\|(\Omega) \leq \varlimsup_{k \to \infty} \|D(u_j - u_k)\|(\Omega) < \varepsilon, \quad j \geq j_0,$$

which completes the proof. □

(3) (Approximation) $\forall u \in \mathrm{BV}(\Omega)$, $\exists \{u_j\} \subset \mathrm{BV}(\Omega) \cap C^\infty(\Omega)$ such that
(a) $u_j \to u$ in $L^1(\Omega)$.
(b) $\|Du_j\| \to \|Du\|$ in the sense of Radon measure.
In particular, $\|Du_j\|(\Omega) \to \|Du\|(\Omega)$.
For the proof, we refer to [AFP], pp. 122–123.

(4) (Compactness) Suppose $\{u_j\} \subset \mathrm{BV}(\Omega)$ satisfies $\|u_i\|_{\mathrm{BV}} \leq M$, then there exists a subsequence $\{u_{n_j}\}$ and some u such that $u_{n_j} \to u$ in $L^1(\Omega)$ and $\|Du\|(\Omega) \leq M$.

Proof By property (3), $\exists \{v_j\} \subset \mathrm{BV}(\Omega) \cap C^\infty(\Omega)$ such that $\|v_j - u_j\|_1 \leq \frac{1}{j}$ and $\|v_j\|_{\mathrm{BV}} \leq \|u_j\|_{\mathrm{BV}} + \frac{1}{j}$.

Since $\|v_j\|_{W^{1,1}} = \|v_j\|_{BV}$, $\{v_j\}$ is $W^{1,1}$-bounded. By the Rellich–Kondrachov compact embedding theorem, there exists a subsequence $\{v_{n_j}\}$ such that $v_{n_j} \to u$ in $L^1(\Omega)$, whence $u_{n_j} \to u$ in $L^1(\Omega)$. By property (1), we have

$$\|Du\|(\Omega) \leq \varliminf \|Dv_j\|(\Omega) \leq \varliminf \|Du_j\|(\Omega) \leq M. \qquad \square$$

In addition, Poincaré's inequality also holds for BV functions.

(5) (Poincaré's inequality) Let $\Omega \subset \mathbb{R}^n$ be a bounded, connected, and extendable region, then the following inequality holds:

$$\left\| u - \frac{1}{|\Omega|} \int_\Omega u \, dx \right\|_p \leq C_p \|Du\|(\Omega), \quad \forall u \in \mathrm{BV}(\Omega), \ 1 \leq p \leq \frac{n}{n-1}.$$

For the proof, we refer to [AFP], p. 153.

Next, we examine the weak-* convergence of $\mathrm{BV}(\Omega)$. Denote $X = C_0(\Omega, \mathbb{R}^{n+1})$. Let

$$E = \left\{ \phi = (\varphi_0, \varphi_1, \dots, \varphi_n) \in C_0^\infty(\Omega, \mathbb{R}^{n+1}) \,\middle|\, \mathrm{div}\, \hat{\varphi} = \varphi_0, \ \hat{\varphi} = (\varphi_1, \dots, \varphi_n) \right\}.$$

Let Y be the closure of E in X, then Y is a closed linear subspace of X.

Consider the linear mapping

$$\begin{aligned} T : \mathrm{BV}(\Omega) &\to \mathcal{M}(\Omega, \mathbb{R}^{n+1}), \\ u &\mapsto (u\mathcal{L}^n, D_1 u, \dots, D_n u), \end{aligned}$$

where $\mathcal{L}^n$ is the n-Lebesgue measure, while $D_i u$ is the ith component of the Radon vector-valued measure Du, then in the sense of distribution, we have

$$\langle D_i u, \varphi \rangle = -\langle u, \partial_{x_i} \varphi \rangle, \quad \forall \varphi \in C_0^\infty(\Omega), \ 1 \leq i \leq n.$$

We point out $Tu \in (X/Y)^*$. This is because $\forall \phi \in E$,

$$\langle Tu, \phi \rangle = \langle u\mathcal{L}^n, \varphi_0 \rangle + \langle Du, \hat{\varphi} \rangle = \langle u\mathcal{L}^n, \varphi_0 - \operatorname{div} \hat{\varphi} \rangle = 0.$$

Thus, $Tu|_Y = 0$, i.e. $Tu \in (X/Y)^*$.

Furthermore, T is surjective, i.e. $\forall (\mu_0, \mu_1, \ldots, \mu_n) \in \mathcal{M}(\Omega, \mathbb{R}^{n+1})$, if

$$\langle \mu_0, \varphi_0 \rangle + \sum_{i=1}^n \langle \mu_i, \varphi_i \rangle = 0, \quad \forall (\varphi_0, \ldots, \varphi_n) \in E,$$

then

$$\langle \mu_i, \varphi \rangle = -\langle \mu_0, \partial_i \varphi \rangle, \quad i = 1, 2, \ldots, n, \forall \varphi \in C_0^1(\Omega).$$

Using techniques from measure theory, it is not difficult to verify that μ_0 is indeed absolutely continuous. Therefore, there exists $u \in L^1(\Omega)$ such that $\mu_0 = u\mathcal{L}^n$.

Since

$$\|Tu\|_{\mathcal{M}(\Omega, \mathbb{R}^{n+1})} = \sup \left\{ \int_\Omega [u \cdot \varphi_0 + Du \cdot \hat{\varphi}] \, dx \,\middle|\, \|\phi\|_{C_0(\Omega, \mathbb{R}^{n+1})} \leq 1 \right\},$$

it follows that

$$\|u\|_{\mathrm{BV}} = \|u\|_{L^1} + \|Du\|(\Omega) \leq 2\|Tu\| \leq 2\|u\|_{\mathrm{BV}}.$$

As a consequence, we can regard $\mathrm{BV}(\Omega)$ as the dual space of the separable Banach space (X/Y). By the Banach–Alaoglu theorem, every bounded set is weak-* sequentially compact.

We now analyze the weak-* topology on $\mathrm{BV}(\Omega)$.

Theorem 20.1

$$u_j \xrightarrow{\text{BV weak}^*} u \iff \begin{cases} u_j \to u, \text{ in } L^1, \\ u_j \text{ is bounded in BV}. \end{cases}$$

Proof "$\Rightarrow$" By the Banach–Steinhaus theorem, $\{u_j\}$ is bounded in BV. By property (4), $u_j \to u$ in $L^1(\Omega)$.

"$\Leftarrow$" By the Banach–Alaoglu theorem, $\{u_j\}$ contains a weak convergent subsequence $u_{n_j}^*$. It remains to show all possible weak-* limits of $\{Du_j\}$ are the same as Du.

Suppose there exists a subsequence $\{u_{n_j}\}$ such that $\lim Du_{n_j} = \mu$, then

$$\int_\Omega u_{n_j} \operatorname{div} \varphi \, dx = -\int_\Omega Du_{n_j} \cdot \varphi \, dx,$$

which implies

$$\int_\Omega u \operatorname{div} \varphi \, dx = -\langle \mu, \varphi \rangle.$$

Thus, $\mu = Du$. $\qquad\square$

The generalized gradient Du of a BV function u in several variables also takes on the Lebesgue decomposition in the sense of distribution:

$$Du = D^a u + D^j u + D^c u,$$

where

$$D^a u = \nabla u \mathcal{L}^n \in L^1(\Omega, \mathbb{R}^n)$$

is the absolutely continuous part, $D^j u$ and $D^c u$ are the jump part and the Cantor part respectively. As usual, $\mathcal{L}^n$ denotes the Lebesgue measure on $\mathbb{R}^n$.

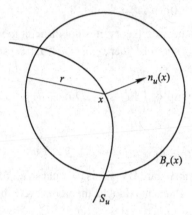

Fig. 20.1

They are defined as follows. Denote $B_r(x)$ the ball of radius $r > 0$ with center x in $\mathbb{R}^n$. Define two functions (see Figure 20.1)

$$u^+(x) = \inf \left\{ t \in [-\infty, +\infty] \,\middle|\, \lim_{r \to 0} \frac{\mathcal{L}^n\{y \in B_r(x) \mid u(y) > t\}}{r^n} = 0 \right\},$$

$$u^-(x) = \sup \left\{ t \in [-\infty, +\infty] \,\middle|\, \lim_{r \to 0} \frac{\mathcal{L}^n\{y \in B_r(x) \mid u(y) < t\}}{r^n} = 0 \right\}.$$

A point $x \in \Omega$ is called a Lebesgue point if

$$\lim_{r \to 0} \frac{1}{|B_r(x)|} \int_{B_r(x)} |u(y) - u(x)| \, dy = 0.$$

In which case, we have

$$u(x) = \lim_{r \to 0} \frac{1}{|B_r(x)|} \int_{B_r(x)} u(y) \, dy$$

and

$$u(x) = u^+(x) = u^-(x).$$

We call the set

$$S_u = \{x \in \Omega \mid u^+(x) > u^-(x)\}$$

the jump set of u. It can be shown that S_u is a countable rectifiable set.

In connection to the $(n - 1)$ dimensional Hausdorff measure $\mathcal{H}^{n-1}$, we can define a normal direction $n_u(x)$ to almost all $x \in S_u$. Consequently, we can define

$$D^j u = (u^+ - u^-)n_u \cdot \mathcal{H}^{n-1}|_{S_u}$$

and

$$D^c u = (Du - D^a u) \llcorner (\Omega \backslash S_u).$$

We refer to [AFP], pp. 187–189 for details.

20.3 The relaxation function

Under certain growth restriction, we know the weak sequential lower semi-continuity of a functional I is equivalent to the quasi-convexity of the Lagrangian in p. In Lecture 13, we mentioned that Bolza gave the counterexample (Example 13.3) of $L(u, p) = u^2 + (p^2 - 1)^2$. It is not convex in p, $I(u)$ has a minimizing sequence $\{u_n\}$ in $W^{1,4}(0, 1)$ which converges weakly to 0, and the corresponding functional values $I(u_n) \to 0$, but $I(0) = 1$.

However, from the viewpoint of extremal values of a functional, in this example, the value of $I(0)$ is not of particular importance, but instead, the minimizing sequence itself reflects how small the values of I may become.

In order to showcase the characteristic of the functional in this regard, we introduce the concept of a relaxation function.

Definition 20.4 Let (X, τ) be a topological space and $f : X \to \bar{\mathbb{R}} = \mathbb{R}^1 \cup \{+\infty\}$. We call $Rf(x) = \sup\{\varphi(x) \mid \varphi : X \to \bar{\mathbb{R}}$ is lower semi-continuous, $\varphi(y) \le f(y), \forall y \in X\}$ the relaxation function of f.

The following are immediate from the definition.

(1) If f is lower semi-continuous, then $Rf = f$.

Proof Choosing $\varphi = f$, then $f \le Rf$. Conversely, by definition, $Rf \le f$. $\square$

(2) $Rf(x)$ is lower semi-continuous. This is because the supremum of a family of lower semi-continuous functions is itself lower semi-continuous.

From this, we see that the relaxation function Rf of f is the largest among all lower semi-continuous functions which are not greater than f.

(3) Let $f : X \to \bar{\mathbb{R}}$ be bounded below. Suppose f has a minimizing sequence $\{x_j\}$ and x_0 is a limit point of $\{x_j\}$, then x_0 is the minimum of Rf and $Rf(x_0) = \inf f$.

Proof Without loss of generality, we may assume $c = \inf f = \lim f(x_j)$. On one hand,

$$Rf(x_0) \leq \underline{\lim} Rf(x_j) \qquad \text{(by (2))}$$
$$\leq \underline{\lim} f(x_j) = c \qquad \text{(by definition)}.$$

On the other hand, $c \leq f(x)$, $\forall x \in X$. Since the constant function c is weakly lower semi-continuous, so $c \leq Rf(x)$, $\forall x \in X$. This means $Rf(x_0) = c$ and x_0 is the minimum of Rf. $\qquad\square$

Theorem 20.2 *Let X be the dual space of a separable Banach space X and $f : X \to \mathbb{R}^1$ be bounded below. If f is coercive, then in the weak-* topology, the relaxation function Rf has a minimum. Furthermore,*

$$\min_X Rf = \inf_X f.$$

Proof Choose a minimizing sequence $\{x_j\}$ of f such that $\lim f(x_j) = \inf_X f$. Since $\{x_j\}$ is bounded in X, it contains a weak-* convergent subsequence $x_{n_j} \overset{*w}{\longrightarrow} x_0$. By definition, the relaxation function Rf with respect to the weak-* topology is weak-* lower semi-continuous, by property (3), x_0 is the minimum of Rf, i.e.

$$\min_X Rf = Rf(x_0) = \inf_X f. \qquad\square$$

Returning to a variational problem, we consider the Lagrangian $L : \mathbb{R}^n \to \mathbb{R}^1$, which is continuous and satisfies the growth condition

$$c_0|p|^q \leq L(p) \leq c_1|p|^q + c_2, \quad 1 < q < \infty,$$

where $c_0, c_1, c_2 > 0$ are constants. Define the functional

$$I(u) = \int_\Omega L(\nabla u(x)) \, dx, \quad u \in W^{1,q}(\Omega), 1 < q < \infty.$$

Noting that

I is weakly lower semi-continuous $\Leftrightarrow$ L is quasi-convex $\Leftrightarrow$ L is convex,

thus, we have

$$RI(u) = \int_\Omega \text{conv}(L)(\nabla u(x)) \, dx,$$

where $\text{conv}(L) = \sup\{\varphi | \varphi(x) \leq L(x), \forall\, x \in \Omega, \varphi \text{ is convex}\}$ is the convex hull of L; that is, the supremum of all convex functions no greater than L.

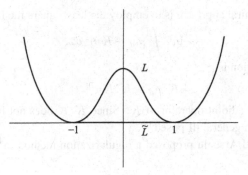

Fig. 20.2

Example 20.3 (The double well) Let $L(t, u, p) = (1 - p^2)^2$ (see Figure 20.2),

$$I(u) = \int_{-1}^{1} L(t, u(t), \dot{u}(t))dt.$$

Define

$$\tilde{L}(t, u, p) = \begin{cases} (1 - p^2)^2, & |p| > 1, \\ 0, & |p| \le 1, \end{cases}$$

which is the convex hull of L, then

$$RI(u) = \int_{-1}^{1} \tilde{L}(t, u(t), \dot{u}(t))dt.$$

In modern calculus of variations, there are abundant literature dedicated to the concrete representations of relaxation functions, for example, we refer to Buttazzo [Bu].

20.4 Image restoration and the Rudin–Osher–Fatemi model

Optic instruments often cause distortions in imaging. Besides noises, instruments themselves can also cause incorrect magnifications of the image functions.

Let Ω be a domain, and let realistic images be represented by functions $u : \Omega \subset \mathbb{R}^2 \to \mathbb{R}$. The optical image of the object is

$$u_d(x) = Ku(x) + n(x) = \int K(x - y)u(y)dy + n(x),$$

where K is the convolution operator derived from the point source dispersion function of the optical instrument itself. It can be viewed as a bounded linear operator on $L^2(\Omega)$. Moreover, $n(x)$ denotes the noise. The so-called image restoration is the process of deducing u from u_d.

The most natural approach is to employ the least square method in finding

$$\inf \int_\Omega |u_d - Ku|^2 \, dx, \tag{20.3}$$

whose E-L equation is

$$K^* u_d - K^* K u = 0, \tag{20.4}$$

where K^* is the adjoint operator of K. Since $K^* K$ does not have bounded inverse, (20.2) is in general ill-posed.

Tikhonov and Arsenin proposed a regularization method in 1977, by adding an energy term

$$T_\lambda(u) = \int_\Omega |u_d - Ku|^2 \, dx + \lambda \int_\Omega |\nabla u|^2 \, dx, \quad \text{where } \lambda \text{ is a parameter}$$

to the target functional in (20.3).

The advantage of this method can not only overcome the ill-posedness of (20.2), but also smooth out the solution u.

If we seek the minimum of the functional on $H^1(\Omega)$, then the E-L equation is

$$K^* u_d - K^* K u + \lambda \Delta u = 0,$$

with the Neumann boundary condition

$$\left. \frac{\partial u}{\partial n} \right|_{\partial \Omega} = 0.$$

The regularity of solution of an elliptic equation insures the smoothness of the solution u. By doing so, some noise can also be filtered out. However, this homogenized smoothing process has its flaw, it inevitably blurs the boundary of the image.

Several attempts were made to remedy this issue. For example, replacing the L^2 norm of the gradient ∇u by the L^p norm, and by decreasing the value of p, one can hopefully retain the image on the boundary. Rudin–Osher–Fatemi suggested to use the L^1 norm, namely, using $\int_\Omega |\nabla u| \, dx$ instead of $\int_\Omega |\nabla u|^2 \, dx$.

Unfortunately, $W^{1,1}(\Omega)$ is neither reflexive, nor is it the dual space of any Banach space. Consequently, it is very difficult to determine the existence of minimum of a functional on it.

We have stated that $\mathrm{BV}(\Omega)$ is the dual space of a separable Banach space. However, a BV solution u may have jump-type discontinuity, while the set of jump-type discontinuity is a rectifiable curve. If we solve the problem in $\mathrm{BV}(\Omega)$, then not only the existence of the solution is relatively easy to establish, but the edge of the body is also easier to meet. The model they proposed is to find the minimum u of the functional

$$E(u) = \frac{1}{2} \int_\Omega |u_d - Ku|^2 \, dx + \lambda \int_\Omega \phi(|\nabla u|) \, dx, \tag{20.5}$$

where $\phi : \mathbb{R}^+ \to \mathbb{R}^+$ is a strictly convex, monotone increasing function satisfying

$$\phi(0) = 0 \tag{20.6}$$

and

$$\exists\, c > 0, \exists\, b \geq 0 \text{ such that } cs - b \leq \phi(s) \leq cs + b, \forall\, s. \tag{20.7}$$

However, on $BV(\Omega)$, we must know how to interpret the second integral in (20.5).

Theorem 20.3 (E. De Giorgi, L. Ambrosio, G. Buttazzo) *In $BV(\Omega)$, the relaxation function of E is*

$$RE(u) = \frac{1}{2} \int_{\Omega} |u_d - Ku|^2 + \lambda \int_{\Omega} \phi(|\nabla u|) \, dx$$
$$+ \lambda c \int_{S_u} (u^+ - u^-) \, d\mathcal{H}^1 + \lambda c \int_{\Omega \backslash S_u} |C_u|, \tag{20.8}$$

where

$$Du = \nabla u \mathcal{L}^2 + (u^+ - u^-) n_u \mathcal{H}^1|_{S_u} + C_u$$

is the Lebesgue decomposition. We refer to [Bu].

Theorem 20.4 *Under the assumptions given above, if we further assume $K \in \mathcal{L}(L^2, L^2)$ and $\|K \cdot 1\| \neq 0$, then the functional $RE(u)$ has a minimum $u_0 \in BV(\Omega)$.*

Proof It suffices to show RE is coercive on $BV(\Omega)$. Let $\{u_j\} \subset BV(\Omega)$ be a minimizing sequence of RE. By (20.7) and (20.8), $\|Du_j\|(\Omega)$ is bounded and $\|u_d - Ku_j\|_2$ is also bounded. It remains to show that $\|u_j\|_1$ is bounded. Decompose

$$u_j = v_j + w_j,$$

where $w_j = \frac{1}{|\Omega|} \int_{\Omega} u_j \, dx$.

By Poincaré's inequality (property (5)), $\|v_j\|_2 \leq C\|Du_j\|(\Omega)$, so we need only to show $|w_j|$ is bounded. Let $\|K \cdot 1\| = r$ and $\|K\| = k$, from

$$\|u_d - Ku_j\|_2^2 \leq M,$$

it follows that

$$r|w_j|(r|w_j| - 2(k\|v_j\|_2 + \|u_d\|_2)) \leq [\|Kv_j - u_d\|_2 - |w_j|r]^2$$
$$\leq \|u_d - Ku_j\|_2^2 \leq M.$$

Since $\|v_j\|_2$ is bounded, so is $|w_j|$.

Finally, by the weak-* sequential lower semi-continuity of the relaxation function RE, there must exist a minimum $u_0 \in BV(\Omega)$. This completes the proof.

$\square$

Bibliography

The main references for Lectures 1–8: [BGH], [LL], [GF], [JJ], [Ka], [Mac], [Mos].

The main references for Lectures 9–14: [Ad], [AE], [CL], [Da], [Ek], [Gi], [JJ], [Mos], [Mor], [Ne], [Po].

The main references for Lectures 15–20:

Lecture 15: [Ch], [JJ], [Ra1], [St1], [MW]
Lecture 16: [BGH], [HT], [Maw], [Mos], [Ra2], [Ra3].
Lecture 17: [Co], [Jo], [Ha], [Hi], [Mor], [Os], [St2].
Lecture 18: [Hu], [Joh].
Lecture 19: [BM], [Mac], [Le], [PBGM].
Lecture 20: [AFP], [AK], [BGH], [Bu], [Fe].

[Ad]   Adams, R. A. (1975) *Sobolev Spaces*, Acad. Press. (Chinese translation by Ye, Q. X., Wang, Y. D., Ying, L. A., Han, H. D., and Wu, L. C., Beijing, People's Education Press, 1983.

[AFP]   Ambrosio, L., Fusco, N., and Pallara, D. (2000). *Functions of Bounded Variation and Free Discontinuous Problems*, Clarendon Press, Oxford.

[AE]   Aubert G. and Ekeland I. (1984). *Applied Nonlinear Analysis*, A John Wiley-Interscience Publication.

[AK]   Aubert, G. and Kornprobst, P. (2002). *Mathematical Problems in Image Processing, Partial Differential Equations and the Calculus of Variations*, Applied Math Sciences **147**, Springer.

[BM]   Brechtken, U. and Manderschild, U. (1991). *Introduction to the Calculus of Variations*, Chapman Hill.

[Bu]   Buttazzo, G. (1989). *Semicontinuity, relaxation and integral representation in the Calculus of Variations*, Pitman Research Notes in Mathematics **207**, Longman Scientific and Technical.

[BGH]   Buttazzo, G., Giaquinta, M., and Hildebrandt, S. (1998). *One-dimensional Variational Problems, An Introduction*, Clarendon Press. Oxford.

[CL]   Chang, K. C. and Lin, Y. Q. (1987). *Lecture notes on Functional Analysis I (Chinese)*, Beijing, Peking University Press.

[Ch]   Chang, K. C. (1993). *Infinite Dimensional Morse Theory and Multiple Solution Problems*, Birkhäuser.

[Co]   Courant, R. (1950). *Dirichlet's Principle, Conformal Mapping, and Minimal Surfaces*, New York, Interscience.

[Da]   Dacorogna, B. (1989). *Direct Methods in the Calculus of Variations, Applied Math. Sciences*, **78**, Springer Verlag.

[Ek]   Ekeland, I. (1979). *Non-convex Minimization Problems*, Bull. Amer. Math. Soc., pp. 443–474.

[Fe]   Federer, H. (1969). *Geometric Measure Theory*, Springer-Verlag.

[GF]   Gelfand, I. M. and Fomin, S. V. *Calculus of Variations*, (English translation by Silverman, R. A., Prentice Hall, 1964).

[Gi]   Giaquinta, M. (1983). *The regularity problem of extremals of variational integrals*, Proc. NATO/LMS Advance Study Inst. on "Systems of nonlinear partial differential equations", Reidel Publ. Co.

[GH]   Giaquinta, M. and Hildebrandt, S. (1996). *Calculus of Variations I The Lagrangian formalism*, Grundlehren der mathematischen Wissenschaften, **310**, Springer.

[GT]   Gilbarg, D. and Trudinger, N. (1983). *Elliptic Partial Differential Equations of Second Order*, 2nd ed. Grundlehren der mathematischen Wissenschaften, **224** Springer..

[Ha]   Hardt, R. (Ed. 2004). *Six Theorems in Variations*, Student Math. Lib. **26** AMS.

[Hi1]   Hilbert, D. (1900). *Uber das Dirichletsche Prinzip*, Jber Deutsch Math. Vere., **8**, pp.184–188.

[Hi]   Hildebrandt, S. (1969). *Boundary behavior of Minimal Surfaces*, Arch. Rat. Mech. Anal., **35**, pp. 47–82.

[HT]   Hofer, H. and Toland, J. F. (1984). *Homoclinic, heteroclinic and periodic solutions for indefinite Hamiltonian systems*, Math. Ann., **268**, pp. 387–403.

[Hu]   Hughes, T. (2000). *Finite Element method, Linear Static and Dynamic Finite Element Analysis*, Dover Publications.

[JJ]   Jost, J. and Jost, X. Li. (1988). *Calculus of Variations*, Cambridge University Press.

[Jo]   Jost, J. (1990). *Two Dimensional Geometric Variational Problems*, John Wiley and Sons.

[Joh]   Johnson, C. (1987). *Numerical Solution of Partial Differential Equations by the Finite Element Method*, Studentlitteratur, Lund.

[Ka]   Kato, T. *Calculus of Variations and its Applications* (Chinese Translation by Zhou, H. S., Shanghai Science and Technology Press, 1961).

[LL]   Lavrentiev, M. A. and Liusternik, L. A. *Lecture Notes on the Calculus of Variations* (Chinese Translation by Zeng, D. H., Deng, H. Y. and Wang, Z. K., High Education Press, Beijing, 1955).

[Le]   Leitmann, G. (1981). *The Calculus of Variations and Optimal Control, An Introduction*, Plenum Press, 1981.

[Mac]   MacCluer, C. R. (2005). *Calculus of Variations, Mechanics, Control, and Other Applications*, Pearson Education, Ltd.

[MS]   Macki, J. and Strauss, A. (1982). *Introduction to Optimal Control Theory*, Springer-Verlag, 1982.

[MT]   Mallianvin, P. and Thalmaier, A. (2006). *Stohastic Calculus of Variations in Mathematical Finance*, Springer Finance, Springer.

[Maw]   Mawhin, J. *Global Results for the Forced Pendulum Equation, Handbook of Differential Equations, Ordinary Differential Equations*, (Ed. by A. Canada, P. Drabek, F. Fonda), Vol. 1, pp. 533–389.

[MW]   Mawhin, J. and Willem, M. (1989). *Critical Point Theory and Hamiltonian Systems*, Applied Math. Sciences, **74**, Springer Verlag.

[Mor]   Morrey, C. B. Jr. (1966). *Multiple Integrals in the Calculus of Variations*, Springer Verlag.

[Mos]   Moser, J. (2003). *Selected Chapters in the Calculus of Variations*, Birkhäuser Verlag.

[Ne]   Nehari, Z. (1961). *Characteristics Values Associated with a Class of Nonlinear Second Order Equations*, Acta Math., **105**, pp. 141–175.

[Os]   Osserman, R. (1986). *A survey of Minimal Surfaces*, Dover Publications.

[Po]   Pohozaev, S. (2008). *Nonlinear Variational Problems via the Fibering Method*, Handbook of Differential Equations, Stationary Partial Differential Equations (Ed. by M. Chipot), Vol. 5, pp. 49–208.

[PBGM]   Pontryagin, L., Boltyanskii, V., Gamkrelidze, R., and Mishchenko, E. (1962). *The Mathematical Theory of Optimal Process*, Interscience Publishers, John Wiley and Sons.

[Ra1]   Rabinowitz, P. (1986). *Minimax methods in critical point theory with applications to differential equations*, CBMS Regional Conference Series Math. **65** AMS Providence, 1986.

[Ra2]   Rabinowitz, P. (1990). *Homoclinic Orbits for a Class of Hamiltonian Systems*, Proc. Royal Soc. Edinburgh, **114 A**, pp. 33–38.

[Ra3]   Rabinowitz, P. (1989). *Periodic and Heteroclinic Solutions for a Periodic Hamiltonian System*, Ann. Inst. Henri Poincaré, Anal. Nonlinearie, **6**, pp. 331–346.

[St1]   Struwe, M. (1988). *Plateau's Problem and the Calculus of Variations*, Mathematical Notes, Princeton University Press.

[St2]   Struwe, M. (1990). *Variational Methods, Applications to Nonlinear Partial Differential Equations and Hamiltonian Systems*, Springer Verlag.

Index

Printed in the United States
By Bookmasters